ALPHONSE CHAPANIS
Human Factors in Systems Engineering

YACOV Y. HAIMES
Risk Modeling, Assessment, and Management, 2/e

DENNIS M. BUEDE
The Engineering Design of Systems: Models and Methods

ANDREW P. SAGE and JAMES E. ARMSTRONG, Jr.
Introduction to Systems Engineering

WILLIAM B. ROUSE
Essential Challenges of Strategic Management

YEFIM FASSER and DONALD BRETTNER
Management for Quality in High-Technology Enterprises

THOMAS B. SHERIDAN
Humans and Automation: System Design and Research Issues

ALEXANDER KOSSIAKOFF and WILLIAM N. SWEET
Systems Engineering Principles and Practice

HAROLD R. BOOHER
Handbook of Human Systems Integration

JEFFREY T. POLLOCK AND RALPH HODGSON
Adaptive Information: Improving Business Through Semantic Interoperability, Grid Computing, and Enterprise Integration

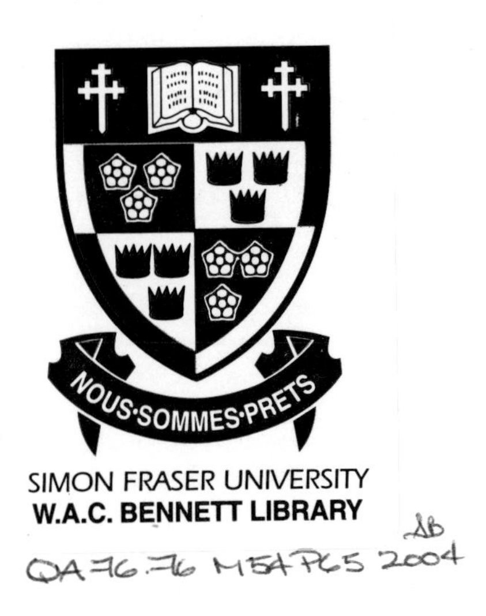

Adaptive Information

Adaptive Information
Improving Business Through Semantic Interoperability, Grid Computing, and Enterprise Integration

Jeffrey T. Pollock
Ralph Hodgson

A JOHN WILEY & SONS, INC., PUBLICATION

Published by John Wiley & Sons, Inc., Hoboken, New Jersey.
Published simultaneously in Canada.

Library of Congress Cataloging-in-Publication Data:

Pollock, Jeffrey T.
Adaptive information : improving business through semantic interoperability, grid computing, and enterprise integration / Jeffrey T. Pollock, Ralph Hodgson.
p. cm.—(Wiley series in systems engineering and management)
Includes bibliographical references and index.
ISBN 0-471-48854-2 (cloth)
1. Middleware. 2. Enterprise application integration (Computer systems)
3. Semantic integration (Computer systems) 4. Computational grids (Computer systems)
5. Business enterprise–Data processing. 6. Information resources management.
I. Hodgson, Ralph, 1945– II. Title. III. Series.

QA76.76.M54P65 2004
005.7′13–dc22 2004040857

Printed in the United States of America
10 9 8 7 6 5 4 3 2 1

Contents at a Glance

Contents

Appendices

List of Illustrations

List of Tables and Other Sidebar Elements

Foreword

For any technology manager (from CIO to technology team leader) or technology specialist within an enterprise that is concerned about the business impact of technology trends, this is a book that covers topics you will want to know more about. You have probably helped your company deal with the use of Internet technologies to enable global connectivity and have multiple projects to use XML to structure your data for information exchange. You may even have begun the development or implementation of projects using Web services technologies. But everything that you may have developed so far is foundational to what is coming next in what these authors describe as semantic interoperability technologies, standards, capabilities and tools.

This book makes a very strong and compelling case for why you need to be aware today of challenges and benefits involved in achieving semantic interoperability. The authors assert that "*the unfulfilled promise of frictionless information exchange is the prime motivation for an array of little known technologies that address semantic interoperability.*"

If your company recognizes that "*competitive advantage for companies in all industries will be largely driven by efficiency in information sharing*," then you are likely not surprised by their assessment that "*most strategic, long-term barriers to efficient information sharing are inadequately addressed by currently popular integration approaches.*" This focus leads to the premise that semantic interoperability is what is needed to enable the level of information sharing that produces a competitive advantage.

To help the reader understand the technology being described, the authors provide extensive examples and background on all of the related terms used in their definition: "*Semantic interoperability is a dynamic enterprise capability derived from the application of special software technologies (such as reasoners, inference engines, ontologies, and models) that infer, relate, interpret, and classify the implicit meanings of digital content without human involvement, which in turn drive adaptive business processes, enterprise knowledge, business rules, and enterprise application interoperability.*"

The authors present in-depth examples of why this technology may soon become a 15 year overnight success. They effectively relate the foundational work on knowledge information frameworks, ontologies and the semantic web.

From the vantage point of oversight of the multiple e-business and web services standards activities, I readily see how the future of semantic interoperability will impact the standards work at OASIS. For instance, the authors describe how Topic Maps (which are being advanced as standards through OASIS Technical Committees) may be executed by a content mediation service. The authors also describe how semantic interoperability is focused on supporting the future technology infrastructures for business. So e-Business and Web services standards need to become more adaptive to incorporate the standards for ontology-based engineering; Web Services need the type of adaptive information capabilities the authors describe to realize their full potential of automatic service discovery and services composition.

Clearly, the time to forge a common framework based on semantic interoperability standards and e-business web services standards is now. *Adaptive Information* puts these requirements into context and gives readers the information they need to understand this development and assure that the promise of Web services extends to support these vital new capabilities.

Patrick J. Gannon
President & CEO
OASIS Open
patrick.gannon@oasis-open.org

Patrick J. Gannon (*patrick.gannon@oasis-open.org*) is President and CEO of OASIS, the global consortium that drives the development, convergence, and adoption of e-business standards. Mr. Gannon also holds a position on the OASIS Board of Directors. He has served as the Chairman of the Internet Enterprise Development Team since it was founded in 2000 by the United Nations Economic Commission for Europe (UNECE), to advise governments with transitional economies on best practices for electronic business. Prior to OASIS, Mr. Gannon was Senior Vice President at BEA Systems, Vice President of Marketing at Netfish Technologies and Vice President of Strategic Programs for the CommerceNet Consortium.

Beginning in 1995 he was involved in catalog interoperability activities and directed research and development efforts in new Internet commerce standards such as XML. While at CommerceNet, he became the first Project Leader for RosettaNet and served as Executive Director for the Open Buying on the Internet (OBI) initiative. He was an active leader in the development of ebXML (electronic business using eXtensible Markup Language), an initiative co-sponsored by OASIS and UN/CEFACT.

Mr. Gannon is co-author of the book: "Building Database-Driven Web Catalogs," and is an international speaker on electronic business and Web Services standards.

Preface

This book is about technology that will enable a new kind of infrastructure. This new infrastructure is dynamic and adaptive—it configures itself and responds to change with little or no human input. No, we are not talking about magic adapters or artificial intelligence. This is an infrastructure driven by models of information, process, and service interfaces. This infrastructure uses models to learn new behavior, find patterns, and rebuild itself for different platforms. It is an infrastructure that can transform data on the fly, establish new communications without programming, and follow process rules without knowing about specific services in advance. No, it is not magic—but it *is* smarter.

Smarter infrastructure is good for business. Business benefits achieved through the use of the semantic interoperability technologies described in this book include:

Increased Profits

- Faster response to market demands
- Better decision making
- Cheaper adoption of new strategies

New Capabilities

- Dynamic information repurposing and reconfiguration
- Adaptive service-oriented networks
- Loosely coupled application connections and information

Reduced Costs

- Faster data analysis and dissemination
- Better quality of service in partner exchanges
- Cheaper maintenance costs for IT systems integration

The authors of this book have watched several technology forces (such as semantics, agents, and services) align for many years. We have helped build software that reaches far beyond commonplace tools. We have interviewed technologists from around the globe so that we can report to you about the widest variety of techniques being leveraged to build this new kind of infrastructure. It is our firm belief that the collection of patterns, technologies, and tools that we describe in this book will inevitably lead toward a future substantially better than today's realities.

Our passion has drawn us into the field of software semantics. Of all the applications for semantic technologies, we believe that the creation of a new enterprise infrastructure framework—semantically aware—will be the foundation upon which all future systems will be built. Our goal in writing this book is partly to share the excitement of the vision that has become so clear to us, but also to provide useful patterns for understanding the scope and capabilities of the software driving this revolution.

HOW THIS BOOK WILL HELP

The primary emphasis of this book will be to provide the reader with a deep understanding of what the future of enterprise infrastructures will look like, while placing special emphasis on describing pragmatic examples of how semantic interoperability is already being used today to save money and enable new capabilities.

After reading this book, you will be able to organize and architect a semantic interoperability project, make decisions about tools, and choose the best architectural patterns and technologies that suit your specific needs. You will benefit by learning from the challenges that have been overcome by a number of early adopter projects discussed in this book.

Architecture and design are two of the most critical ingredients in a successful semantic interoperability project. Therefore, significant attention is given to the subject of middleware architecture—covering traditional Enterprise Application Integration (EAI) architectures, emerging semantic interoperability architecture, and design patterns.

ROAD MAP

This book is organized in three parts. Part One, consisting of Chapters 1–3, provides a detailed look at the problems that plague the EAI industry and some of the emerging semantics-based infrastructure that will remedy those problems. Chapter 1 gives special attention to the business problems that many industries are facing, connects those business issues to the integration technologies that should be able to help solve them, and discusses strategic barriers preventing their success. Chapter 2 introduces a new approach for thinking about enterprise middleware architectures—splitting the architectural views into two broad categories: connections and information. In Chapter 3 we examine trends in middleware development and start to envision how they may be manifested by considering a hypothetical intelligence information-sharing solution.

Part Two, a semantic interoperability primer, is intended to provide the reader with enough information to understand what semantics are and how metadata, ontologies, and software tools are used to provide semantic interoperability. Chapter 4 provides a foundation in the roots of the semantics discussion by giving a history of some of the questions of meaning that have been handed down by great thinkers over thousands of years. In Chapter 5, a detailed look at the kinds of semantic conflicts is provided alongside the solution patterns that have emerged to solve some of these conflicts. Chapter 6 is a high-level discussion of metadata layers intended to show how current architectures neglect crucial aspects of metadata architecture. The subject of ontology and its usages, patterns, and formats are at the core of Chapter 7—including a discussion of ontology representation and transformation. Chapter 8 is about the components and configurations required to enable semantic architecture runtime environments. We close Part Two with Chapter 9, which summarizes the knowledge from the previous chapters (about semantics, semantic conflict, metadata, ontology, and architectures) and describes the infrastructures that can be deployed in support of various E-business models.

Part Three is dedicated to grounding the reader in pragmatic examples of how to manage and deploy semantic interoperability solutions. Chapter 10 offers a substantive look at industry case studies and the ways that business is already deriving value from semantic interoperability solutions. The emphasis of Chapter 11 is to provide guidance to the manager who needs to build a plan for adoption and implementation of semantics-aware solutions. Finally, in Chapter 12 the future of semantic interoperability is examined to identify special business requirements, emerging solutions, and barriers to adoption as interoperability technology con-tinues to evolve.

GUIDE TO THE READER

There are two kinds of people who should be reading this book: senior technologists and technology managers. The senior technologist will benefit from the patterns provided for ontology, infrastructure, and semantic approaches. The technology manager will benefit from the implementation guide, the capability cases, and the vendor tools survey. Because the first part (Chapters 1–3) of the book is essentially a business case, any reader who is not familiar with the semantic technologies business space, or who isn't sold on the vast potential of these technologies, would greatly benefit from reading this part.

There are a few different ways to read this book. To thoroughly understand why the semantic interoperability technology space is important, Chapters 1, 2, 3, 4, and 10 are recommended. For business managers, and others who are interested in what semantic interoperability is and how it is used, this may be sufficient.

More technical managers, software architects, software developers, and those who want a better understanding of semantic interoperability technologies should additionally read Chapters 5, 6, 7, 8, and 9 to really understand how they are used and deployed. After reading these parts, the reader will be well prepared to examine the case studies in Chapter 10.

The last option would be to read the whole book straight through. This is recommended for system architects, integration architects and others who want a thorough understanding of the technology and implementation of semantics-based interoperability solutions.

Book Iconography and Picture Key

This book uses consistent iconography wherever possible.

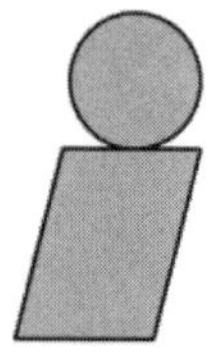
Agent/Actor

Working Software User

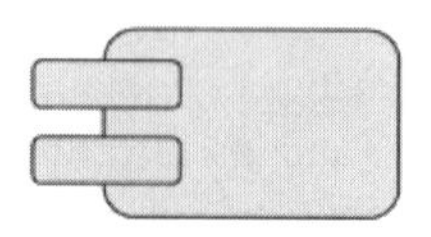
Software Component

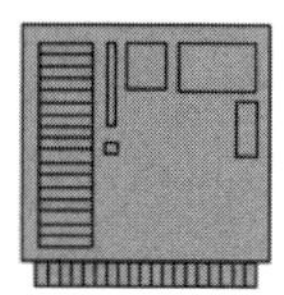
Enterprise Server

Commercial Enterprise

Manufacturing Factory

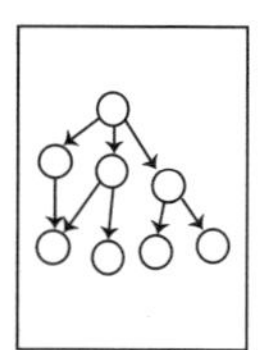
XML Schema

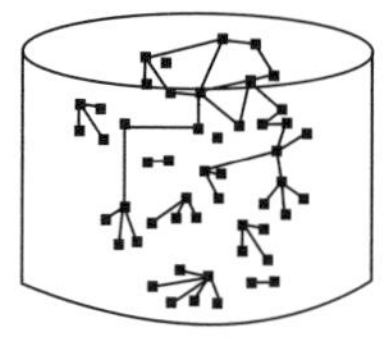
Database Schema

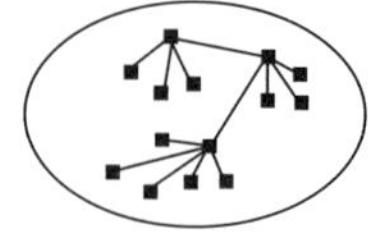
Schema (ontology)

XML Instance

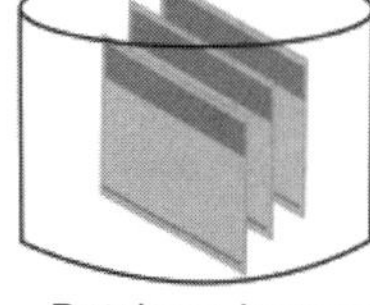
Database Instance

Semantic Maps

Logics

Software Code

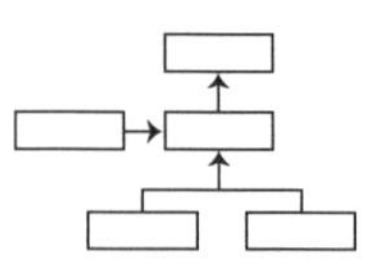
Object Model

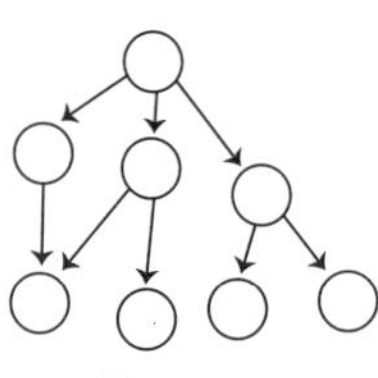
Taxonomy

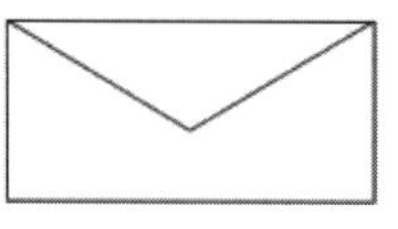
Message

PEDAGOGICAL ELEMENTS IN THE TEXT

Many sidebar elements are included throughout this book to assist the reader in understanding the central text.

Insider Insights

These are "guest spots" where industry luminaries voice their thoughts about an important topic of interest. These sidebars are inserted throughout the text wherever an "insight" overlaps with the content being discussed in the main text. In this way, the reader should enjoy a diversity of ideas surrounding many topics of interest in the book.

Perspectives

These are real-world experiences and/or author opinions that are presented as an excerpt where they apply to a given discussion in the main body of work. Often these perspectives will be tangentially related to the text, but they usually provide some interesting thoughts that otherwise wouldn't have been included.

Definition

Inserted at strategic points, a Definition may help the reader get through a section of text more efficiently. Usually it is helpful to have access to basic definitions for words like ontology, inference, autonomic, and so on.

Data Table

A Data Table is usually a simple matrix that summarizes narrative text in the main body of the book.

Soapbox

A Soapbox is an impassioned position statement about a subject that could be seen as controversial. Soapbox entries tend to be in the first person.

Code

The code element is space where code samples are provided to support nearby text.

Counterpoint

When the book is moving an idea along in a certain direction based on assumptions, a Counterpoint element will be included to help the reader achieve a more balanced view of the particular issue being discussed. Often, these Counterpoints will make an assertion and identify supporting rationale, but the main text will continue to move along.

Acknowledgments

A book such as this cannot be written without a substantial team effort. Although only two names will appear as the authors of this book, its underpinnings are connected to so many more to whom I am greatly indebted.

First, the staff at John Wiley & Sons, Inc., George Telecki, and Danielle Lacourciere have made the process easy for this first-time book author—without their support and hard work this project would have stalled out several times. Also, my co-author Ralph Hodgson contributed superhuman effort in the midst of many other important commitments to produce great work—significantly improving the book from end to end. Louise Bartlett and Cheryl Roder-Quill provided invaluable assistance with concepts and design of graphics included in this book.

Thinking back to a time in my life when semantics seemed to belong more in a philosophy book than in a computer trade book, I am astounded at the degree to which my own thinking has changed. Without friends like Julian Fowler and William Burkett, who constantly engaged me in new ideas about semantic computing, I would have never developed the deep passion for this subject, which now consumes me.

Without support from critical reviewers, a book like this could be destined for the trash heap. Michael Daconta has been an invaluable source of feedback and suggestions that reflect his deep experience with both semantic technologies and book authoring. Jack Berkowitz consistently challenged me to simplify the book structure and to clarify the architecture designs so that they are useful to a wide range of technologists. Dave Hollander's thoughts and engaging conversation always resulted in new insights and clarity that I hope to have passed along to the reader. Ken Fromm and Martha Kolman-Davidson are two people who strove to help me make this complex subject simpler—any confusing text is my fault despite their best efforts.

My passion for software design existed before I was even aware of the notion of semantic technologies. As such, I arrived at the problem already greatly influenced by a number of people who shaped my own internal schema. The Three Amigos (Grady Booch, Ivar Jacobson, and Jim Rumbaugh) and the Gang of Four (Erich Gamma, Richard Helm, Ralph Johnson, and John Vlissides) shaped the way I think about software problems—their influences can be felt in both the structure and organization of technical content in this book. More recently, Tim-Berners Lee, Amit Sheth, and Jim Hendler have all guided many of my views on semantic technology. John Sowa enabled me to believe that the words "semantic technology" even belonged together in the first place.

Without the help of many former colleagues from Modulant, this book would not have been possible. Yuhwei Yang introduced me to the work at STEP, which served as a genesis for much of the semantic mapping technology we employed.

Craig Wingate, Curtis Olson, Steve Bastasini, and Andrea Bruno all provided much-welcomed support, education, and encouragement for this book.

And finally, I would like to express a deep debt of gratitude for all the authors of Insider Insight pieces for this book—their contributions have made the book more interesting and eminently more valuable for each and every reader. Thank you Jeff Dirks, Julian Fowler, Dr. Jürgen Angele, Mike Evanoff, Rick Hayes-Roth, Dr. Rod Heisterberg, Sandeep Maripuri, Stephen Hendrick, and Zvi Schreiber.

Jeffrey T. Pollock
San Francisco, CA
January 2004

There have been two software technology loves in my life: Object Technology and Logic Programming. When Object Technology emerged in the 1980s, there followed much excitement about how software systems could mimic the real world. Object models were seen as ways of ensuring fidelity of the software system to the concepts and phenomena of the world in which they were used. The software industry gained a lot from what happened with objects.

Without objects, perhaps there would not have been a patterns movement, or component-based development, or the developments in modeling methods and development processes. With semantic technology I firmly believe we are at the start of just as significant a change in how we will build systems in the future. I have the hope that standards like OWL will help us overcome the ambiguities and imprecisions in the semantics of the Unified Modeling Language (UML). For Logic Programming, I have the hope that its power will finally be realized in more commercial systems that employ semantic technology. In many ways, semantic technology brings together these two loves of mine—the love of abstraction and the love of the elegance and precise semantics of logic.

Writing this book with Jeff Pollock has been an exciting and challenging venture. Jeff's fortitude and unrelenting drive kept my energies sustained. Despite the many other activities that surround building a consulting company in the Semantic Technology space, we managed to get the job done. Thank you, Jeff, for your trust and patience during this journey we took together.

I have to say that this journey would have been impossible without the contributions and support of my colleagues in TopQuadrant. Especially I want to thank Irene Polikoff. She remains, in my estimation, unequaled in her ability to take a complex field, research it well, and make sense of what is important at unrivaled speed. Many times I needed Irene's memory of what we had learned in our consulting assignments in semantic technology. Irene continues to impress us all with her ability to discover knowledge and make important insights and predictions. Without Dean Allemang's keen sense of journalism and critical inquiry, our solution stories and capability cases would be much impoverished. Thank you, Dean, for your expertise in the semantic technology field and your unrivaled ability to see the essence of things. Finally, I thank Robert Coyne for his encouragement and many hours of constructive dialog on the issues surrounding the adoption of semantic technology.

Many vendors contributed to the collective knowledge of the capability cases and their solution stories. I would especially like to thank Amit Sheth of Semagix, Michael Ullrich and York Sure of Ontoprise, Danielle Forsyth of Thetus Corporation, Loren Osborn of Unicorn Solutions, and Mathew Quinlan and Paul Turner of Network Inference. I am also very appreciative for the interactions that I have enjoyed with thought leaders in the semantic technology field, in particular, Professor Jim Hendler and his staff at the University of Maryland's MINDlab and Professor Nigel Shadbolt of the UK-based AKT Consortium.

Ralph Hodgson
Alexandria, VA
January 2004

Part One

Why Semantic Interoperability?

The first part of this book makes three primary assertions. First, contrary to some pundits' claims, information technology can still matter to business. The third wave of economic fortunes depends squarely on business' ability to maximize the utility of corporate information. However, several inherent, largely nontechnical issues with the way digital information is created and stored block substantial improvement in the corporation's ability to share information across internal and external boundaries.

Second, entrenched approaches to implementing information sharing infrastructures continue to rely on high degrees of human involvement—leading to systemic barriers preventing both agility and automation. Despite the hype, service-oriented architectures (SOA) and business process management (BPM) do little to improve the overall situation. The purpose and foundation of SOA and BPM approaches is information sharing between discrete content owners, but neither will address fundamental problems with the interpretation of information in different contexts.

Third, the unfulfilled promise of frictionless information exchange is the prime motivation for an array of little-known technologies that address semantic interoperability. Organic middleware infrastructure capable of highly autonomous behavior will dramatically reduce the effort required to connect, configure, and communicate meaning among disparate information resources. Semantic interoperability is a required underlying foundation of these new capabilities.

Adaptive Information, by Jeffrey T. Pollock and Ralph Hodgson
ISBN 0-471-48854-2

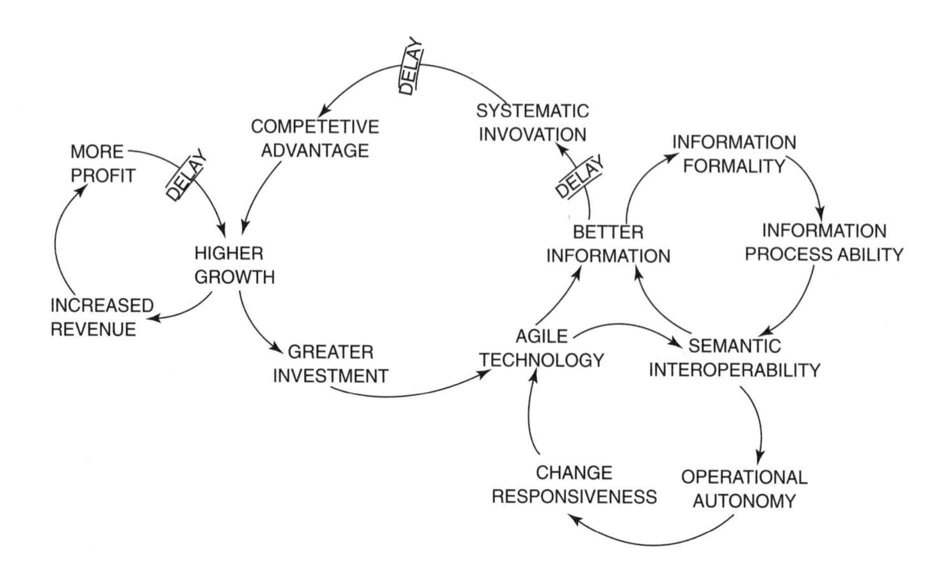
MORE
PROFIT
DELAY
COMPETETIVE
ADVANTAGE
DELAY
SYSTEMATIC
INVOVATION
DELAY
INFORMATION
FORMALITY
HIGHER
GROWTH
BETTER
INFORMATION
INFORMATION
PROCESS ABILITY
INCREASED
REVENUE
GREATER
INVESTMENT
AGILE
TECHNOLOGY
SEMANTIC
INTEROPERABILITY
CHANGE
RESPONSIVENESS
OPERATIONAL
AUTONOMY

Chapter 1

Semantic Interoperability Gives IT Meaning

KEY TAKEAWAYS

- Semantic interoperability is an enterprise capability derived from the application of special technologies that infer, relate, interpret, and classify the implicit meanings of digital content, which in turn drive business processes, enterprise knowledge, business rules, and enterprise application interoperability.
- The third wave indicates that competitive advantage for companies in all industries will be largely driven by efficiency in information sharing.
- Most strategic, long-term barriers to efficient information sharing are inadequately addressed by currently popular integration approaches.

Innovation in information technology (IT) is often represented as a series of waves. Going back to the 1950s, various technology waves appear with large amplitudes. These waves include the mainframe, the network, the mini-computer, the personal computer, and the Internet. Then there are technology waves with smaller amplitudes—the disk drive, the computer terminal, the relational database, the graphical user interface, the laser printer, markup languages, and component software.

Viewing technical innovation in this manner is useful, but its simplicity often obscures much of the real story. Rarely does just one technology make for a revolution. More often it is a combination of technologies that create a new platform that then creates fertile ground for an even larger set of innovations. Often, these technologies are not even new. Famous authors and musicians often joke that they became overnight successes—over the course of ten years. The same is true with technology. Almost all "breakthrough" technologies are typically preceded by years of research and prior iterations. It is typically the combination of technologies that creates the large-amplitude waves that set off a host of other advances. These advances come not only within technology but also in technical development with the advent of new design and engineering processes; in business with the development of new business models and new products and services; and—when technology touches the consumer—in social dynamics and interactions.

Adaptive Information, by Jeffrey T. Pollock and Ralph Hodgson
ISBN 0-471-48854-2

The most recent period of innovation, the Internet revolution, featured at its foundation the Internet with its IP protocol and network topology, which had been in existence for over 30 years. And although the HTTP protocol was a big step forward from other file and data transfer protocols, the use of the World Wide Web only really took off when low-cost web servers and free graphical browsers became available.

Even then, however, technical advancement was aided by a host of other enabling technologies—many of which were independent efforts that happened to be at the right time and in the right place. These technologies included innovative search algorithms that provided immediate utility to the web and advanced computer languages such as Java that enabled greater platform independence for component software. Additional progress came through of the use of rapid development techniques that fused information and interaction design closely with engineering and programming disciplines. The open source movement contributed to the pace of advances by providing thousands of innovators with the basic tool sets for working and creating new products and services. The Internet revolution was not one technology or one protocol, nor was it just about technology. New thinking was applied to almost all areas of business and many areas of social interaction. And although it is amusing to look back at all the talk about the "New Economy," business and society have been forever changed by a simple and yet revolutionary set of component technologies and protocols.

The personal computer revolution was a similar story. It started with a relatively simple and programmable operating system and extended to computer screens, disk drives, floppy drives, and laser printers. Novel consumer programs such as spreadsheets, word processors, and databases appeared, soon to be followed by desktop publishing. A vast number of companies sprang up to create and sell add-on hardware components that further extended PC capabilities. New distribution systems emerged, first in the form of small computer stores, then evolving into mail order, office supply stores, and computer megastores. Ethernet networks and file sharing were key components in making personal computers standard office equipment in the workplace and fueled the second and larger trajectory of the PC wave.

Whereas consumers' lives may not have changed as much as they did with the Internet, no one can ignore the change that personal computers brought about in the workplace. New business structures arose, and roles and responsibilities were drastically redefined—all of this because of a revolutionary platform that increased access to information and to information tools.

DOES IT MATTER ANYMORE?

The Harvard Business Review published a controversial article entitled "IT Doesn't Matter" in May 2003.[1] In this article Nicholas Carr made the bold claim that America's great IT industry no longer had any strategic value to offer corporations.

[1] *Does IT Matter? An HBR Debate.* Carr, Nicholas, June 2003.

Noting that the "data machines" of the past 50 years are limited to the core functions of "data storage, data processing, and data transport," he builds a case that software will no longer be a source of competitive advantage in business.

He is mostly right. That is, about current popular tools and capabilities.

Traditional integration tools (EAI, B2B, BPM), and packaged applications (ERP, CRM, PLM) for that matter, fall into Carr's category of "data machines." However, he misses the important point that software capabilities are a moving target—constantly evolving in response to changing business demands. As you will see throughout the course of this book, emerging technology approaches like semantic interoperability will drive new opportunity for business advantages by enabling companies to strengthen their own unique differentiators in the marketplace.

Because semantic technologies—technologies that enable data, process, and service meaning to become machine processable—offer a new paradigm for managing information inside the corporate infrastructure, not just offering new technical function points that can quickly become commodity, the application of these technologies will deliver highly tailored competitive advantages among companies who adopt them.

So, although Carr may feel that each technology offers the same advantages to companies who invest and adopt it, semantic technology is not just another me-too software application. *Rewiring the infrastructure that delivers information throughout the enterprise is sort of like enhancing the human body's central nervous system—it doesn't turn everybody into clones with the same capabilities, it enhances the characteristics of who you already were.*

However, a significant portion of Carr's thesis is dead-on accurate. Today, the vast majority of popular packaged software applications, off-the-shelf integration and middleware products, and information technology approaches are largely commodity products that offer incremental value to corporate consumers. Indeed, the IT industry had reached a point of diminishing returns because nothing was changing the rules of how technology worked. IT was simply about function points, interfaces, and custom coding requirements.

Actually, the whole notion of *information* technology was a misnomer to begin with; as Carr implies in his paper, we probably should have called it *data* technology!

Carr, like other skeptics, is standing at point A in Figure 1.1 and forecasting the inevitable commoditization of data processing technology,[2] which will lead to point B. However, the rise of new technologies and approaches has already resulted in new rules and a new curve—information processing. Emerging technology approaches discussed in this book, such as the application of semantic tools, technologies, and methodologies, will deliver us into the technology world beyond data—toward a real information technology industry.

[2] This S-curve example is similar to Clayton Christensen's (Innovator's Dilemma) concept of sustained technical change within a value net—in this case we are referring to the constant innovation in the IT industry as a whole, and suggesting that strategic competitive advantage remains possible by out-maneuvering competition and adopting new technology at the right time.

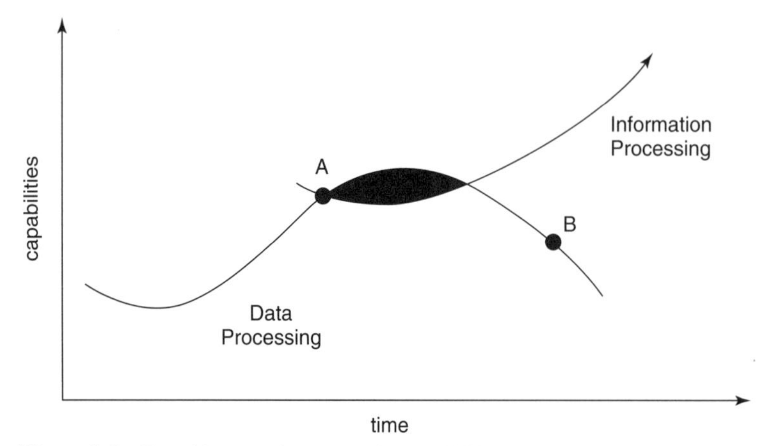

Figure 1.1 Sigmoid curve of IT processing capabilities

SEMANTIC INTEROPERABILITY: 15-YEAR OVERNIGHT SUCCESS

Semantic interoperability may initially sound downright foreign to some casual readers. However, it is grounded in long-established concepts of information theory and enterprise software practices. It will become the next 15-year overnight success in the information technology industry. The discipline of data semantics has broad and far-reaching applications in a number of software solutions. From work on the Semantic Web to emerging database structures, the concepts underlying the use of semantics in digital systems are having an immense impact on businesses industry wide.

> ***Definition: Semantic Interoperability***
>
> Semantic interoperability is a dynamic enterprise capability derived from the application of special software technologies (such as reasoners, inference engines, ontologies, and models) that infer, relate, interpret, and classify the implicit meanings of digital content without human involvement—which in turn drive adaptive business processes, enterprise knowledge, business rules, and software application interoperability.

Semantic Interoperability vs. the Semantic Web

The term "semantic interoperability" is not interchangeable with the term "Semantic Web." Much of the work on the Semantic Web is focused on the ambitious goal of allowing relatively ubiquitous and autonomous understanding of information on the Internet. Semantic interoperability, on the other hand, represents a more limited or constrained subset of this goal. More immediate—and many would say sufficient—returns can be gained by using semantics-based tools to arbitrate and mediate the structures, meanings, and contexts within relatively confined and well-understood domains for specific goals related to information sharing and information interoperability. In other words, semantic interoperability addresses a more discrete problem set with clearer defined end points.

It may be an overgeneralization to say that this problem subset fits the classic 80–20 Pareto principle, but nevertheless, the benefits in solving challenges within this subset are immediate and cannot be denied. A more apt differentiation between the two might be that semantic interoperability represents a 3- to 5-year vision whereas the Semantic Web represents a 5- to 10-year vision. The concepts surrounding semantic interoperability share many common terms, technologies, and ideas with the Semantic Web, but they also contain principles that have been developed and proven separately from Semantic Web initiatives. As semantic interoperability gains visibility, though, it is quite likely that there will be further blending of the concepts, protocols, and tools of the two.

Insider Insight—Stephen Hendrick on Why Semantics Now

Semantics are important because they help us cope with diversity. Diversity runs rampant in the Information Technology (IT) industry where new languages and standards are continually being introduced but rarely retired. Until recently, most vendors and organizations weren't concerned with this mushrooming diversity. However, as IT has progressively crept into most dimensions of our lives, the software asset base has expanded rapidly creating a software complexity crisis due to the large number of disparate standards, applications, languages, and datasets that were created without an eye toward interoperability and integration. The burden created by the software complexity crisis now means that most organizations spend more time maintaining and integrating existing applications than building or implementing new applications.

XML and Web services are considered an important first step in addressing interoperability. While there is no doubt that these technologies and their descendants are constructs that will aid in solving the software complexity crisis, their existence has simply brought us face to face, and far sooner than we expected, with the hard problems posed by diversity and the challenges of integrating disparate applications.

Addressing and resolving issues of diversity and integration involve building meaningful relationships between objects. Building relationships applies to both data objects and process objects. The value in building these relationships is the ability to coherently link together all aspects of the value chain that previously were isolated by virtue of their age, architecture, or implementation. The need for more structure in IT development tools is a phenomenon that has been taking shape for two decades. The widespread adoption of relational and object-relational databases as well as the pervasive shift to object-oriented programming languages specifically identifies the importance of being able to define intricate object-based relationships and hierarchies. Recent industry experience with objects and relationships is however simply a prerequisite for addressing the far more complex task of building granular relationships between complex objects comprised of many heterogeneous databases and systems. The quid pro quo is that our legacy

Continued

of widely disparate heterogeneous software assets in their current form can only be transformed into a coherent integrated array by first establishing what relationships exist in a semantically structured way. However, the power which results from a large scale unification of IT assets—first intra-company and then inter-company—will revolutionize IT and expand B2B and B2C relationships by several orders of magnitude. As this unification continues, it will dramatically simplify the way each of us relates to this far more expansive and semantically rich IT environment, laying an extensible foundation for building even better semantically structured relationships in the future.

Stephen D. Hendrick
Vice President, Application Development and Deployment
International Data Corporation (IDC)

Differing Uses of the Term "Interoperability"

The use of the term "interoperability" has become increasingly popular over the last several years, and its seemingly sudden appearance in the nomenclature has led to several differing uses and definitions. *Interoperability at its base level means using loosely coupled approaches to share or broker software resources while preserving the integrity and native state of each entity and each data set.* Common variants in use include data interoperability, Web Services interoperability, and information interoperability.

The term "data interoperability" is commonly used in conjunction with the communications frameworks used for connecting differing sets of emergency workers, such as first-responders. First responders are typically defined as the public service agencies—specifically local police, fire, and emergency medical, but also including federal and state law enforcement, emergency management, public health, National Guard, hazardous materials, search and rescue, and other specialized disaster recovery agencies or units. Although the information discussed in this book will have some relevance to this issue, specifically with respect to the access to various data sources, the primary problem set is one of mediating between data spectra and radio and data communication protocols and not necessarily addressing semantic conflicts.

"Web Services interoperability" primarily refers to ensuring compatibility between various Web Services protocols and frameworks. The rapid pace of innovation in the Web Services space reflects the efforts of companies to gain a competitive advantage over others which means that the ubiquity promised by web service interactions is under constant risk. As with data interoperability, this book may have some relevance but the issue regarding transport interoperability is more one of market forces as well as the correct implementation of web service specifications such as SOAP, WSDL, UDDI, XML, and XML Schema.

Information interoperability is the core subject of this book: Data alone is insufficient for successful software application interoperability—semantics turn data into information, and, in turn, the interoperability of semantics enables widespread interoperability of software components, processes, and data. At its core, information interoperability is the ability to address and mediate the logical nature and context of the information being exchanged, while allowing for maximum independence between communicating parties. This means addressing syntactic, structural, and semantic differences within structured, semistructured, and unstructured data. The larger definition encompasses greater transparency and more dynamic communication between information domains regardless of business logic, processes, and work flows.

Throughout the course of this book we will examine how this emerging discipline will drive business innovation through strategic value, and we will provide a detailed analysis of the technologies and techniques embodied in the emerging discipline. To make a complete case as to why this technology is important, we must explain the need, present forces at play in business, and identify the major barriers to strategic progress in information technology.

THE THIRD WAVE

We are at the cusp of what has been called the "the third wave"—an information society. Ten thousand years ago, the first wave, the agricultural revolution, launched a society no longer based on hunting and gathering, moving to one based on farming and a less nomadic lifestyle. The second wave, based on manufacturing and production of capital, resulted in urban societies centered around the factory. The new era, based on information, brainpower, and mediacentric communications is taking shape now. The future will reveal the profound social, cultural, political, and institutional effects this shift will have. The foundation of this new wave rests squarely on our capabilities to make information sharing ubiquitous and meaningful.

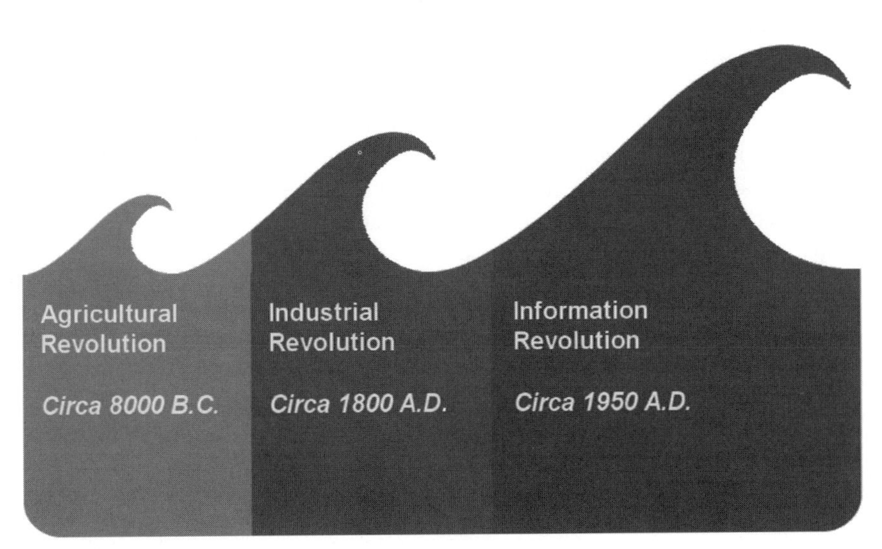

Figure 1.2 The third wave, an information revolution

Competition has been a crucial part of society for our entire history. There is no reason to think that the forces and motivation for competition will diminish along with the rise of the third wave. However, there is growing evidence that the fundamental *drivers* of competition will change as the basis for productivity changes. The decisive factors of production have shifted. In the past, land and capital were the dominant factors of production. Today, knowledge (of patents, processes, formulas, skills, technologies, customer information, and materials) is as important to businesses as land and capital were two hundred years ago.

IT can be used to improve competitiveness in a myriad of ways. Unfortunately, many of the applications of IT are quickly becoming commodities. E-mail connects us. Enterprise resource planning (ERP) systems monitor and store our company financials, human resource data, and supply chain information. Product data management (PDM) systems enable us to produce cars, airplanes, and other heavy equipment with greater efficiency and higher quality. The ubiquity of IT has reduced its impact on corporate strategy because companies have access to the same technologies, making it ever more difficult to use it in differentiated ways. Nonetheless, information used in computers is what keeps the world of business moving.

Strategies for Competitive Advantage

Today's marketplace is the most competitive and dynamic in all of history. Businesses are vying for competitive advantage, indeed survival, on many fronts. Systems in the value chain make businesses more efficient and reduce operating costs when they work together collaboratively—eliminating labor-intensive work and speeding responsiveness. Change is the only certainty in business. Changing environments create the need for businesses to adapt more quickly in order to stay ahead of their competitors. One way to meet this challenge is collaborative IT systems, which can enable businesses to change processes and procedures on the fly—thereby further differentiating themselves against their competitors. These differentiating factors can be leveraged to drive customer loyalty. By making a business's outward-facing demand-based systems more collaborative, corporate leaders can ensure that their customers always have the best service, information, and insight possible. Finally, innovators across industry will leverage collaborative technology to drive market expansion—growing the opportunity with new services and offerings that can expand their customer base.

Improved competitiveness can be accomplished in three primary ways:

- Improve cost savings by driving down expenditures
- Improve products, services, and organizational structures
- Create strategic advantage by exploiting new opportunity

Information technology can be a crucial enabler for each of these approaches. Despite IT's relative ubiquity in the marketplace, businesses can learn to leverage IT in unique ways that add to a core differentiation of the business.

The IT industry is in a slump. Recent innovations in technology, particularly technology available to business through IT vendors, have not offered innovations that shift businesses core capabilities in differentiated ways. Too often IT offers "me-too" functionality that is not substantially different from what competitors already have applied.

When competitive pressures drive business decisions, technology is more often a competitive barrier than a competitive enabler. Take, for example, three common business tactics used to drive competitive advantage:

- Merger and acquisition activity
- Corporate consolidation
- Accelerated IT spending

Far from sure things, these tactics are rife with challenges for businesses seeking competitive advantage. As the following sections point out, there is more than a little irony in the fact that businesses are both a master and a slave to the forces, successes, and barriers that information technology presents.

Insider Insight—Dr. Rod Heisterberg on E-Business Transformation

In 1995 something wonderful happened, something quite remarkable—a true phenomenon. During that year the whole world unilaterally adopted the same set of Information Technology (IT) infrastructure standards to share business data, the Internet. No governmental edict or channel master command forced this technological paradigm shift. In 1994, the Internet was still an arcane technology primarily known and used by academia, computer scientists, and the military. By 1996, businesses were being established and were generating revenue from commercial buying and selling transactions by leveraging the enabling technology known as the World Wide Web. This seminal event has forever changed the way we do business—now, electronic business (e-business).

Circa 1999 market share leaders in aerospace, consumer packaged goods, and high tech electronics industries had transformed their e-business models beyond electronic commerce (e-commerce) into Collaborative Commerce (c-commerce). The *Internet Encyclopedia*, recently published by John Wiley & Sons, defines Collaborative Commerce as a strategy for the next stage of e-business evolution with business practices that enable trading partners to create, manage, and use data in a shared environment to design, build, and support products throughout their lifecycle, working separately to leverage their core competencies together in a value chain that forms a virtual enterprise.

Sustainable competitive advantage may be realized by adoption of c-commerce strategies and business models. Rather than simply exchanging procurement transactions as with e-commerce practices, leading enterprises are executing c-commerce strategies to share intellectual capital with their trading partners working as a value chain that provides a competitive advantage for the development and distribution of their products. These collaborative business practices are enabled by semantic interoperability, which is fundamental to realizing e-business transformation. The convergence of business process reengineering and Internet technology that spawned e-business during the 1990s has set the stage for reengineering the resulting management decision-making processes to promote c-commerce as the dominant business model of this decade of the 21st century.

Rod Heisterberg, PhD
Senior Consul for ManTech Enterprise Integration Center

Mergers and Acquisitions

Irrational exuberance and hyperinflated expectations during the 1990s drove capital markets to new highs and increased the desire of business leaders to acquire new profit centers. Even though the number of announced transactions has been lower since then, activity is still at a comparatively high 1996 level and projected to remain fairly steady in the years to come.

Business activity surrounding mergers is important to the discussion of semantic interoperability. M&A activity constitutes a great deal of impact on IT systems for all the parties involved in the transaction. Frequently, mergers have financial plans that set out a projected savings based on efficiencies derived from the synergy of the combined entities. These efficiencies rely on the successful integration of a diverse set of IT systems that typically include financial systems, human resource systems, manufacturing systems, and other mission-critical enterprise systems. Rarely are these business or technical synergies achieved.

When the companies do merge, there is a period of time during which they are legally one entity but actually operate internally as two or more discrete businesses. Typically this situation is maintained for only a few quarters while the business units work through the logistics of integration.

Frequently a number of difficulties arise with the process of integrating IT systems between the newly wedded companies. Solving these issues typically presents problem solvers with one of three IT-centric solution choices:

- **Systems Integration**—Massive spending goes toward the integration of disparate systems so that they can effectively use each other's data and processes.
- **Systems Consolidation**—Extensive effort is expended to move mission-critical data from one system to another. Sometimes, such consolidations are not entirely possible when there are significant differences in the information stored in two or more systems.
- **Systems Fragmentation**—Let each system continue to function in the manner in which it was designed: If it ain't broke, don't fix it. This usually results in unnecessary duplication of human resources and system resources and significant complications with data access and knowledge dissemination

Perspectives: Dozens of ERP Systems per Enterprise?

Consider a large electronics manufacturing company that underwent five merger and acquisition transactions in 18 months, and several more over three years. Its information technology infrastructure was a target-rich environment for consolidation and integration. With over eight financial packages, five human resource systems, and four major product management systems it was having serious trouble realizing the efficiencies that had been promised during the M&A transactions. Different brands, such as Oracle, SAP, PeopleSoft, MatrixOne, Agile, and version numbers R3 vs. R5) further convoluted its system architecture.

Situations like this are more common than many would think.

The bitter truth behind these expensive alternatives is that they are repeated for every new merger or acquisition. Constant change is the speed at which business operates, yet current technologies require months of time and massive expenditures to keep up!

Businesses that can respond to changes more effectively will most certainly earn a competitive advantage in their marketplace.

Corporate Consolidation

Corporate consolidation—sometimes called downsizing or right-sizing—is a business reality with tangible effects on IT infrastructure. Large organizations often turn to business integration as a way to streamline processes, cut costs, and provide better value to customers.

Consolidating IT systems requires extensive analysis and development effort to unify disparate business rules, work flow, and data definitions. Today this is a very labor-intensive process that cannot be automated because of the rigid nature of most legacy systems—human experts must be involved to make sense of poorly managed technical infrastructures.

Businesses often grow in fits and starts. One consequence of this is the fragmented nature of their IT infrastructures. Consolidation efforts are frequently required to impose order on a long legacy of chaos. These efforts can work on smaller scales, but major problems can occur if too much is taken on at once. Significant systems consolidation efforts can stall out, and other solutions—like starting from scratch—are pursued.

Perspectives: To Consolidate or Rebuild?

A large personals and dating website grew rapidly in the early years of its formation—resulting in a fragmented and chaotic architecture for its systems. When undertaking the effort to consolidate systems, as a mature business, it encountered an entrenched architecture that could not be modified in a cost-effective way. A separate team was formed to begin the process of architecting new systems from the ground up—in parallel with its live systems.

Sometimes rebuilding is the best alternative.

Although the reasons for consolidation will no doubt change throughout the years, business drivers that encourage executives to choose consolidation will be with us for quite some time. Unfortunately, today's popular integration and consolidation technologies do little to make this process easy or repeatable.

Increased IT Spending, Increased Agility?

Popular wisdom during the late 1990s had executives spending more than 50% of their annual budgets on technology[3] hoping for increased productivity. A recent survey shows that only 10% of industries show a significant correlation between IT spending and productivity.[4] Yet despite evidence to the contrary, business leaders still envision information technology as a key enabler for achieving long-time goals of becoming a real-time, event-driven organization that can keep pace and even lead industry changes. Faster availability of information on market conditions, customer satisfaction, and product performance are still enticing goals.

In the past, typical cycles for market intelligence and other trends analysis took months and sometimes quarters to process and report on. But as information technology is increasingly connected to remote information providers (customers, products, market indices) businesses are taking advantage of the new information supplies to create real-time views into the health of their organization. The speed at which a company can acquire, process, and understand information about itself, its customers, its suppliers, and the market is in direct proportion to its relative advantage over its competitors. In other words, the better you get at understanding information in real time, the bigger your advantage can be over your competitors.

Ultimately, a commitment to using innovative information technology can be a key enabler for the agile business. For example, the 10% of businesses that showed positive correlation between IT spending and productivity effectively used information technology to drive significant innovations within their industry.[5] An agile business is a business that creates new market opportunities and first outs of downturns. Commitment to using IT innovatively is embodied in a variety of forms including staff training, technology initiatives, and corporate culture—not just a measure of IT spending. But today, the most significant barrier to effective use of IT in the business is nearly insurmountable: end-to-end development speed.

End-to-end development speed is the total time it takes from executive sign-off to deployed functionality. It encompasses the time used for analysis, design, development, and several levels of testing. For most medium and large initiatives, development speed is measured in months, not weeks.

Soapbox: The "My Product Does It All" Syndrome

One of our biggest peeves with the EAI industry is the insistence of some vendors that their software is the right solution for all applications. But EAI and Web Services tools rely on a software platform that requires development of new configurations, adapters, and message formats for every new "node" in their proprietary networks.

[3] *Technology Spending.* CIO Insight, June 2002.

[4] US Productivity Growth, 1995–2000. McKinsey Global Institute, October 2001.

[5] *Does IT Matter? An HBR Debate.* Hagel, Seely-Brown, June 2003.

For example, if the FBI becomes aware of a threat to the port of Sacramento, California it might need nearly instantaneous connectivity to a wide array of state and local IT systems that track shipping and persons information. If the FBI doesn't have access to that data in its own systems, making that data available could take weeks or months—hooking into various protocols, data formats, vocabularies, and query structures.

EAI vendors need to understand where their solutions fall short. Only something radically different from the same old EAI or Web Services solutions will make a difference for these kinds of real-time needs.

Suppose a major manufacturing company has a new business partner. The manufacturer wants to streamline processes—enabling each person in the supply chain to view and modify invoices inside his own systems, rather than having buyers, sellers, assemblers, and engineers work over the phone, fax, or E-mail. This is a pretty straightforward scenario that could take 6–12 months to accomplish inside most legacy supply chain environments.

For executives this delay often represents the hidden costs associated with partnerships, new suppliers, and outsourcing. This limitation can severely hinder the organization's competitive advantage by delaying the creation of new capabilities and economies of scale.

The bottom line is that the value of IT is not in the size of an IT budget. No matter how innovatively a business approaches IT spending, the barriers to rapid deployment of new capabilities remain a significant impediment to the business's ability to execute its core strategies.

BUSINESS AND APPLICATION IMPACTS

The impacts of the third wave will be both broad and deep. As more businesses go global and more businesses go on-line, information access and timeliness are crucial to successful strategy.

Key Industries and Vertical Markets

Industries affected by information technologies are not limited to software or computer manufacturers. Today, nearly all businesses are information intensive. From the automotive industry and its reliance on manufacturing information and the supply chain to shipping businesses with sophisticated logistics tracking programs, big companies use IT systems for all aspects of the business.

Information-intensive industries include:

- Automotive
- Aerospace and defense
- Heavy and light manufacturing
- Life sciences

- Finance
- Professional services
- Government (federal, state, and local)

Key Processes and Application Areas

Within a given industry, indeed within a given business, several processes operate more or less independently of one another. They may deal with customer management, supplier management, or product development. Each process can have dependencies on other related processes, but in and of themselves they usually span multiple, geographically disperse software applications that are united by a common domain space.

Information intensive systems and processes include:

- Supply chain management (SCM)
- Demand chain management (DCM)
- Customer relationship management (CRM)
- Collaborative product commerce (CPC)
- Collaborative planning, forecasting, and replenishment (CPFR)
- Product data management (PDM)
- Enterprise resource planning (ERP)
- Decision support systems (DSS)
- Industry exchanges and portals

It is an understatement to say that information is central to these processes. In fact, these business processes rely on the digital information that flows within their boundaries. However, current technologies that apply to these processes cannot provide real-time mechanisms for handling dynamic changes in the information. This means that most current IT infrastructures cannot fully cope with the complexity of their business environments or the volume of information within them.

THE INFORMATION EXPLOSION

The world we live in today is unlike anything ever experienced in the past. Our global economy runs on no fewer than 200 billion lines of legacy code,[6] and our world population produced over 5 exabytes of content in 2002.[7] This is equivalent to adding half a million new libraries the size of the United States Library of Congress full of printed information in one year: simply staggering. In 2002, so much content was created that it was as if every person on the planet generated over 800 megabytes of content herself.[8]

[6] Rekha Balu, "(Re) Writing Code," Fast Company, April 2001.

[7] How Much Information, Berkeley, 2003 (www.sims.berkeley.edu/research/projects/how-much-info-2003/).

[8] How Much Information, Berkeley, 2003 (www.sims.berkeley.edu/research/projects/how-much-info-2003/).

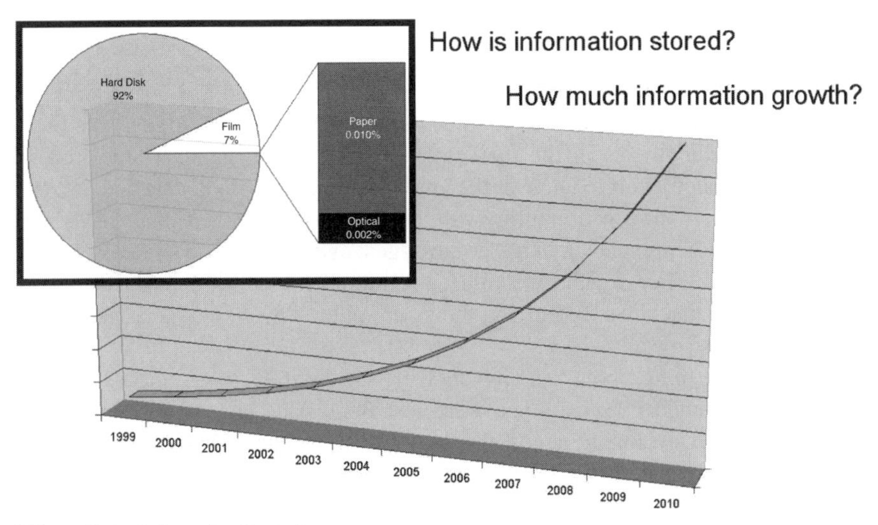

Figure 1.3 Information Explosion

In the three years from 2002 to 2005 more data will be created than in all of recorded history up to 2002.[9] Humanity's economic progress seems inexorably tied to how we create, disseminate, and use information about the world of business. Meanwhile, our tools for making that new information meaningful are severely outpaced.

Information Matters More Than Ever

As a response to this incredible growth, businesses are driving the evolution of software to support data *and* semantics together—which is actually the beginning of information. This capability is crucial to survival and growth. Businesses that effectively use the data they already have will reap the rewards, while those who flounder will be left behind. The interesting things about data and information that matter to most businesses are visibility and configurability of processes, business rules, and key data values. This need will drive development of newer kinds of dynamic information infrastructures. To fill this gap, a richer method for working with enterprise information must be created based on enabling the semantic interoperability of network services and components.

Ultimately, technology by itself doesn't matter to the business—but information does—and as long as information matters, the evolution of *information* technology will, too.

[9] Information Integration A New Generation of Information Technology, M.A, Roth. *IBM Systems Journal*, Volume 41, No. 4, 2002.

Data Rich, Information Poor

Nicholas Carr's article on the death of data processing should serve as a wake-up call to vendors and proponents of the same old middleware technologies. The focus of tools and technologies in the marketplace today is still on data and process. Without a dedicated effort to uncover the implicit meanings behind data and process, finite limitations to information utility will always exist. IT professionals have been too far down in the trenches looking at messages, events, and data to really see the forest for the trees. For decades we've been selling IT middleware products that are basically simple machines for moving things around. As they say, "garbage data in, garbage data out."

The hardest thing about moving on will be admitting to customers that the products they bought five years ago were just Band-Aids on a wound that really needs penicillin. Developing new technologies to move away from the datacentric software paradigm will be easy; many of them already exist. In fact, much of the software behind the semantic interoperability evolution has been developing in labs around the world for more than a decade.

Information Dominance

It is clear that information technology does still matter. But to really get competitive gains from information technology, businesses must also invest in technologies that make information more useful. In effect, this is an investment in information dominance over competitors. This goal can be difficult to justify to cash-strapped financial officers.

A delicate balance is needed between more traditional technology investments (like hardware, desktops, servers, ERP systems) and newer investments (like semantic interoperability, ontology development, and modelcentric development). A successful balance requires a targeted and creative application of IT budgets unique to individual corporate entities.

Balancing these priorities will lead to significant IT innovations and ultimately to competitive gains for the companies who invest in them. Information dominance is truly the key strategic objective for businesses. The struggle to achieve and maintain information dominance will play out as the crucial battle in capital markets. However, longstanding barriers to IT innovation will have to be overcome to really see substantial progress. It is these barriers, which stand in the way of this next great wave of technology improvement, that go unaddressed in the vast majority of vendor products on the marketplace today.

SYSTEMIC BARRIERS TO IT INNOVATION

The next great hurdle in enterprise computing is interoperability among the multitudes of core systems that enable businesses to function. In addition to the many known technical issues with certain middleware technologies, there are several strategic issues that cut across all approaches to solving the interoperability challenge. These big issues are the most significant barriers to achieving an information technology infrastructure that will create competitive advantage indefinitely into the future. The same old approaches will run into the same roadblocks unless fundamentally new concepts are introduced for handling enterprise information.

The Speed of Change Barrier

The speed of change barrier is the gap between the rate of change in business and the ease of changing the IT infrastructure. The more quickly market conditions and organizational structures are changing, the more difficult it becomes to change the IT infrastructure. Therefore, when the cumulative effects of IT on a business enhance its agility, it becomes more difficult for that business to gain additional flexibility and agility. The closer an IT infrastructure comes to the actual "speed of change," the more effort is required to make additional gains. This behavior is crucial to understanding the business's ability to rapidly scale vertically and horizontally, meet new demands from customers, partners, and suppliers, and manage IT spending.

More than ever, business and governmental organizations are asked to move very quickly. Whether it is federal agencies responding to crisis or corporations forming ad hoc alliances to capture new markets, the demand for rapid, fluid, and frequent alignment of organizations is on the rise. Accordingly, IT infrastructures must evolve to meet these new demands, and a number of enterprise architectures have emerged as candidate solutions for this problem. Business pundits call these visionary solutions by a litany of different terms, but a common characteristic for all of them is their heavy reliance on technology as the key enabler for managing rapid changes in a business environment.

However, IT's actual capabilities leave business leaders wanting. In practice, the amount of effort required to take on even the most trivial of tasks is monumental. This is so because of a number of factors such as different corporate stake-holders with different interests and the sheer volume of customization and code-writing that must be done to accomplish any given enterprise technology solution.

One way to look at the problem is to attach figures to the costs associated with providing interoperability of information systems—and to determine a traditional cost/benefit trade-off. But, as implied above, a more accurate measure of the cost is to analyze how much opportunity is lost by *not* doing the integration—now and in the future.

This notion of real-time scalability is best understood by imagining the opportunities a large business has every day to streamline its value chain, connect to partners, listen to customers, and create new value. Hundreds of small opportunities get overlooked each day because middle managers know that building a business case and lobbying decision makers is a time-consuming effort with an unlikely payback—executive approval and a secured budget.

Now imagine that the business had an infrastructure that dynamically adapted to new requirements for communication, connections, and integration. Imagine that it only took minutes to look up, connect, and establish interoperability, rather than taking months to go through design, development, and testing phases of most integration efforts. Costs per integration would be drastically reduced, and partners, customers, and suppliers could be brought closer to the business.

Experienced technologists reading the previous paragraph are likely to be laughing right now, because even if a technology existed that could dynamically adapt itself and maintain a real-time network infrastructure, there is another significant barrier to integration nirvana that is often the cause of long development cycles—entrenched enterprise system infrastructures.

Entrenched IT Infrastructures

Most large businesses do not have the luxury of modern packaged applications that already speak Web Services and XML. The dominant landscape today, and for the foreseeable future, is that of large legacy systems with thousands of business rules, data formats, and program architectures. These systems were not designed to share information.

Often called "silos of information," these legacy systems are written in a wide variety of languages including COBOL, C++, Fortran, and proprietary languages. More significantly, the processes and data definitions that these systems use are widely divergent and frequently understood by only a few within an organization.

These entrenched IT systems can be neither replaced nor connected together with ease. Inside large manufacturing businesses it is common to have dozens of incompatible systems, each responsible for a different product family. Sometimes the only cost-effective way to accomplish a basic level of integration is to perform swivel chair integration techniques over a sneaker net.

Definition: Analog Integration Techniques

Swivel Chair Integration—integration that is accomplished by a human being who sits in a chair that swivels between different computer terminals and keyboards and manually enters information into the different systems

Sneaker Net—the network between computer systems that is connected by human beings in sneakers, or some other footwear, running between them with new information

These entrenched IT systems represent tangible technical barriers to widespread availability of truly dynamic information sharing. However, even if these technical challenges are solved, there are still more significant barriers.

Fuzzy Data, Fuzzy Process

Even for businesses that have the luxury of modern systems that speak Web Services and XML, a serious and expensive barrier to dramatic improvement exists:

a severe lack of explicit knowledge about business processes and associated corporate information. Process and corporate knowledge is frequently intangible, difficult to locate and associate value to. This slipperiness of enterprise information and processes is the most difficult problem to understand and fix.

The specific meaning of business processes and information inside a company remains black magic to outsiders. Employees steeped in their own culture and practices are responsible for creating and maintaining the entrenched IT infrastructure of the business by using what they know of the folklore of the organization to determine how data is structured, processes are defined, and business rules are implemented.

To illustrate this, consider that there is no single "best" way to model data for IT systems. Given a parts inventory system for the same shop floor and the same parts, fifty different data modelers would model it in fifty different ways. The modeler's own background, combined with his assumed understanding of the organization, results in differences with the models. Although many patterns for good data modeling are understood, even the smallest of subtle differences in the models can result in absolute incompatibility among them.

Even with packaged applications, culture and folklore of the company influence how they are used. Supply chain systems, financial systems, product data management systems, and all other ERP packages require significant amounts of customization to the business rules, data models, and application code. ERP vendors encourage this customization because they know that every business, every organization, every process is different. However, the result of this customization means that even identical versions of an ERP application installed at nearly the same time for two different organizations will not automatically interoperate—customized integration is required.

Perspectives: The Customized Packaged Application

By the late 1990s the federal government had made significant steps toward modernizing its IT systems. Multiple instances of SAP R3 were deployed by several Department of Defense organizations. However, like many large businesses, the government quickly realized that it had another significant challenge on its hands—none of the customized SAP systems could natively talk to one another.

Despite the fact that the versions were the same, the amount of customization required to implement each SAP system made it impossible for them to communicate with each other.

This spawned additional federal programs.

It is difficult to overestimate the severity of this little-recognized truth. Every engineer who participates in the creation, configuration, and deployment of an application contributes to that application's uniqueness. Yet each bit of uniqueness must be accommodated for that system to communicate effectively with other systems.

Humans deal with fuzzy logic very well. We can walk into new situations, find patterns, react, and adjust as we go without completely halting the entire communication process. Unfortunately, IT systems cannot do this. Their architectures are based on the precision of 1s and 0s in expected sequences. This is why an SAP R5 ERP package cannot understand an XML document with an unknown schema. It's why an application developer might use the SAP IDE to write an XSL/T script to convert the unknown type into a known type.

Writing code to accommodate these kinds of situations is the standard response from IT vendors. Sometimes they might include graphical tools, perhaps with some drag-and-drop functions, to assist with writing that code—but the final result is always some type of custom script or compiled object to accomplish an integration task.

Each company's folklore, culture, fuzzy process, and data combine to create the single most challenging barrier to IT evolution.

The "H" Factor

Humans themselves are the single largest bottleneck to adaptive IT systems. Humans must be involved in the finest of details because of the hands-on nature of resolving data and process meanings. We are still the bottleneck for time, costs, and quality. The foundational elements of computing are still driven procedural behaviors that are codified by human programmers. This means that software will continue to answer the questions of how data gets processed and what it will do with the data—but it will not answer the question of *why* it is performing a given set of logic or *why* it should convert data this way instead of that way.

Handling the question of why is crucial in computing because it gives the IT system the opportunity to automate some things on its own. Shifting the burden of programming enterprise interoperability routines from humans to machines can make a dramatic impact on all of these barriers to strategic innovation in IT. For software to handle the questions of *why* it must begin to interpret the semantics of the data it is processing. Semantics are the foundation of all other substantial improvements in information technology and the key to unlocking information from data.

SUMMARY AND CONCLUSIONS

IT matters now more than ever. Persistent challenges to business demands greater flexibility and dynamism. Challenges ranging from the volume of new enterprise data, to the inherent rigidity of custom-coded applications and the increasing demand for real time business change require new ways to manage the information inside the technology. We've seen how key industries and business mergers, acquisitions, and consolidation continue to demand more agility and more flexibility at greater speed. In short, the need for an adaptive enterprise infrastructure persists. Thus far in the book we have examined the systemic business pain points that drive the need for better technology. In the next chapter we will take a look at some traditional solutions that, despite the hype, fall well short of solving these business needs.

Chapter 2

Information Infrastructure Issues and Problems

KEY TAKEAWAYS

- Traditional middleware solutions are built on outdated technology.
- Today's businesses are wising up to the importance of the information contained in the messages—as opposed to the technology surrounding the messages.
- Plumbing, or integration, is necessary but not sufficient. Without it, enterprise software communication would not be possible. With it alone, communication is inefficient, expensive, and lacking focus on what matters most—information.
- The real challenge for enterprises today is to build an infrastructure that enables information, in all formats, to be utilized freely at the right time and place.
- Dynamic, adaptive middleware architectures will rely on semantic interoperability.

Defining holistic middleware architecture may not be the first order of business for most CEOs, but cutting costs and improving business relationships surely are. So, although the question of semantic interoperability may be technical geek speak for most, software architects and CIOs should take heed of this emerging vision. It could make you a corporate hero.

In this chapter we will examine the roots of application integration solutions and discover the reasons why they so frequently become money pits. Key technical components of typical EAI solution sets will be discussed, highlighting specific areas that lead to high maintenance costs and problematic configuration management. It will be made clear why enterprise information, in the form of structured, semi-structured, and unstructured content, should take primacy over all other technical middleware concerns. Finally, semantic interoperability will be introduced as a necessary complement to typical integration solutions because it can alleviate the cost of deploying existing integration solutions.

Adaptive Information, by Jeffrey T. Pollock and Ralph Hodgson
ISBN 0-471-48854-2

LIMITATIONS OF CURRENT TECHNOLOGY APPROACHES

Failed integration projects are common today. The sad fact is that the vast majority of enterprise efforts to link trading partners, unify corporate data, extend business processes, and enable IT system collaboration end with unsatisfactory results. Analysts report that over 80% of all data integration projects have failed or significantly overrun budgets.[1] These numbers must surely be taken with a grain of salt, but they do indicate that something is amiss.

Off-the-Shelf Integration Products

Enterprise application integration (EAI) and business process management (BPM) companies offer integration platforms that provide adapters, transport, work flow, and transformation capabilities to companies seeking to integrate disparate applications.

EAI approaches have been successful because they create a centralized management paradigm for controlling information flows throughout the enterprise. This is vastly improved from the era before EAI, when no alternatives to spaghetti integration existed at all. Additionally, the clear separation of architecture layers in most EAI tools enables these vendors to consistently create new value-added services such as B2B, I2I, A2A, supplier networks, analytics, etc.

However, EAI-type approaches are struggling. IT managers have found problems that become impediments to widespread adoption. The typical EAI solution requires a very tightly coupled environment, which severely restricts the flexibility, agility, and adaptability of the integration framework. This disadvantage negatively impacts the portability of the business and process data contained within the framework.

In addition, EAI solutions include adapters that require significant amounts of custom code to facilitate any integration that is not supported by prepackaged templates—which turns out to be most integrations. For EAI or BPM solutions to work properly, custom code must be written to connect every application within the system, at the data layer, the adapter layer, or both. This custom code is extremely expensive to implement and maintain.

Furthermore, when a company selects an EAI application it is virtually forced to remain with that one vendor to continue integrating new applications—which usually means buying more services, tools, and adapters. Total cost of ownership (TCO) is very high for most EAI solutions because the solutions are tied so heavily to a business's processes. Every time an application or business process is modified, changes must be made to the integration framework.

After years of buying EAI solutions, Fortune 1000 corporations and government agencies have created an entrenched industry that cannot be easily displaced. Alhough EAI technologies are entrenched, they are not the final word for linking

[1] The Big Issue, Pollock, *EAI Journal.*

incompatible business systems. The ongoing focus of most EAI vendors on proprietary process-based solutions detracts from their overall utility, and future promise, precisely because TCO customer concerns often derive from other causes.

Web Services and Service-Oriented Architectures

The Web Services framework has recently taken the computing industry by storm. Sun, Microsoft, IBM, and most other large software vendors have embraced the concepts and languages that underlie the Web Services model. The combination of UDDI, WSDL, and SOAP forms a triad of technologies that will shift the entire market toward service-oriented architectures (SOA). Together, these technologies provide directory, component lookup, and exchange protocol services on top of an HTTP or SMPT network protocol. This capability translates into a loosely coupled physical communications channel for moving messages around.

However, Web Services are not without shortcomings. From a business perspective service-oriented architectures do not yet solve several crucial aspects of the integration problem such as robust transactional support, adequate security features, improved directory services, and sufficient architectural soundness to provide mission-critical performance.

Perhaps the most significant improvement opportunity for Web Services is in the area of information management and schema transformation. Fundamentally, Web Services technologies handle messages in a loosely coupled manner, but they will not enable the recipient to understand the message that has been sent. With Web Services this part of the exchange relies on custom-coded solutions or widespread community agreement (rarely achieved) on some kind of document exchange standard.

Therefore, the recent and rapid adoption of Web Services continues to highlight the pressing need for semantic interoperability technologies.

Data Warehouses and Metadata Management

Performing systems integration with a data warehouse involves creating a central database that is the primary owner of enterprise data—and the unified data source for client applications. As a way of enabling applications to share information, the data warehouse appears conceptually sound, because it puts a strong emphasis on collecting and unifying enterprise data.

However, implementing this approach leads to a series of challenges that undermine this solution. For one, creating a unified data model for the central repository is extremely time-consuming and generally not adaptable to change. For another, homogenized data is frequently not as valuable because context and relationships are typically lost. Finally, updating and maintaining the data in a central repository on a real-time basis is extremely difficult because of the diversity and complexity of layering multiple business process rules and the cleanliness issues of the data. Because of these shortcomings, data warehouses are predominantly used for archiving historical data, mining data, and performing trend analysis, and not as an application integration solution for operational data.

Some companies are beginning to provide metadata management capabilities that involve moving, viewing, and extracting data from databases. Typically, such software solutions offer some query capabilities and the ability to mediate requests from a number of different data stores. In addition, many of these companies focus directly on analytics and decision support systems.

From a technology perspective, these kinds of solutions are not ideal for a generalized integration platform in the enterprise for a number of reasons. Metadata management tools tend to be database-centric: built on relational database systems, usually ignoring the need for API-based integration and sometimes even ignoring robust XML support. Their vision of metadata is frequently limited, two-dimensional, and fails to include the much richer form of environmental metadata that a robust integration solution must accommodate.

Portals

A portal is sometimes used within an organization, or across organizations, as a single entry point for a number of systems behind the scenes. Leading edge companies that provide portal-based federated search and retrieval capabilities typically have serious limitations with regard to data variety and deployment options. Technologies that do exceptionally well at searching, pattern recognition, and taxonomy generation on unstructured data don't work very well with structured sources. Generally speaking, these tools lack sophisticated ways to account for embedded contexts (by way of the inherent structure of RDBMS, object, or hierarchical data sets), native metadata, or precision in structured query formatting. Aside from technical limitations, concerns with vendor lock-in and handling machine-to-machine interoperability (as with traditional EAI) still go undressed with the portal-based approach.

Systems Integrators—Custom Solutions

Systems integrators get nearly any IT job done. Major consultancies like BearingPoint, CGEY, AMS, IBM Solutions, and Lockheed Martin Consulting are the masters of brute force approaches to solving technical challenges. In fact, they have a vested interest in solutions like this because they often help create one-off point-to-point integrations—which they might be contracted to maintain for years to come. This is possibly the least efficient approach for a technically advanced, elegant, and easily maintainable technology solution, but, sadly, it is probably the most common IT integration solution.

Standard Data Vocabularies

As an approach to application integration, standards-based interchange involves agreeing on a specific standard data format for recording and storing information. One of the earliest and best known attempts at standards-based interchange is EDI, used to exchange information among trading partners. More recently, many existing and newly developed standards have shifted into the more modern technology of XML. Whereas EDI was used almost exclusively for intercompany data exchange, XML standards are evolving as a way to exchange information both internally and across multiple enterprises.

Using a standards-based approach means that each standard develops based on input from a wide community and typically has cross-organization collaboration and support. Such a standard represents advance agreements on the syntax and meaning of a given set of exchanges. However, standards have a difficult time responding to business and environmental changes that occur rapidly in most industries. Furthermore, the politics surrounding standards creation often undermine their ubiquity and effectiveness. Perhaps the largest problem with many standards is that local business context may be lost because each system that uses the standard has to convert its information to a common, specialized "view of the universe." Historically, the standards-based formats either have proliferated almost out of control (XML-based vocabulary standards) or have been very narrow in focus (X12, EDIFACT), creating a problem for IT managers attempting to decide on one or only a few.

Many people mistakenly believe that XML will solve all application integration problems. The reality is that XML is simply a mark-up language that enables information to be "carried" from one incompatible system to another. The use of XML alone will not resolve differences in how information is processed by different systems (i.e., how the "semantics of the information" are interpreted). XML is simply one piece of the puzzle, not a complete solution. Standards have their place in the IT toolkit, but standard vocabularies will not be the primary vehicles of information between systems that need a flexible and robust method to communicate with each other.

TRADITIONAL APPLICATION INTEGRATION

In the early 1990s MRP/ERP vendors were making a killing by selling the promise of integration-proof monolithic enterprise systems. The idea was that if the enterprise would just buy all its software from one vendor, there would be no need to integrate anything. But a threat to their plan for world domination was looming on the horizon. Message-oriented middleware (MOM) arose out of the telecommunications industry with the kickoff of the Message Oriented Middleware Association (MOMA) in 1993. Even as early as 1985 commercial work was beginning on DACNOS (IBM and the University of Karlsruhe), software that utilized an asynchronous message-driven communication model.

As initially conceived, the MOM was intended to solve the messaging problem with a simpler approach, document-like messages, than CORBA—another key mid-

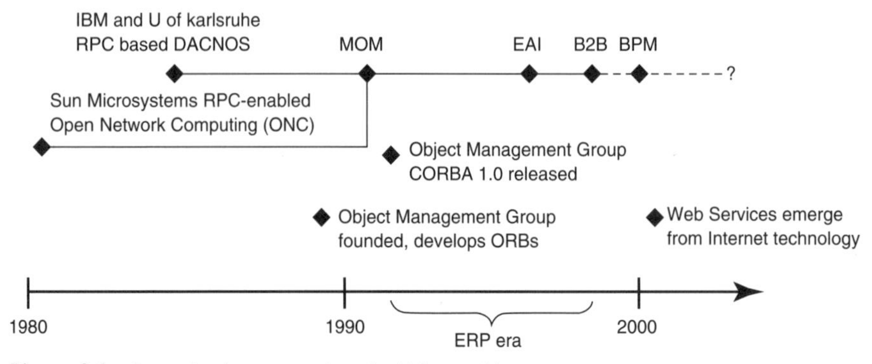

Figure 2.1 Synopsis of message-oriented middleware history

dleware technology of the day. Early MOM technologies eventually adopted many extract transform and load (ETL) capabilities and spawned a new service offering that, in 1997, was renamed enterprise application integration (EAI) software. EAI companies began to offer more comprehensive end-to-end platforms with adapters, transport, and transformation capabilities to companies that needed to extend the reach of their existing systems. More recently, EAI companies started to layer even more management tools, like partner, process and standards management, on top of their transport systems in an attempt to rebrand as B2Bi or process integration vendors.

Modern EAI approaches have been successful because they have adapted to the needs of exchanging data within business processes and created a centralized way to control information flow from a single point—instead of through distributed code and granular interface, as with CORBA. As mentioned above, the clear separation of concerns in the typical EAI architecture allows for the easy layering of new value-added services such as B2B, I2I, A2A, supplier networks, and analytics.

Key Components

Most integration infrastructures are built around two key concepts: events and messages. This is the logical equivalent of a letter and the action of dropping it in a mailbox—with dozens of variations on that simple theme. Users can define messages that are broadcast to everyone or messages that go to specific individuals. They can route all messages via a central hub, like a post office, or they can bypass the hub and go straight to the recipient. Services such as translation routines can rewrite the messages' content into a different "language" for the recipient or convert values to different scales and formats . . . and so on and so forth.

EAI tools have already been well written about in many books, and the authors have no intention of providing deep technical insight into common EAI architectures here. However, because so many concepts are important to grasp, it will be useful to briefly cover the following major components of the typical EAI tool set:

- **Adapter**—small or large units of code that reside physically close to the participating source application to convert its native protocols, vocabulary, and languages to formats that are known by the information bus or central hub
- **Transport**—the over-the-wire protocol that is responsible for the physical movement of data, message and content, from one system to another
- **Message**—instructions, which accompany the data from one physical location to another, that can contain work flow rules, metadata, and queries
- **Process Controller**—sometimes known as an event manager, depending on the topology that is employed, that listens for messages and can determine the appropriate actions to take with them—such as sending them along to other systems or launching other subroutines to work on the data

Many other components can be included in integration packages, but most of them include some variation on these components. Many integration solutions also include a transformation controller. Most EAI tools use application code to perform data transformations. Usually an integrated development environment (IDE) is provided to augment a simple graphical user interface for building these data transformations. Code is then deployed in the adapters, the hub, or both. In practice, some tool vendors opt not to provide sophisticated transformation capabilities because they deliver highly specialized software targeted to specific industry verticals and deem sophisticated transformation capabilities unnecessary.

Disadvantages and Concerns

However, EAI-type applications have struggled in their own right because IT managers are wary of several key disadvantages that impede adoption. EAI solutions create tightly coupled integration environments that eventually restrict the flexibil-

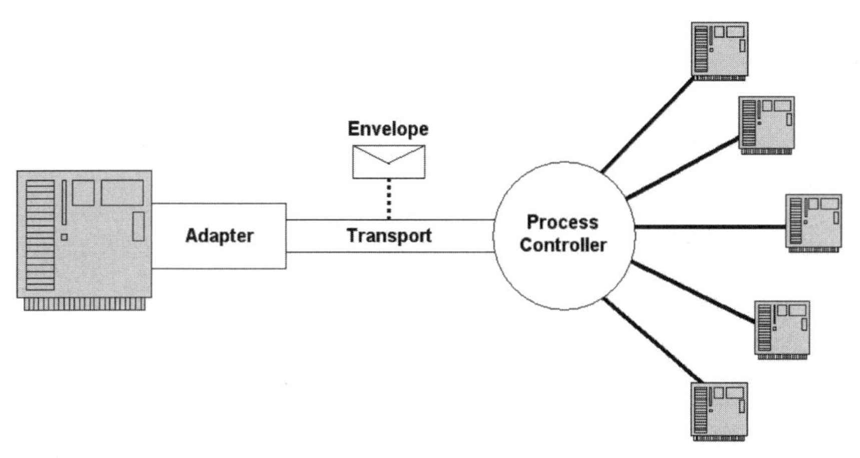

Figure 2.2 Simple EAI architecture view

ity, agility, and adaptability of a enterprise. Tight coupling impacts the bottom line in both lost opportunity cost and high maintenance costs. Integration vendors would have IT managers view the enterprise as a single monolithic system of subsystems. In this view the middleware itself is the enterprise operating system—managing processes and messages to subprocesses. But like a computer operating system, this EAI paradigm only works well if every subsystem implements the same vendor's software.

Other significant concerns and disadvantages include the amount of custom code required for the application adapters used to facilitate integration. To summarize, the two most significant concerns are not technical; instead, they are business issues:

- Vendor lock-in
- Total cost of ownership (TCO)

However, these concerns stem widely known EAI technical limitations such as:

- Proprietary interfaces for adapters, process control, and sometimes even messaging and event protocols
- Tightly coupled architectures resulting from hard-coded interfaces between adapters, hubs, and message buses
- Clumsy interfaces for customization of process and data transformations
- Static, precompiled, or prescripted, routines for event management, data transformation, and resource management

These are technical legacies of the EAI's roots grounded in the remote procedure call (RPC) style of distributed computing. RPC epitomized the tightly coupled interface, and the tightly coupled interface is what makes responding quickly to business changes so difficult.

Application Integration Trend: Target the Vertical Markets

A visible trend in the EAI and B2B industry is the specialization of solutions to specific verticals. For example, a single vendor may repackage the same tool set to target health care, automotive, or customer relationship management market segments. This gives an illusion of a tight match between customer requirements and the vendor's tool set. For better or worse, the behind-the-scenes rationale for this kind of specialization has less to do with actual technology innovation or benefits than with the sales team's quarterly numbers.

However, the vertical specialization of these integration solutions also indicates a growing trend with integration customers—they are beginning to care more about their data and the processes in which the data participate. Therefore, companies are beginning to demand that middleware vendors offer solutions that are specifically targeted toward their industry verticals.

These customer demands stem from highly publicized failure rates of EAI and the strong desire to manage risk, under the natural assumption that if a solution is specialized for a particular kind of data and process it will have better odds of success. However, the technical causes of failures in the past—tight coupling, poor data management, and proprietary interfaces—are still present in the vertically targeted systems. Eventually, the high total cost of ownership and vendor lock-in issues will resurface and further disrupt the EAI industry.

Integration's Core Issue

In reality, and from a technical perspective, when it comes to middleware—the black box with no face running between lots of gray boxes—data is just data and process is just more data. Middleware is truly a horizontal infrastructure solution that requires few, if any, capabilities that are specific to a given industry. If some middleware solution works really well in one industry it has really strong odds at working well in another.

This is true because most middleware components are not aware of details about specific industries, such as data and process labels—they're busy looking at messages and putting stuff on queues. The rest of the middleware components that do operate on industry-specific data or processes can't inherently discern the difference between XML tag that says `<customer_number>` and `<rfid_part_num>`—the software engineers who wrote the programs or configured the middleware are the ones who really know the difference!

This sort of reliance on human factors is what leads to integration's inefficiencies. The "verticalization" of the EAI industry is sure to sell more products, but it will only add minimal value to the end users. Thankfully, those who have become aware of these facts are beginning to refocus on the importance of the data, concepts, and information that actually move through integration's pipes and fittings. It has been the unfortunate lack of focus in these areas that has caused the further popularization of the idea that integration middleware is just a commodity.

APPLICATION INTEGRATION AS PLUMBING

Plumbing is a good thing. Plumbing provides the infrastructure to connect key services in a house. But plumbing alone does not make a house. Plumbing is a commodity. However, plumbing is also a necessary requirement in a home.

Although many integration, B2B, and process integration vendors will tell you that their software is much more than plumbing, if you look closely at what value they actually provide, chances are it will probably primarily involve the movement (and management) of message, documents, or data between different physical instances of enterprise software. In other words, plumbing.

Why Connecting Systems is Necessary

Physical movement of messages and transactions among software systems has been accomplished in a wide variety of ways. File transfer protocols like gopher, FTP, and HTTP along with network and transport protocols like DECnet, SNMP, TCP, and others are among the more widely deployed connection protocols. Each of these technologies' role in the enterprise highlights the importance of the software plumbing. Without the core layers of the standard communications stack[2] and the management services provided by integration tools, the volume and scale of information transfer today would simply not be possible.

However, as with household plumbing, not all pipes and fittings are created equal.

The Coupling Question

Software coupling is the style and characteristics of how two different software components are connected. The easiest way to think about the kinds of connection techniques for integrating systems is to divide technologies into two categories:

- Tightly coupled plumbing
- Loosely coupled plumbing

This breakdown makes it clear that—despite the complexity and rhetoric in the application integration industry—there are not really many significant differences among EAI alternatives.

Tightly coupled integration systems use rigid control mechanisms that require each system to know specific details about other systems, or about the middleware itself, for the system to effectively communicate. To use the integrated interfaces, control flows, process and work flow languages, vocabulary transformations, management and monitoring schemes, and other elements of a network community each participant must have a tight coupling with some aspect of the network.

Tight coupling is not inherently bad. However, some consequences of tight coupling drive up total cost of ownership and drive down return on investment inside large enterprises. These negative consequences usually only become apparent over time and at a large scale. Because technology must always keep up with changing business environments, tight coupling causes a myriad of expensive maintenance nightmares.

On the other hand, loosely coupled integration techniques, which have only recently emerged, form the future of most integration styles. Loose coupling is characterized by indirection—always using an intermediary step to get to a final destination—between networked systems. This indirection establishes connections without predefined or explicit knowledge of the details behind participant nodes or middleware. Loose coupling insulates the nodes on the network from change,

[2] For additional information about the OSI network model, see Chapter 6.

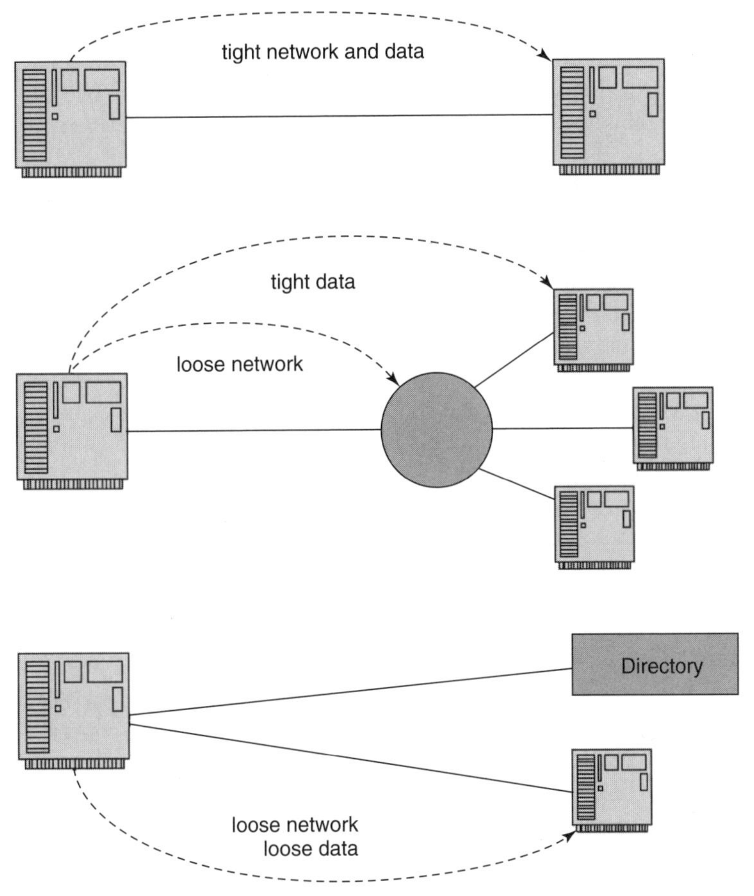

Figure 2.3 Tight-coupling and loose-coupling components

while delegating more of the "knowledge" about the system interfaces to the network itself.

Loose coupling is not inherently good. Early implementations of loose coupling suffered from poor performance, lax security, and inadequate error recovery. As these concerns are addressed, loosely coupled enterprise systems will overcome many of the persistent barriers presented by tight coupling.

Although a significant proportion of current integration systems could benefit from loosely coupled technology, let us not forget that a number of specialized application areas will continue to be best served by tightly coupled integration frameworks.

Business Process Plumbing

Advocates of business process management (BPM)—which is process-centric integration—believe that the critical challenge in enterprise integration can be solved by better management of the transactional work flow among participating nodes on the network. On the surface, this is a compelling argument. The single greatest factor of change in the enterprise is business process, which cuts across organizational and system information silos. However, these advocates are missing one key point with far-reaching implications: To computers, business process models are just data.

Business process consists of instructions, or business rules, that define the timing of specific events, the treatment of messages, and what to do with messages during error flows. Execution of business process is delegated to software components that can reside in different physical locations on the network. For business process execution to take place, these units of code must interpret the messages, proprietary formats that can only be understood by the software that created the message to begin with.

Today's business process management solutions have had mixed results at best. Their tight coupling of interfaces and process rules create brittle environments that require constant attention to run effectively.

Because business process vendors build their tools around process data, rather than process semantics, the connections between enterprise systems are just as static as they always were.

Ironically, attempts to eke more value out of BPM approaches have led to the adoption of a loosely coupled middleware approach: service-oriented architectures (SOA). Even with an embedded SOA underneath the BPM solution, BPM will still be the plumbing. Fundamentally, the business of managing messages, work flow, and transactions in a message centric middleware solution are plumbing concerns. Because the focus of BPM is on the process, as opposed to the information contained in the message, it exists at a technology layer several abstractions removed from the content of the message itself.

Service-Oriented Plumbing

Service-oriented architectures (SOA) have emerged as the next great alternative to legacy middleware technologies and a key enabler for loose coupling. The SOA framework is the first widespread, viable solution for loosely coupled integration. SOA builds on the architectural patterns of component-based development and enables IT groups to build services and connect them on demand as they are needed. Unlike traditional plumbing, SOA gives users the ability to establish dynamic "pipes" that are initiated on demand with little overhead or prior knowledge about existence of the service. An infrastructure built on this approach can support integrations without a confusing collection of different protocols, interfaces, and conduits into various applications.

Nevertheless, the SOA paradigm is just plumbing. Because it is concerned with the physical nature of connections, interfaces, protocols, transport, and control mechanisms for the movement of messages, it is just plumbing.

Why Plumbing is Insufficient

Plumbing, or integration, is necessary but not sufficient. Without it, enterprise software communication would not be possible. With it alone, communication is inefficient, expensive, and lacking focus on what matters to the business—information.

In many cases, technologists responsible for specifying and deploying integration-based solutions do not fully understand how important the contents of the messages being transmitted is. If they did, they would spend more time on the underlying information architecture. Instead, today's most common solution is to just create a new XML schema that is adequate for describing the data that needs to go over the wire.

Unfortunately, this haphazard care for enterprise information has created an environment in which business's decision makers rarely have a complete view of the business and they pay dearly, in money and chaos, for each operational change they make.

ENTERPRISE CONTENT, NOT PLUMBING, IS KING

Okay, so it might have been 1999 when you last heard that "content is king," but here it is in reference to the content of enterprise messages—not Internet web pages. The invoices, repair schedules, financial information, patient data, sales figures, and customer data inside those messages are what comprise enterprise content. Isn't the whole point of integration to move content around so it can be used in new ways?

Crack open a Web Services SOAP envelope or an EAI message and you can learn something about the business. The XML documents and text files that fly around at the speed of light contain the records of importance for process flows and business-to-business exchanges. Even for process-centric applications, the representation of process in the middleware is the information that enables business logic to control the process flow events. The rules and structure that define a process are content that look and behave just like data.

Unfortunately, the amount of attention given to enterprise content, information, and data inside technology circles is minuscule compared to sexier subjects like Web Services and programming frameworks like. NET and J2EE. This is because it is difficult to understand and articulate the enterprise data that represents the foundation of every enterprise system.

Enterprise Information Data Structures

Present understanding of enterprise data structures is quite sophisticated. Not only are enterprise data different in internal binary formats, such as the difference between a text file and an object, but the information is also organized within a particular structure and representation. Basically, the continuum of data structure formats ranges from very unstructured to very structured. This structural continuum is commonly broken into at least three parts:

- **Structured Data**—This is the most organized of this continuum. Typically, it will have definitions stored as metadata, such as type, length, table name, and constraints. Examples of structured data include relational databases, object models, and XML documents.
- **Semistructured Data**—This type of data is less clear; no specific delineations exist on where to draw its boundaries. For the most part, it is acceptable to think of semistructured data as data that are organized but not explicitly defined in a self-contained way. Traditional examples of semistructured data include positional text messages, such as EDI, tokenized, and comma-separated value (CSV) data files.
- **Unstructured Data**—This is simpler to understand: data that possesses no inherent organization intended to communicate its own meaning. Essentially it is like free-form text that is completely arbitrary in presentation and contains no metadata or structure that can be useful in describing its concept structures.

Fundamentally, the organization or structure of the data does not inherently make it more meaningful, only easier to process. The work of understanding what the data means, its semantics, is still a task that has to be undertaken with lots of effort in understanding the meaning behind the data.

Continuum of Knowledge

Understanding data content still requires human interaction. Without a programmer to encode the rules inside the software, the software is incapable of discerning meaning. One way to understand the reasons behind this is to consider the relationship between data and knowledge inside the continuum of knowledge.

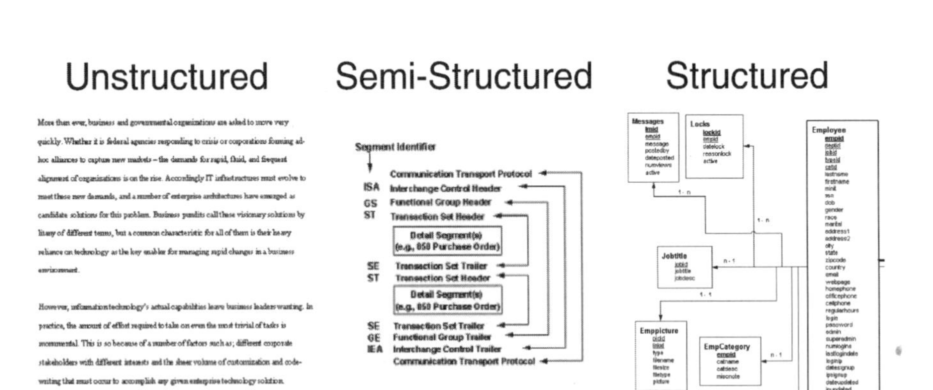

Figure 2.4 Structured data continuum

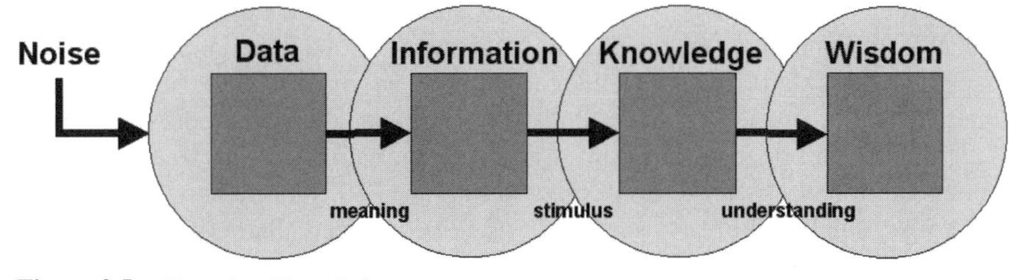

Figure 2.5 Hierarchy of knowledge

Although philosophers, scientists, and other academics do not have a comprehensive understanding of what knowledge actually constitutes, the above representation is commonly accepted as a basis for understanding the relationship between data, information, knowledge, and wisdom.[3] Consider the following example:

Data: "1234567.89"

Information: "Your bank balance has jumped 8087% to $1,234,567.89."

Knowledge: "Nobody owes me that much money."

Wisdom: "I'd better not spend it because I will surely get jailed."

Basically, information is data plus meaning. The discussion of what constitutes meaning is covered in Chapter 4, Foundations in Data Semantics, and is too broad to cover here.

The real challenge for enterprises today is to begin to build an infrastructure that enables enterprise information, of all formats, to be utilized freely at the right time and place. These capabilities will transition the industry away from commodity enterprise plumbing solutions and uncover new enterprise value enabled by seamless information interoperability.

FINAL THOUGHTS ON OVERCOMING INFORMATION INFRASTRUCTURE PROBLEMS

All problems of information interoperability can be solved, but brute force methods are still the most popular ways to effectively deal with these problems. However, brute force efforts result in inefficiencies and lost opportunities and have likely cost integration customers trillions. Results like these have soiled the reputation of the EAI category and prompted pundits like Nicholas Carr, as mentioned in Chapter 1, to write *Harvard Business Review* essays entitled "IT Doesn't Matter."

Carr's essential point is that IT doesn't enable companies to distinguish themselves in a meaningful way from their competitors. He concedes that IT is essential for competing but successfully argues that it is inconsequential for strategic advantage.

[3] *Experience Design*, Nathan Shedroff.

Strategic advantage is won when companies can build unique differentiators valuable to their customers. Building unique differentiators is precisely what happens when enterprise *data* evolve to enterprise *information*, because information interoperability enables businesses to better adopt and adjust their core capabilities. This kind of information capability transcends the argument that technology is a commodity precisely because information interoperability results are unique to an enterprise, thus enabling it to strengthen its existing differentiators—or pursue new ones.

Semantic Information Interoperability

The next great step forward in the computing industry will be its shift toward capabilities that can discern meaning inside digital systems. These new capabilities will enable enterprise content visibility and interoperability of unprecedented scales, which will in turn drive businesses' ability to establish strategic competitive advantages in both established and new ways.

Semantic interoperability emphasizes the importance of information inside enterprise networks and focuses on enabling content, data, and information to interoperate with software systems outside of their origin. Information's meaning is the crucial enabler that allows software to interpret the appropriate context, structure, and format in which the information should reside at any given moment and inside any given system. This information ubiquity is the beginning phase of a truly information-driven organization.

Integration	*Interoperability*
• Participant systems are assimilated into a larger whole	• Participant systems remain autonomous and independent
• Systems must conform to a specific way of doing things	• Systems may share information without strict standards conformance
• Connections (physical and logical) are brittle	• Connections (physical and logical) are loosely coupled
• Rules are programmed in custom code, functions, or scripts	• Rules are modeled in schemas, domain models, and mappings
• Standard data vocabularies are encouraged	• Local data vocabularies are encouraged

The technology components to enable interoperability of this kind already exist. Universities, companies, and standards bodies throughout the world have been working on semantic technologies for many decades. Widespread adoption of these technologies is now possible because of the rising maturity of these approaches, increased R&D funding, and poor performance of more popular traditional techniques.

Chapter 3

Promise of Frictionless Information

KEY TAKEAWAYS

- Frictionless information flows are the inevitable future of information technology.
- Models, independent of platforms and data, will drive semantic interoperability.
- Component-driven service-oriented architectures (SOA) will provide flexible and dynamic connectivity—once fully enabled by semantic interoperability.
- Autonomic computing concepts will drive strategic technology development in a number of industries and software solution spaces.
- US intelligence sharing problems are a useful context in which to examine the strengths of flexible, dynamic and loosely coupled semantic interoperability architectures.

In an environment in which business leaders freely discuss the end of IT innovation, it may seem overly optimistic to claim that semantics-aware middleware is the strategic future of technology, much less of business as a whole. However, it is true—the evolution of enterprise middleware is not only the strategic future of information technology, it provides the single greatest potential for businesses to gain strategic differentiation in the foreseeable future.

More common beliefs about what the future may bring include:

- A belief that many other technologies, not just middleware, will have strategic impact on the future of technology
- A belief that some other individual technology (e.g., wireless, handheld, or enterprise packages) is the strategic future of IT
- A belief that no technology can be strategic, because IT is only a tactical commodity

The central problem with either of the first two positions is that information sharing software, like middleware, is a trump card. Software that facilitates information sharing among physically or logically diverse enterprise applications is a

Adaptive Information, by Jeffrey T. Pollock and Ralph Hodgson
ISBN 0-471-48854-2

critical path technology for every other enterprise information technology's success. Middleware is the enterprise cardiovascular system that carries an organization's lifeblood to its many parts. It is the path of least resistance to information contained in distributed enterprise applications such as enterprise resource planning (ERP), customer relationship management (CRM), knowledge management (KM), network appliances, and supply chain radio frequency identification (RFID) applications. Middleware is responsible for connecting islands of information within the enterprise—thereby enabling other information technology to reach its fullest potential.

Middleware can be any technology that fulfills some significant portion of machine-to-machine collaboration, such as CORBA, Web Services, EAI, B2B and custom-written solutions in the programming language du jour. Most middleware solutions currently in production are custom. Perhaps the most significant argument for middleware's position as the strategic future of software technology is the gap between what it does today and what we know it can do in the future. It is true that middleware's past failures leave room for doubt regarding software's strategic value to the enterprise. Skeptics use middleware's failed promises, and a myriad of other examples, to argue that IT cannot generate any strategic value at all.

The problems with the idea that IT is a tactical commodity and offers little, if any, strategic value are multifaceted:

- It assumes that all companies use technology the same way.
- It assumes a static and level competitive playing field.
- It assumes that IT is mature and ignores the existence of disruptive technologies.

For example, even though golfers have nearly equal access to high-performance clubs, their individual training and styles contribute to how well they use those clubs.

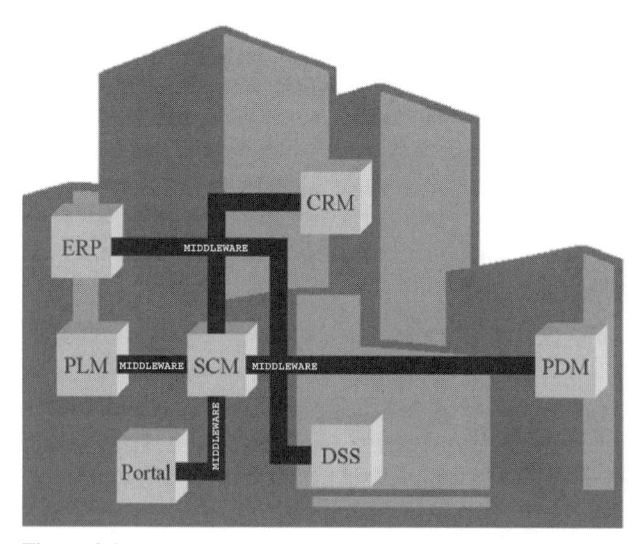

Figure 3.1 Software packages keep things running inside the enterprise

Likewise, although information technology is a tool that everyone has access to, not everyone uses it in the same way achieving the same results. However, unlike the rules of golf, the rules of business change all the time—information technology can exploit new rules and provide advantages to savvy users who are on top of the game. Finally, and like Schumpter's economics, creative destruction is constantly generating higher-quality, cheaper technologies to solve persistent business problems. The bottom line is that—within traditional corporations and the IT industry as a whole—innovation marches on.

IT is indeed a strategic asset to innovative companies, and the most strategically important part of IT in the coming years will be the adaptive and dynamic infrastructures enabled by semantics-aware middleware.

ORGANIC MIDDLEWARE: "SOFTWARE, INTEGRATE THYSELF!"

We are on the cusp of significant changes in interoperability technology. These changes will drive substantial improvement in capabilities and value. Paradoxically, they will also reduce the complexity and expense of systems maintenance—although they will have far more complex underlying architectures and functions.

Various analysts, reporters, and business strategists have all seen these changes coming, calling it by different names—one of which has been the word "organic." Organic middleware means that the network will evolve, adapt, and change on its own, with minimal input from programmers, to make systems interoperable. This adaptability stimulus will drive efficiency, cost reduction, and emergent new behavior that can exploit new business opportunities.

Insider Insight—Zvi Schreiber on Semantic Interoperability ROI

How do companies measure their return-on-investment for capturing the formal business meaning of their data? At Unicorn Solutions we introduce our customers to a methodology that measures value in four main ways:

- IT Productivity—Semantic tools can automate the generation and maintenance of data translation scripts. For an organization with many IT professionals, this creates a substantial and measurable ROI.
- IT Efficiency—Many organizations buckle under the cost of maintaining redundant databases and applications. Capturing data semantics based on an agreed ontology allows redundant systems to be decommissioned with measurable savings.
- Business Agility—Ontology provides a conceptual business view which may adapt to reflect business change quickly—this will result in quicker integration to new supply chain partners, quicker realization of savings after M&A, quicker optimization of business processes and quicker adoption of new products.

Continued

- Information Quality—Information quality problems, which according to TDWI cost US businesses $600 billion per year, are frequently the result of semantic inconsistencies among heterogeneous data sources.

Additionally, other soft benefits may include less dependence on specific vendors and their proprietary formats for capturing your data and business rules. Experience has shown that when real pain points are selected for semantic projects, payback takes a few months with massive ROI over the first year or two.

Zvi Schreiber
CEO, Unicorn Solutions

A Tough Road Forward

Middleware, and interoperability in particular, is one of few technologies that can offer rising returns on efficiency for existing enterprise systems. Unlike most packaged enterprise applications, middleware provides infrastructure and infrastructure provides foundation. But middleware has no face, no fancy GUI for every employee to log on to and check it out. Instead, middleware quietly hums away all of the time, making sure that applications get the data they need at the right time.

At least that's how it should work.

The reality is that, just moving into its second decade of existence, information sharing software is a very young technology. Middleware solutions, both custom and off the shelf, tend to be inconsistently implemented with mixed results. Nagging fear, doubt, and confusion plague IT buyers who doubt its ability to deliver on vendors' inflated promises.

Integration approaches today rely on vast amounts of human intervention even after initial implementation. Managing change has emerged as the most significant challenge in the IT sector as a whole. Middleware fails miserably at achieving long-term savings and consumes vast proportions of annual corporate budgets with every change.

The fundamental technology barriers identified in Chapter 1 prevent today's approaches from achieving higher levels of success. Traditional middleware approaches discussed in Chapter 2 focus on the transport and orchestration of message delivery and not on its content—business information. Data is a second-class citizen in these common middleware integration frameworks. To enable truly organic and dynamic network capabilities a greater level of semantic awareness is a mandatory requirement. This new-generation middleware must be built on much more sophisticated concepts of information and process to enable technology to drive strategic business advantage.

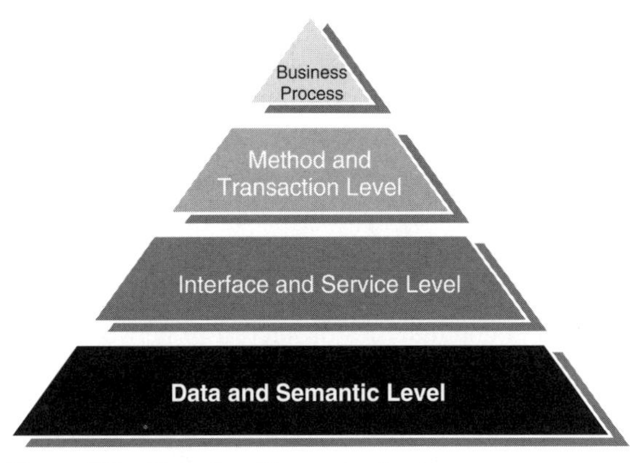

Figure 3.2 Hierarchy of integration levels

ORGANIC MIDDLEWARE HINGES ON SEMANTIC INTEROPERABILITY

Data have always been the foundation of any information-sharing programs. Next-generation systems rest upon an expanded view of data—information. In Figure 3.2, interface-level integration, method-level integration, and process-level integration have all developed on top of a foundation of data. With semantic interoperability, the expanded notion of data includes semantics and context, thereby turning data into information. This transition both broadens and deepens the foundation for all other integration approaches—enabling new organic capabilities to emerge.

With a robust foundation of information, data, and semantics as a baseline, new capabilities for interoperability can emerge:

- **Data Interoperability**—Semantic interoperability of data enables data to maintain original meaning across multiple business contexts, data structures, and schema types by using data meaning as the basis for transformations.
- **Process Interoperability**—Semantic interoperability of process enables specific business processes to be expressed in terms of another by inferring meaning from the process models and contextual metadata and applying it in a different process model elsewhere or outside the organization.
- **Services/Interface Interoperability**—Semantic interoperability of services enables a service to look up, bind, and meaningfully communicate with a new service without writing custom code in advance.
- **Application Interoperability**—Semantic interoperability of applications enables the granular interactions of methods, transactions, and API calls between heterogeneous software applications to be platform independent.

- **Taxonomy Interoperability**—Semantic interoperability of taxonomy enables any category to be expressed in terms of other categories by leveraging the intended meaning behind the category definitions.
- **Policy Interoperability**—Semantic interoperability of policies and rules enables businesses to protect valuable resources regardless of what technologies their security mechanisms have been implemented in or how complex the rights management issues have become.
- **Social Network Interoperability**—Semantic interoperability of social networks enables people in different communities of interest to network, infer, and discover meaningful connections through previously unknown contacts and interests.

When these components can interoperate with either an automated or a human agent, network configuration can become emergent. Emergent behavior implies a level of dynamism and organic growth that can enable a pervasive autonomic environment with features like:

- Self-configuring interface and schema alignment across data vocabularies, directory taxonomy, service descriptions, and other components
- Self-optimizing transactions, routing, queries, and transformations
- Self-healing error flows that can recover from otherwise dead-end use cases
- Self-cleansing data validation that can scrub instance values from various sources

FRAGMENTED INDUSTRY EFFORTS AND ORGANIC COMPUTING

Despite common goals and objectives, there are many competing and complementary efforts under way that all rely on innovative uses of data semantics. The following section will briefly introduce a few key technology movements that are driving momentum toward the future of semantic computing.

Autonomic Computing

IBM first popularized the notion of autonomic computing in its manifesto published in October of 2001.[1] This manifesto clearly established the needs and goals for the next evolution in information technology. Research and development of autonomic concepts has been a priority of IBM's research arm ever since and should ultimately lead to further innovation in IBM's software products.

Microsoft has also embraced the autonomic concepts and applied them, not in middleware, but to data center technology. By late 2003 their public ad campaign of "automated systems recovery" was in full swing.

[1] For more information, see: www.research.ibm.com/autonomic/manifesto.

Definition—Autonomic Computing

Autonomic Computing[a]—Autonomic computing derives its name from the autonomic nervous system and denotes its ability to free our conscious brains from the burden of dealing with the vital, but lower-level, functions of the body. As used in the software industry, autonomic computing refers to self-managing systems that are comprised of four core characteristics: self-configuration, self-healing, self-optimizing, and self-protection.

[a] The Dawning of the Autonomic Computing Era, A.G. Ganek. *IBM Systems Journal*, Vol 42, No 1, 2003.

More of a technology vision, autonomic computing doesn't necessarily represent any specific technologies or set of technical components. In fact, the visions of autonomic computing are somewhat different among some giants of software: IBM, Microsoft, Sun, and BEA. However, all of these companies are using technology to move toward more organic and change-resilient infrastructures.

Semantic Web

The Semantic Web has become an umbrella term applied to everything related to processing semantics in data. Initially conceived and articulated as an expansion of Internet capabilities to enable more meaningful searches among the vast amounts of data living on the Web, it has since morphed to include other notions of semantic processing on more traditional structured enterprise data as well. Indeed, the topics covered in this book—regarding the semantic interoperability of middleware and enterprise systems—are rightly considered part of the Semantic Web.

Definition—Semantic Web

Semantic Web[a]—The Semantic Web is an extension of the current web in which information is given well-defined meaning, better enabling computers and people to work in cooperation.

[a] *The Semantic Web*, Berners-Lee, Hendler, Lassila. May 2001.

The World Wide Web Consortium, an international standards body directed by Tim Berners-Lee, controls the Semantic Web's vision and direction. Initial efforts with the Semantic Web have resulted in significant activity with knowledge representation markup languages (RDF and OWL) as well as the architectures and logic required to implement them alongside existing data sets. A wide range of companies, many discussed in this book, have adopted the Semantic Web vision and are actively pursuing technology strategies that advance it further.

Semantic Web Services

Semantic Web Services are the intersection of the Semantic Web and Web Services. The Semantic Web Services Initiative (SWSI),[2] an international committee of visionary technologists, is defining how convergence will occur between these two technologies. A significant objective for the Initiative is to create an automated and dynamic infrastructure for Web Services provisioning and usage. A secondary objective for the Initiative is to assist in the coordination of research efforts in commercial and academic programs.

Definition—Semantic Web Services

Semantic Web Services[a]—Semantic Web Services are Web Services implementations that enable greater adaptive capabilities through rich semantic metadata at the interface layer. The Web Ontology Language Service specification (OWL-S) provides a mechanism towards this end. OWL-S supplies Web service providers with a core set of markup language constructs for describing the properties and capabilities of their Web Services in unambiguous, computer-interpretable form. OWL-S markup of Web Services will facilitate the automation of Web service tasks including automated Web service discovery, execution, interoperation, composition and execution monitoring. Following the layered approach to markup language development, the current version of OWL-S builds on top of W3C's standard OWL.

OWL-S is not the only mechanism under development to support Semantic Web Services. The Web Service Modeling Ontology (WSMO) is being developed within the Digital Enterprise Research Institute (DERI) to solve many of the Semantic Web Services challenges. Unlike OWL-S, the WSMO specification focuses on a workflow ontology for web services that can capture rich semantics about interfaces and exchanges—it will not require the use of inferencing technology for discovering semantic metadata. The WSMO work, in conjunction with other efforts from DERI, has already produced working prototypes of Semantic Web Service implementations. This research area is rapidly becoming commercialized through the efforts of DERI, W3C, and IEEE.

[a] Definition adapted from: www.daml.org.services.

Service Grid

A service grid is a concept that represents a specific set of technology components. The Open Grid Service Infrastructure (OGSI) specification extends Web Services WSDL specifications to enable more dynamic usage patterns for service-oriented

[2] For more information, see: www.swsi.org.

architectures. The Global Grid Forum (GGF) is a technical community process, also focused on processorcentric computing grids, that drives the OGSI and Open Grid Service Architecture (OGSA) technical specifications.

Definition—Service Grid

Service Grid[a]—A service grid is a distributed system framework based around one or more grid service instances. Grid service instances are (potentially transient) services that conform to a set of conventions (expressed as WSDL interfaces) for such purposes as lifetime management, discovery of characteristics, notification, and so forth. They provide for the controlled management of the distributed and often long-lived state that is commonly required in sophisticated distributed applications.

[a] *Grid Service Specification, Draft 3 2002*. The Globus Alliance.

Both IBM and Sun Microsystems have already made substantial investments in the grid computing arena. They both participated in the specification process, each sponsoring its own proposed specifications for various portions of the grid computing challenge. With continued support, some combination of OGSA efforts and the Semantic Web Services efforts will produce a standards-based approach that finally delivers on the initial promise of Web Services—dynamic service discovery.

Model-Driven Architecture

Originally focused more at the application level than at the enterprise network level, model-driven architecture (MDA) concepts and specifications will drive the next evolution of code generation for enterprise information systems. Consisting of platform-independent models and platform-specific models, the MDA insulates business and application logic from technology evolution. Using MDA, software engineers spend more time with enterprise models than with application code. A step beyond CASE (computer-aided software engineering) technology, code generation is the foundation of MDA.

Definition—Model-Driven Architecture

Model-Driven Architecture[a]—Model-Driven Architecture is an approach to system development that emphasizes the use of models to separate the specification of software application independent from the platform that supports it. The three primary goals of the MDA are portability, interoperability, and reusability.

[a] *MDA Technical Guide Version 1.0*. Object Management Group, 2003.

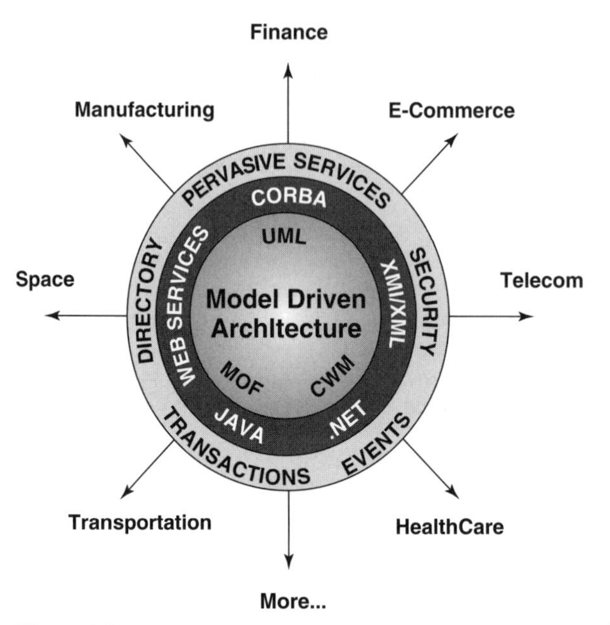

Figure 3.3 MDA vision depicted by the Object Management Group[3]

The Object Management Group (OMG), which has brought us many innovative standards (CORBA and UML), now offers the MDA. Adopted in 2001, the ongoing efforts of the OMG in this regard will continue to drive the industry toward a modelcentric method of software development. More on the MDA is covered in Chapter 6, Metadata Archetypes.

Intelligent Agents

Agent technology has had a long, but anticlimatic, history. For nearly a decade, agent technology was supposed to have brought us all closer to IT nirvana. However, multiple barriers to agent capabilities have prevented this from happening. To begin with, there is no single definition of what an intelligent agent is. This book uses the definition provided by Dr. Nicholas Jennings, a leading researcher in agent technology.

Definition—Intelligent Agent

Intelligent Agent[a]—An agent is an encapsulated computer system that is situated in some environment and that is capable of flexible, autonomous action in that environment to meet its design objectives.

[a] N. R. Jennings, "On Agent-Based Software Engineering," *Artificial Intelligence*, vol. 177, no. 2, 2000, pp. 277–296.

[3] Image is property of Object Management Group, www.omg.org/mda.

Agent technology will not offer widespread benefits to the enterprise in and of itself. Instead, it requires an infrastructure to support agent activities. When coupled with other technologies, like service grids, model-driven architecture, and inference engines, agent technology will become a crucial aspect enabling dynamic, adaptive, and autonomic enterprise computing.

ACHIEVING SYNTHESIS

Each of the previously mentioned technology innovations would accomplish significant new breakthroughs in IT on their own, but the synthesis of these approaches is what will drive momentum toward a holistic, dynamic, and organic middleware framework. Likewise, each of these technology innovations relies on an improved information framework that enables the full scope of their capabilities.

Let's take a look at how the synthesis of these technologies might be applied in a hypothetical scenario involving security and intelligence information sharing.

INTELLIGENCE INFORMATION SHARING IN THE TWENTY-FIRST CENTURY

Homeland defense is a pressing problem that highlights many of the barriers faced by current information interoperability software. Nontechnical issues such as organizational culture barriers, decentralized IT management, politics of negotiation, and the sheer complexity of policy and change management are quite significant across federal government agencies. These challenges are extreme examples of what all information sharing programs face—albeit of lesser scale and with fewer worst-case consequences. The scope of these challenges for intelligence information sharing and homeland defense offers a poignant scenario for examining some of the requirements that next-generation middleware should meet and points toward the kind of technical architecture needed to solve them.

The remainder of this chapter will explore these intelligence information sharing problems and explore high-level technology capabilities that may provide a solution. As you read the following scenario, remember that the synthesis of emerging technologies will be greater than the sum of its parts. This example highlights only the portions of the overall architecture that depend on semantic interoperability for information sharing.

Information Sharing Imperative

Although it is impossible to pinpoint the precise beginnings of an information sharing network for the United States, Paul Revere's ride in the spring of 1775 marked one of the earliest examples. From the famous cries of "The British are coming! The British are coming!" the demand was already apparent for an efficient mechanism to relay information when national security is threatened.

Today, much has changed. We no longer need to hang lanterns in church windows and ride swift steeds to pass emergency messages among our communities. Computer networks and telephones shrink even the greatest of distances.

However, much is also the same. Despite significant advances, computer technologies are only 30–50 years old, which makes them immature compared with other modern technical luxuries, and they are generally incapable of performing even the most basic communications without being assisted by skilled human gurus.

To see how important better information sharing is, consider a report by the Markle Foundation Task Force on National Security in the Information Age entitled "Protecting America's Freedom in the Information Age," which provides a sobering analysis of the dangers we face.

It is impossible to overstate the importance of swift, efficient, and secure information sharing among our local, state, and federal agencies. At the same time, to say that this challenge has been difficult to accomplish is an understatement.

Information Sharing Conundrum

A seemingly simple task, information sharing has proven to be a treacherous sinkhole of time and money for our government. And lest you think that these woes are the fault of bureaucracy, it is important to note that large business enterprises routinely fail at the same task—upwards of 88% according to some analysts.[4]

The reasons for our problem in achieving swift and efficient information sharing in homeland defense are intricate and intertwined. These problems also cast a bright spotlight on some issues that are usually seen as minor problems in commercial enterprises, but have enormous consequences.

- **Technical Incompatibility**—Usually the most cited problem, compatibility issues tend to be black-and-white in nature. Typical compatibility issues will boil down to application protocol, programming language, or data definition problems that can be fixed, but sometimes only at significant cost.
- **Operational Limitations**—In the not-so-distant past this has been perhaps the most significant barrier to joint efforts on information sharing. Differences in agency objectives have led to differences in processes, equipment, and organizational authority. Along with cultural issues, operational limitations have greatly prevented substantial progress.
- **Budgetary Constraints**—The nature of budget constraints has been twofold. Both inadequate funding and the lack of specific plans for coordinating different programs have limited the possibilities of finding ways to share required information.
- **Cultural Differences**—Often overlooked, this is frequently difficult to spot because it is not a technical problem. Agency culture is embedded in the way one government agency chooses to operate or the way differences in termi-

[4] Gartner Group Report, December 2001.

nology evolve among organizations. Culture is subtle, culture is folklore, and culture is pride in differences—which often leads to difficulty when sharing information. Considering the historical differences between the State Department and the Defense Department, various agencies in the intelligence community (for example, CIA, NSA, DIA and other intelligence agencies), and agencies tasked with overseeing the nation's borders (for example, U.S. Customs, Border Patrol, etc.) can highlight examples of differences along these lines.

- **Political Wrangling**—Like it or not, people in power often wield that power in order to secure even more. Information is power. This persistent power struggle leads to fiefdoms of control and dangerous silos of isolated information. Even where power is not the prime motivation, many agency leaders worry that ceding operational control of information could compromise their primary mission of their agency—this mentality leads to information hoarding.

It is not impossible to overcome these challenges. As the Markle Task Force working group "Connecting for Security" has suggested, there are at least 10 major characteristics that the next-generation homeland security information network should meet. In addition, the Federal Enterprise Architecture (FEA) Working Group has specified a minimum set of frameworks that should unify agency development programs. Other groups, such as the Regional Organized Crime Information Center (ROCIC), have installed information sharing systems that mostly consist of Internet portal type applications (such as RISSnet) that enable specialists to search by keywords within various systems for data.

Perspectives—Markle Key Characteristics:[a]

- Empowerment of Local Participants—empowerment to access, use, and analyze data on the network's edge
- Funding and Coordination—decentralized, but adequately funded, coordinated, and empowered participants
- Safeguards to Protect Civil Liberties—Guidelines should be implemented to prevent abuse.
- Elimination of Data Dead Ends—Stovepiped data should not lead to a dead-end, with no accountable authority.
- Robust System Design—Minimize possible points of failure.
- Network Analysis and Optimization Capabilities—Detecting patterns in the security network may trigger alerts.
- Solid Growth and Upgrade Plan—The use of good architectural patterns ensures maximum longevity.

Continued

- Support for Existing (legacy) Infrastructures—A successful network will exploit existing information resources.
- Create Network Aware Scenarios—Complex models of security scenarios can assist analysts in the field.
- A Connected Culture—Being connected and participating must become second nature to participants.

[a] *Protecting America's Freedom in the Information Age*, Markle Task Force on National Security in the Information Age. October 2002.

There are two primary demands for information sharing: (1) enable visibility into disparate systems, preferably with analytic capabilities to identify trends that develop over time and across systems, and (2) enable interoperability among these systems so that local alert systems and agency software applications can operate on the data directly, from within their own environments. Early steps have been taken to accomplish both of these goals, but mostly through uncoordinated smaller efforts with little cross-mission planning.

This problem, when viewed in its entirety, is exceedingly complex. Federal agencies already have volumes of information that can legally be shared, so interagency exchange is technically possible—in fact, most consulting companies would be happy to work on specific intersystem communications. The core challenge today is to imagine an infrastructure that can be built, and built upon, in the years to come—while enabling government agencies and commercial entities access to required security information on demand.

Given enough money and technology staffing, any software vendor or systems integrator could propose a reasonable plan. However, the future of such an infrastructure must also account for the human factors.

Human elements are frequently assumed to be a nonreducible factor of technology deployments. Technologies for middleware and information sharing lack contextual sensitivity in their deployment architectures—which leads to custom-written code that solves the same problems over and over again. No matter which vendor-supplied tools an end user works with, the result of software deployments inevitably requires that a programmer be proficient in Java, XSL/T, XML, SQL, or some other language in order to manipulate and give meaning to the raw data moving through the system.

At a small scale this is not a problem. But consider the scale of a broad-based cross-agency intelligence sharing solution. Potentially hundreds of thousands of unique data sources may need to contribute or tap into relevant information. This is a middleware solution of a scale that has never been considered before. It is a middleware problem that, based on person-hours alone, could never be solved with today's technologies.

Today, software's connectivity and information architecture concerns are mixed into the same physical components—creating tightly coupled data and communication protocols. In environments that change frequently this brittleness results in expensive maintenance scenarios.

To reiterate, these issues have been overcome in the past because most middleware ties together dozens (not hundreds of thousands) of applications—and humanpower is much more cost effective and manageable at those levels.

In summary, there is much agreement on the core functional characteristics that a successful solution must have. The human factor and scalability of possible solutions must be a key consideration in finding a reasonable answer. Unfortunately, popular notions about integration technology—and the vast amount of disinformation that surrounds commonly available high-profile vendor offerings—confound these goals.

Toward a Pragmatic Solution: Semantic Interoperability

Thinking about a solution for a problem of this scope and this complexity is extremely challenging. To frame it in terms that are policy oriented or business focused would result in a solution that fails to meet fairly specific technical goals. To frame the solution in terms that are technology focused would result in tactical solutions that are missing the forest for the trees. A balance of these forces is required.

A balanced solution will enable us accomplish both the technical and nontechnical requirements of an acceptable solution. These requirements derive from the characteristics and goals put forth by the Markle Foundation task force and succeed at addressing problems at multiple levels such as cultural, political, technological, operational, and budgetary.

Because of the need to balance diverse and loosely defined concerns while still addressing multiple layers of complexity, it is pretty clear that what we need from technology in the future is vastly different than what is generally available today. We will need it on a scale that has not yet been seen before. We will require it as a matter of safety—not just for generating return on investment. We will need to apply new thinking to old problems to determine an ideal course of action.

We start with a few core premises:

- For any solution to succeed, a high degree of organizational autonomy among federal agencies should be supported. In other words, few, if any, vendor-based, proprietary, or narrow vocabulary standards should be required.
- Organizations must be able to support a wide variety of historical, legacy applications and processes. It is not practical to assume any baseline of technical competency when planning for the infrastructure.
- The solution must be infrastructure, which is supported by appropriate tools, methodologies, and verification to allow for nondeterministic deployment options (it must be self-contained and relatively easy to maintain and deploy in a decentralized operating environment).

A fundamental aspect of a successful solution is application of the architectural pattern of separating concerns. In this proposed architecture data semantics are purposefully separated from the connectivity technologies as a way to divide and conquer the inherent complexity of the system.

Some publications[5] have referred to the distinction between logical and physical components of middleware. For the purposes of this solution, the logical layer encompasses logical data models, conceptual models, business rules, process models, and logical query structures. The physical layer traditionally encompasses connectivity protocols, work flow brokers that execute process and routing models, data persistence mechanisms, data exchange formats (instance data), physical security rules, and various tools to manage systems participating in exchanges.

SEMANTIC INTEROPERABILITY FRAMEWORK CHARACTERISTICS

One way to describe a system is with a set of buzzwords. A standard set of these has been used to describe the framework. The focus of the rest of this section is on explaining what is meant by those buzzwords and the problems that are being addressed.

Definition—Semantic Interoperability Framework

Semantic Interoperability Framework—A highly dynamic, adaptable, loosely coupled, flexible, real-time, secure, and open infrastructure service to facilitate a more automated information-sharing framework among diverse organizational environments.

Dynamic

The essence of being dynamic hinges on the infrastructure's ability to support the configuration of information assets in previously unforeseen ways without the intervention of humans. To say it another way, sometimes an interaction between two sources of information needs to take place, but development teams have not predicted that particular interaction and therefore no code has been written to facilitate the transaction. This leaves the organization with an option to either expend development person-hours to get the data required (if they are even aware of the request at all) or to do without the information.

The semantic interoperability framework makes use of four core technologies that enhance the dynamic nature of the final solution and reduce the need for human intervention:

- **Dynamic Data Transformations**—By creating an architecture whereby data schemas are linked via context-sensitive maps, the underlying run time com-

[5] The Big Issue, Jeffrey Pollock. *EAI Journal*, April 2002.

ponents can generate the transformation code on the fly—without having to have prior knowledge of the specific interaction desired.

- **Dynamic Query Generation**—By providing rich conceptual models (ontology) that intelligent agents and/or dedicated inference engines can navigate with the aid of description logics, the software can dynamically assert truth statements in the form of queries against previously unknown schema types.
- **Dynamic Map Generation**—By providing human users, or intelligent agents, the ability to assign relationships among concepts, basic semantic equivalence can be asserted among schema entities, thereby enabling the software to make intelligent guesses at relationships among schema types.
- **Dynamic Discovery of Information and Services**—By providing agents and humans the ability to look up new items on registries or via taxonomy. This is a fairly well understood capability that has driven much of the initial excitement around Web Services. Although it is not trivial, we are much farther along in this area of dynamic capabilities than in the others.

These dynamic capabilities are the key enablers for eliminating the low-level involvement of programmers and application experts in the day-to-day operations of information sharing. The ability to initiate and complete round-trip transactions and intelligence operations on the fly and without prior knowledge about specific outcomes is essential to a sustainable long-term vision of information sharing at the scale we require for homeland security initiatives.

Real Time

People coordinating the day-to-day requirements for information sharing should be able to establish previously unforeseen data sharing requirements rapidly, without months of development time, without months of analysis, and without the costs associated with all of that effort. Real time is the capability of establishing new modes of communication among services and systems on demand and in real time.

Of course, the more technical meaning of the term "real time" must also be a core capability: to facilitate transactional message exchanges over different network topologies (i.e., publish-and-subscribe and request/reply) without significant latency. Closely associated with real-time messaging is queuing technology—resulting in guaranteed delivery for mission-critical data delivery when lives depend on it.

The bottom line is speed. Speed to respond to new requirements, and speed to get the data to the right place.

Loosely Coupled

A basic tenet of systems design and architecture is indirection. Indirection is a concept that is used to plan for future uncertainty. Simply put, indirection is when two things need to be coupled, but instead of coupling them directly, a third thing is used to mediate direct, brittle connections between them. A simple example of indirection is one that you probably use every day. When you type a web address, such as www.yahoo.com, into your web browser it will not go to Yahoo directly. First it is directed toward a DNS (domain name server) that tells your computer what IP (internet protocol) address the simple name www.yahoo.com is mapped to. These numeric addresses, like phone numbers, identify the correct location of the Yahoo website.

Indirection is used to insulate separate concerns from being too tightly connected with each other. For instance, when Yahoo changes its IP address, which it may do all the time, your computer system doesn't break when you try and connect to it—because of indirection via the DNS.

In object-oriented software systems, indirection is used in software patterns such as proxy, facade, and factory, among others. In a semantic framework indirection is a fundamental aspect of ensuring loose coupling for the purposes of efficient, dynamic, and automated information sharing.

Abstraction is also a primary enabler for this kind of architecture. If you could imagine the above scenario of using a pivot data model without abstraction, it would require the aggregation of all of the data elements in a particular community. The result would be a community of 500+ applications, each application with approximately 100 data elements, requiring a pivot model with about 50,000 data elements. An abstracted model would conceivably be capable of representing this information in far fewer than 100 data elements.

By creating loose coupling in the fundamental aspects of the technology, the semantic interoperability infrastructure is built for change. It's this built-in capability to manage change, adapt, and be flexible that makes this infrastructure concept more powerful than other approaches.

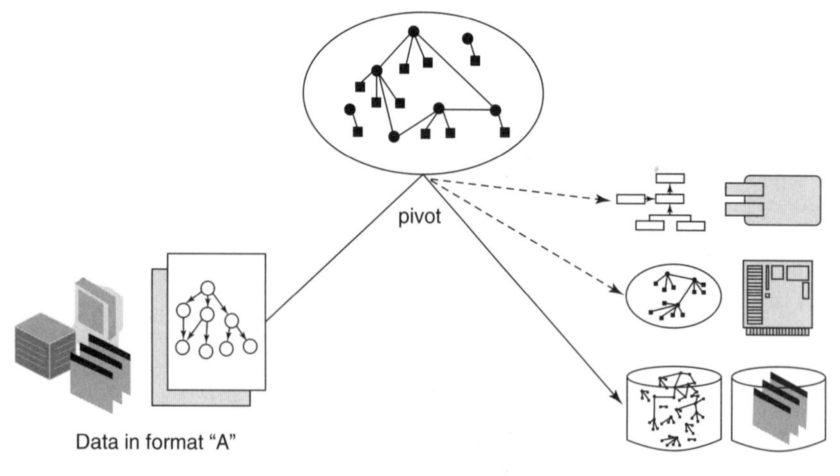

Figure 3.4 Indirection with data schema and ontology

Highly Flexible

Each enterprise system or information resource should be capable of participating in configurable and adaptable interoperability architectures—despite different core connectivity platforms.

A semantic interoperability framework must be pluggable. Pluggability means that different core technology platforms can make use of the framework by simply connecting and publishing some basic metadata about it. Functionally speaking, the framework should be able to support many different kinds of information technology resources. Resources such as:

- Applications (software that provides functions to users)
- Middleware (software that connects other software)
- Databases (software that stores data)
- Portals/hubs (software that distributes functions broadly)

Flexibility also implies an ability to work with various kinds of technical transaction modes such as publish-and-subscribe events as well as request/reply calls. These transaction modes often define middleware deployment topologies—at the network and message levels. Typical deployment topologies include:

- **Hub-and-Spoke**—First popularized with early deployments of NEON middleware, the hub-and-spoke deployment architecture is textbook request/reply message transactions typical of many popular vendor offerings.
- **Message Bus**—The message bus architecture is usually employed with a publish/subscribe event management scheme. This architecture was popularized and most frequently associated with Tibco middleware offerings.
- **Web Services Grid**—The newcomer on the block, the Web Services grid is the registration of many application and utility interfaces and service descriptions in an open network cloud. This allows other members to use their services dynamically, so long as it is consistent with appropriate security policies.

The framework needs to unify data across diverse topology ecosystems by making data of all types—including unstructured, semistructured, and structured—interoperable. The semantic interoperability framework makes middleware architectures more flexible by embedding machine-processable metadata required for interpretation of the meaning of operational data.

In today's technical terms, the framework must also be accessible via industry-standard APIs including Web Services, Java and J2EE, COM and DCOM, CORBA, and others.

Secure

Security, like semantics, is a technical component that is difficult to draw a single box around. In fact, security needs exist at every layer of a complete technology

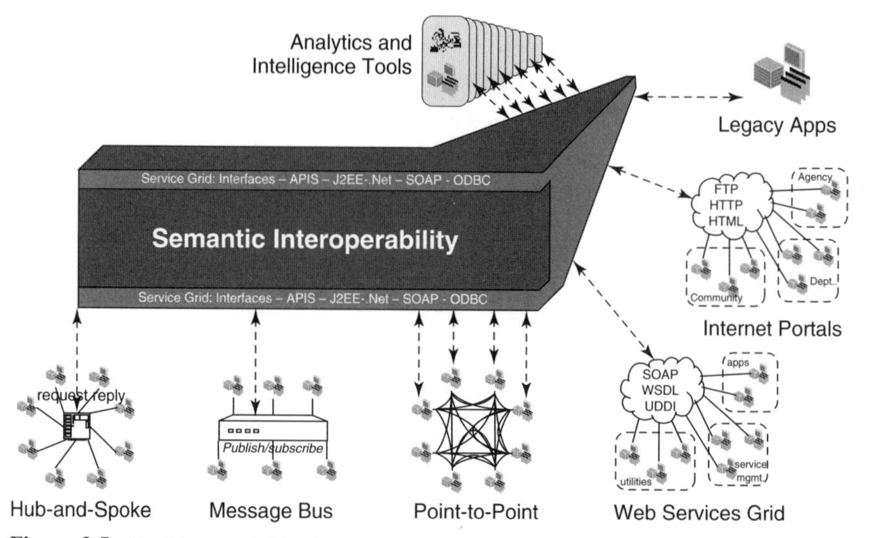

Figure 3.5 Flexible accessibility from enterprise architectures

architecture and are inherent in the tools we use during deployment. Network security is a different kind of mechanism than application security—and application security is a different kind of security than data security.

At the network level, mechanisms such as network fire walls are part of a successful infrastructure. At the application level, the application fire wall is the cornerstone tool that provides security to the deployed information framework. At the data level, mechanisms should exist that separate different realms and hierarchical access points to sensitive information. Various models of access levels can be applied from the CIA, the FBI, law enforcement, and the Department of Defense.

Without going into excessive details here, common information technology security components would typically include:

- Fire walls
- Application/network authentication/authorization
- Private/public key encryption
- Malicious attack protection
- Certificates and signature support
- Encryption and hash
- Auditing and reconciliation

The semantic interoperability framework should make use of each of these proven, industrial-strength security precautions. Most importantly, this entire proposal is about security—helping us gain confidence in our nation's ability to respond before and after potential threats to its citizens.

Open

A solution built to stand the test of time will meet the criteria of the openness test:

1. A solution should not depend on proprietary products or fail to accommodate open standards in order to best meet the test for openness.
   - Platform neutral services (web service, J2EE, .net)
   - Schemas (OWL, RDF, UML, Express, XML)
   - Registries (UDDI, LDAP)
   - Communication protocols (HTTP, RMI, SOAP, FTP, SNMP)
2. A solution should remain neutral with regard to core business differences to appropriately meet the test for openness.
   - Data vocabulary mediation—agree to disagree
   - Query mediation—write once, query everywhere
   - Process mediation—loosely coupled process sharing
3. A solution should not require information consumers to use the brand of application software or middleware that the information provider uses to best meet the openness test.
   - No adapters required
   - Insulate the business process from other organizations
   - Insulate the internal technology from other infrastructures

Service Oriented

A robust information framework is an "on the wire" service that is always waiting to be used. It is a nonfunctional (in terms of providing business functions to end users) part of the overall systems infrastructure providing a utility to the rest of the applications, systems, and processes within the security community.

The infrastructure service ensures that applications and processes can get on the grid and plug into a broad service utility. It is the kind of software component that gets built and then forgotten about. Once in place, the services continue to be used, but the malleable aspect of the solution is maintained through the schemas, business rules, and domain models registered within the framework.

Information-Centric

Throughout the history of computing, information has taken a backseat[6] to other computing aspects. Whether code, messages, or processes, it seems that people just want to assume that data will take care of itself. This lack of information-centricity has created a massive problem with the way data is stored and represented. The semantic interoperability framework challenges this central problem through recently emerging approaches for working with data semantics. Four core principles are applied to a computing environment to formalize the description, management,

[6] Smart Data for Smart Business: 10 Ways Semantic Computing will Transform IT, Michael Daconta. *Enterprise Architect*, Winter 2003.

and interoperability of information assets: mediation, mapping, inference, and context.

By separating the management of information from processes and software applications—through the use of context—a pluggable environment may be created that dramatically reduces new solutions' development time and excessive maintenance costs. In this way, the semantic interoperability framework can become the chassis for the next generation of government information technology.

Autonomic

Solutions that have a high degree of human involvement, because of a lack of automation, experience much higher costs, much longer cycle times, and higher maintenance. In the integration world it has been notoriously difficult to automate key processes of linking systems together. Because of differing technologies, cultures, and organizational processes specialized resources are almost always required to be hands-on when integrating systems. In fact, a great deal of integration is still done with "swivel chair" methods in which people literally type information from one system to another.

For a next-generation solution to change the rules of how involved people need to be in creating middleware solutions, we need to introduce new technologies that can learn more about businesses. The framework should leverage these technologies to automate the process of several key areas of middleware solution development:

- **Query Generation**—rules and constraints associated with querying databases, XML documents, and custom APIs
- **Data Transformation**—algorithms and structures for manipulating instance data in a variety of forms
- **Schema Mapping**—relationships and entity associations among a community of related schema
- **Schema Generation**—structure and concepts of information in both broad domains and specialized applications, for unstructured, semistructured, and structured data

Part of the overall goal of automation is to build in capabilities for the framework to evolve on its own. Steps in this direction have been partially implemented in other systems; for example, data mapping tools use thesauri to associate synonyms to data elements—these thesauri can store new associations that help to further refine the semantics of the thesauri each time a new map is completed. Other capabilities that exhibit elements of evolutionary growth can be found in the commonly used ranking systems employed by websites such as eBay, where community participants rank the value of the information and services they receive from others.

Next-generation tools will incorporate these kinds of ideas to form a type of artificial intelligence (AI) within middleware. We will see these capabilities highlighted by software that can infer data meanings, alter data semantics dynamically over time, establish community rules for interactions, and reason about the information and services that are widely available. These strengths will ensure that humans are not required to facilitate the programming of rules at the very lowest levels—and thereby ensure that we can more rapidly make use of the information distributed across agencies, organizations, and commercial entities.

DEVELOPING A SEMANTIC INTEROPERABILITY SOLUTION ARCHITECTURE

It is no secret that a solution with all the characteristics presented in the previous section does not yet exist. However, tomorrow's semantic interoperability capabilities will fulfill these promises with a combination of several technologies that are already available.

Tomorrow's information sharing platforms will contain key technologies that enable loosely coupled, dynamic, and more automated information sharing:

- **Ontology Tools**—tools that create, manage, and link conceptual models, taxonomy, and canonical models to actual enterprise data, process, and business rules schemas
- **Dynamic Mediation Tools**—tools that operationalize the ontologies within or across organizations to mediate conflicting processes, policies, data, and business rules without custom-written programs for the purposes of information sharing

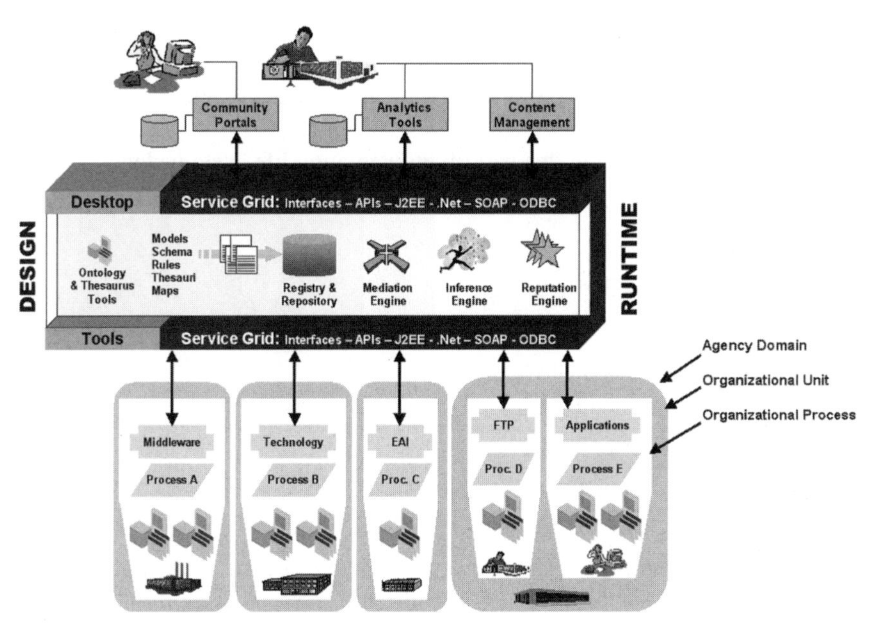

Figure 3.6 Plugging into the semantic framework across agency domains

- **Inference Tools**—tools that dynamically navigate ontology and schema to facilitate the generation of new schemas, queries, and business rules in a loosely coupled model-driven environment
- **Thesaurus Tools**—tools that allow nonexpert information consumers to identify basic semantics for data, processes, and business rules, thereby enabling automated relationship building among schema, ontology, and business rules

These four core information sharing mechanisms will complement many of the now standard protocols and data sharing techniques that are commonly understood, such as:

- Web Services
- EAI/B2B/MOM Messaging
- J2EE/DCOM APIs
- Application and network security
- Registry and repository services
- Other communications protocols

With the proper architecture utilizing these core information-sharing tool sets, a wide array of flexible interfaces, and a number of possible network data exchange topologies, a complete solution may be constructed.

Design Time

One way to differentiate the views of the architecture is to consider the differences inherent in design time activities vs. run time activities.

Design time activities are those that are typically performed before any physical attempt to connect, integrate, or otherwise make interoperable a set of two or more applications.

Typical activities done in this part of an integration life cycle include:

- Requirements analysis
- Object modeling
- Logical data modeling
- Conceptual modeling
- Business rules identification
- Data dictionaries and thesauri
- Work flow and process analysis
- Interface API descriptions

As Figure 3.7 indicates, the intent of the semantic framework is to make the results of all of this design time work storable and executable within a shared platform.

Once we have the opportunity to drive the run time processes from the actual artifacts that are created during the design time portion of a project, many efficien-

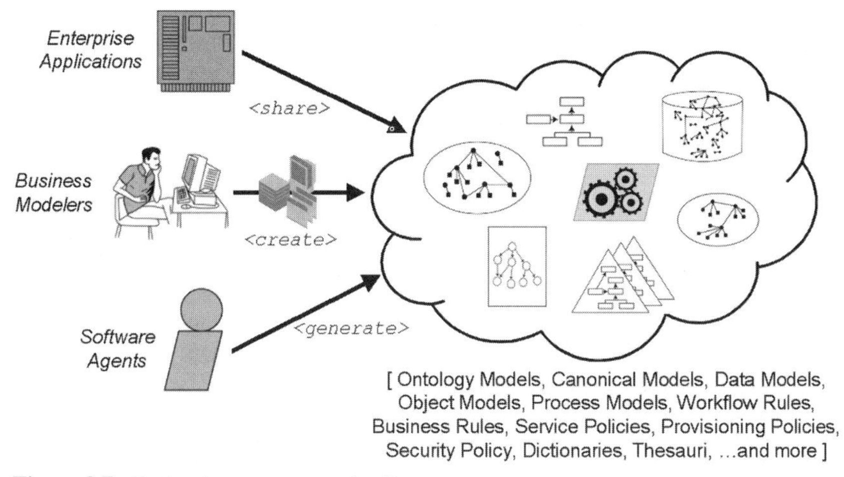

Figure 3.7 Design time processes and artifacts

cies will be realized. These efficiencies will translate into fewer person-hours spent in the design and development phases of integration efforts. Traditionally, programmers are asked to interpret the design intent of numerous design deliverables into executable code. Instead, the design deliverables (models, work flow, business rules, dictionaries, etc.) will be made available to the run time layer directly, to enable the dynamic configuration of a wide variety of possible data and process interactions—without attempting to predict the precise nature of a data or process exchange.

Run Time

The persistent challenge to integration has been the amount of human involvement required to get data, processes, and systems that have been developed at different times, in different places, and by different people to interoperate. The subtleties in business rules and semantics associated with these challenges (data, processes, and systems) have always been more of a gating factor than any specific technical issue of connecting to a new application. It is these factors that mandate significant human involvement in the process of software development. Software development processes guide engineers through a formalized process of understanding and analyzing a problem and then designing, developing, and testing a solution for it.

This is one core reason why today it takes months, instead of minutes, to make one set of databases at a local sheriff's office to communicate with legacy applications maintained by an international crime database.

A semantic interoperability architecture would change the rules of the game.

Semantic interoperability succeeds because it reduces the amount of human intervention in the acquisition of data from previously unknown data sources. The premise of the framework's run time is to accept any data, any time, over any pro-

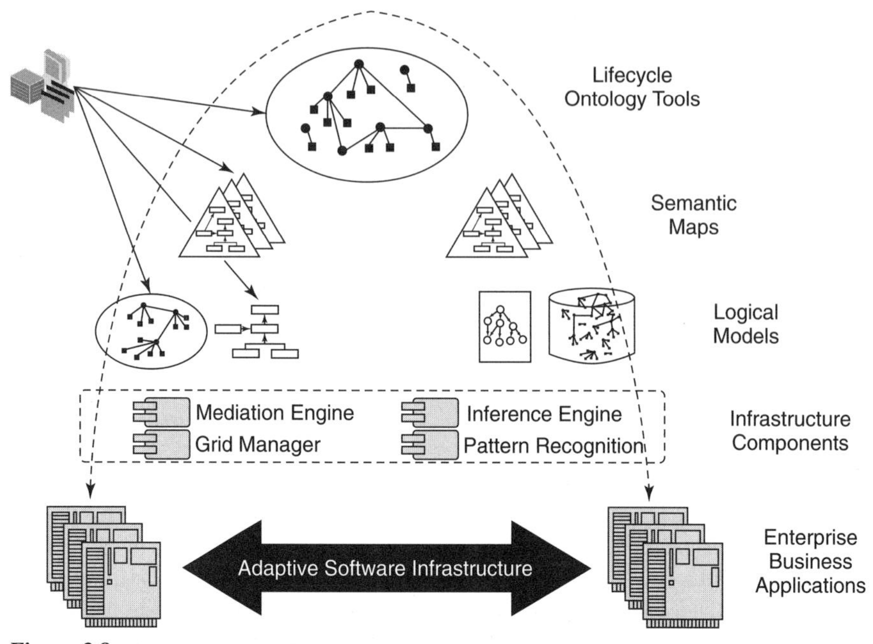

Figure 3.8 Any to any, dynamically, without a priori knowledge or code

tocol. Run time service components would be responsible for deriving the meaning of data. Once the meaning was established in a platform neutral way, it would make the data queryable and transformable into the context of any other system that is also plugged into the framework.

If we start with the baseline assumption that a source will have access to the framework's architecture over a communications channel—which can vary by application or security requirements—then the process of sharing information begins with registering process and data models. This process can be automated in ways that reach far beyond what we presently do when designing and writing programs to accomplish tasks—and can provide a multitude of benefits over time that custom-written code cannot.

Once these steps are completed, the framework would have enough "knowledge" about the data, process, and business rules to dynamically query, transform, and share the local application's contents with any other framework subscribers without having to predict the precise nature of interactions in advance.

Of course, this scenario is a simplification for the purposes of presenting a high-level flow of events; some nontrivial assumptions include:

- The existence of one or more community models (process or data) that are sufficiently abstract to serve the purpose of a mediating schema
- The existence of tools to infer popular community usages and help direct the appropriate mapping into the common (pivot) formats

- The linkages of diverse local query syntax to a common neutral syntax that operates on ontology (i.e., the mechanisms to map description logics to traditional query interfaces)

So, although this architecture may sound complex and uninviting today, the benefits of what it delivers far outweigh the trepidation in attempting to use it to solve our pressing problems. There are no other possible solutions, save humans coding queries, transformations, and business rules over many months of development time.

FINAL THOUGHTS ON FRICTIONLESS INFORMATION

Although we used homeland defense as an illustrative example, the issues and requirements in commercial businesses closely mirror the demands of agency-to-agency interactions. As difficult as it may be to admit, a ready-made out-of-the-box solution for this kind of advanced information sharing does not exist today. What we have described in this chapter is still largely just a vision. However, the next generation of tools that will meet advanced intelligence sharing requirements does exist—some assembly required!

Fortunately we are in a strong position to draw on expertise from several key disciplines to enable this convergence to occur. Experts in the fields of ontology engineering, Semantic Web, Web Services, EAI/B2B, inference engines, data and schema mapping, work flow, and process modeling are guiding the next generation of information sharing platforms each and every day. Part Two of this book, The Semantic Interoperability Primer, focuses more deeply on the nature of these emerging technology programs and begins to identify some patterns of their usage.

Part Two

Semantic Interoperability Primer

The second part of this book is explanatory in nature and takes on three essential questions. The first question, covered in Chapters 4 and 5, is What is semantics? This deceptively simple subject is shown to be more complex than many pundits would claim. Semantics are shown to exist along a continuum that roughly mirrors the form data can take in different layers of enterprise architecture—with significantly different conflicts that could occur. In these chapters, an overview of semantic solutions is provided and the role of context is shown to be a crucial factor for solving semantic conflicts of different types.

The second question, covered in Chapters 5 and 6 is, is What is metadata? Metadata is shown to be a critical element for capturing both semantics and context, but to fully understand metadata, each of its many layers must be clarified. Metadata is far more than just data about data. Five horizontal layers of metadata are decomposed with one vertical layer of rules-oriented metadata that cuts across the others. Chapter 7 takes the discussion of metadata further by focusing on ontology and ontology design patterns. A crucial aspect of semantic metadata, ontologies are shown to vary in their structures and in their possible uses.

The third question, covered in Chapters 6, 7, and 8, is How would I use semantic technology? Semantic architectures are shown to be both nontrivial and multimodal. For software to effectively solve problems of the complexity required, its architecture must support different modes of operation for solving different aspects of the semantics problem. Different

Adaptive Information, by Jeffrey T. Pollock and Ralph Hodgson
ISBN 0-471-48854-2

views of a sample architecture are discussed in depth and applied to a hypothetical PLM interoperability problem. Chapter 9 concludes Part Two by identifying key infrastructure deployment patterns and assessing their applicability to commonly understood E-business models.

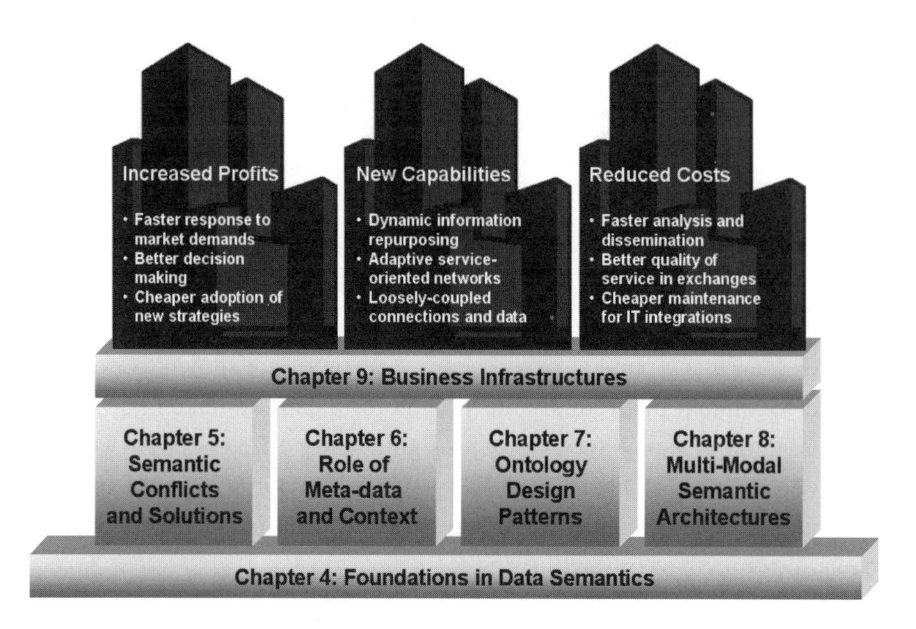

Chapter 4

Foundations in Data Semantics

KEY TAKEAWAYS

- The foundations of semantics lie in a 3000+-year debate of philosophy, scientific method, and mathematics.
- Understanding meaning is inherently fuzzy, paradoxical, and context dependent.
- Semantics in digital systems can be discovered through multiple avenues including pattern analysis, thesauri, inference, semantic mapping, and data nets.
- Most information technology contains inherent, but only implicit, semantics.
- Semantics are evolutionary—data meanings change over time.

A great debate involving some of humanity's most influential thinkers throughout history, on subjects like truth, logic, knowledge, and wisdom, has raged on for over 3000 years. This discussion, as the study of knowledge, has in large part shifted during the past hundred years—toward subjects like linguistics, natural language, and semiotics.

As much as this extended debate might be interesting for philosophy enthusiasts, one should certainly question how relevant it is to what matters in day-to-day business activities. Today, the topics explored in this debate should have great relevance to the reader more than ever. For many reasons these ideas have particular relevance to a modern organization, including business processes, human relations, organizational structures, and most importantly in an information-based economy, applications of information technology.

One engineering friend said that he could not imagine an engineer who would not want to understand how ideas are connected and related. However, many others have said that the concepts and philosophy underlying a discussion of semantics are uninteresting to practical people, who just want to know how to implement software in more effective and efficient ways.

The authors strongly believe that it is crucial to understand the conceptual underpinnings of a new technology in order to appreciate its utility in the enterprise—especially when the new technology requires a significant shift in approach and mind-set. Perhaps in several years semantic technologies will become as

Adaptive Information, by Jeffrey T. Pollock and Ralph Hodgson
ISBN 0-471-48854-2

commonplace and easy to use as relational databases or object-oriented programming, and thus require very little explanation of their inner workings. But today there are still significant questions surrounding the validity of the semantic approach, the theoretical limits of its utility, and the very nature of meaning in information technology, and so it is necessary to lay a conceptual foundation.

Therefore, this chapter is intended to provide some answers for those who are skeptical about semantic technologies, to demonstrate its theoretical foundations and the limits of its applicability. We will utilize the first third of the chapter to discuss some of the more ephemeral ideas, history, and social implications of what meaning means. The middle third of the chapter will focus on explaining the different conceptions of semantics that can be applied to information technology, with a focus on providing a pragmatic taxonomy for understanding what the terms mean when IT people talk about semantics. The final third of the chapter will explore how these different conceptions of semantics are actually embodied in digitally stored information.

INTRODUCTION

Even the most advanced IT is inherently incapable of the kinds of analog processing that the human brain is capable of. Because IT is built on digital structures, patterns of ones, zeros, and absolutes, it is limited to a rules-based system at its processing core. Therefore, we are forced to implement technology in relatively unsophisticated ways. Historically, improvements to the processability of digitally based knowledge have usually occurred in Kuhnian-like fits and starts. Like the Internet a decade ago, a set of disruptive technologies are emerging that may forever change our conceptions of IT. And like the Industrial Revolution, these disruptive changes will manifest themselves at an ever-faster rate and find their way into the everyday lives of millions.

These new information-centric technologies will make semantics explicit, rather than leaving it to the eyeballs of programmers and the implicit meanings in their heads. This will allow us to imbue the system with a kind of intelligence. But, unlike AI of the 1970s, this newer "AI for middleware" optimizes decentralized network and information-centric communications, not monolithic, self-pondering machines like HAL. However, along with the increasing importance of semantics in IT, new ideas are emerging that challenge long-held belief systems and prompt us to reevaluate old questions in a new light. One simple core question of this discussion leads to some uncomfortable answers: What exactly is communication?

Our current notion of software-based communication is broken. Software-to-software communication is broken because the IT community mind-set has not yet fully adjusted to a distributed world. Generally speaking, we still tend to build software as though it existed in a closed world. These closed systems result in the construction of software that does not maintain links between data and shared, but usually implicit, background knowledge that is necessary for two or more parties to have the same understanding of the meaning behind shared data. Take, for example, how the cornerstone of current IT communications theory (the OSI communications stack) basically ignores the role of *understanding* in communication and systems interconnectivity. This utter lack of attention to the role of data meaning shows that a new model, which makes explicit a necessary and sufficient formalization of data's background knowledge required for complete communication, should be considered.

Today, a definitive approach does not yet exist. Several efforts with the OMG (Object Management Group), W3C (World Wide Web Consortium), and other industry groups are building models to support these capabilities. Leading computer scientists, researchers, commercial software architects, and software engineers across the globe are participating in this coming revolution by contributing their ideas and software.

They all stand on the shoulders of giants.

It is hard to imagine, but many of the very things researchers discuss today, in the context of software, have been discussed in other communities for thousands of years. In so many ways, we owe our thoughts and worldviews to brilliant minds long gone. But in particular, the subjects of semantics, context, and knowledge have a longstanding intellectual heritage. This heritage must be understood to fully appreciate where we are going with enterprise software in the future.

THE GREAT DEBATE

A great debate is taking place among history's greatest thinkers, occupying minds a thousand years apart. This debate has centered on seemingly esoteric ideas on truth and knowledge, yet these esoteric issues are of immense importance to the human race. Thinkers in the western tradition, such as Plato, Aristotle, Augustine, Aquinas, Hume, Locke, Bacon, Kant, Schopenhauer, Nietzsche, and Russell, all have recorded their commentary for others to ponder throughout the ages.

The very foundations of western civilization rest upon their shoulders. They've spoken to each other regarding subjects that still perplex us and offer us new frontiers today.

Some of the topics they have explored and debated include:

- Truth
- Knowledge
- Logic
- Wisdom
- Causality
- Scientific method
- Mathematics
- Aesthetics
- Physics
- Relationships
- Universal and particular
- And many more . . .

Concerns such as these are of crucial relevance to competitive businesses at several levels of analysis. Business processes, and the complex dynamics that occur within them, are inescapably tied to the uncertainty underlying the great conversation. Patterns of human relations, at home as well as in the office, offer many areas for further study. Organizational structures, in most circumstances, are founded upon management theories that were formed decades ago. And, most relevant to the topic of this book, applications of technology are heavily influenced by many topics of the great conversation.

Of particular relevance to information technology, struggles with digitization of process, information, and idea circle around the larger transcendental issues of truth, logic, mathematics, relations, idea, and abstraction like moths around a flame. The necessity to embody intangibles in the tangible drives the codification of these eternally fuzzy notions in a variety of ways. Ephemeral ideas surface in seemingly mundane concerns like:

- The logic of truth in business rules (by methods and exceptions)
- Abstraction and data modeling (designing software classes and objects)
- Relationships and interoperability of closed systems (communication of meaning)
- Handling variables and constants (determining global and local subjectivity)
- And many, many more . . .

The ways in which our software systems process these ephemeral ideas is the key factor in how new capabilities can manifest themselves in information technology. Understanding the philosophical context underlying these issues is a necessary first step in overcoming the currently limiting factors in information technology.

Plato and Aristotle Argue About Truth

Over three thousand years ago, Plato and Aristotle framed the core debate about knowledge and truth that continues to this day. Does the essence of reality and truth lie in the ideal realm or in the tangible realm? Plato was an ardent believer in the truth of the idea and abstract concepts, whereas Aristotle defended the view that more concrete and measurable elements of the world are what comprise truth and knowledge.

Plato's lifelong contention was that the idea is the ultimate reality; his famous cave allegory is a poignant example of this. He was interested in the ideal form of reality; his realism was the world of form and idea. In his theory of forms Plato believed that there exists an immaterial universe with perfect aspects of everyday things such as a table, a bird, and ideas/emotions, joy, action, etc. The objects and ideas in our material world are mere shadows of those forms. This solves the problem of how an idea, such as the idea of a chair, unites all the various instances of a chair in the physical world. For example, Socrates, in asking for a definition of piety, says that he does not want to know about individual pious things, but about the idea itself, so that he can judge "that any action of yours or another's that is of that kind is pious, and if it is not that it is not".[1] Plato concludes that what we look upon as a model is not an object of experience, but some other real object.

[1] Using the G.M.A. Grube translation (*Plato, Five Dialogues, Euthyphro, Apology, Crito, Meno, Phaedo*, Hackett Publishing Company, 1981, pp. 6–22) [6e, G.M.A. Grube trans., Hackett, 1986].

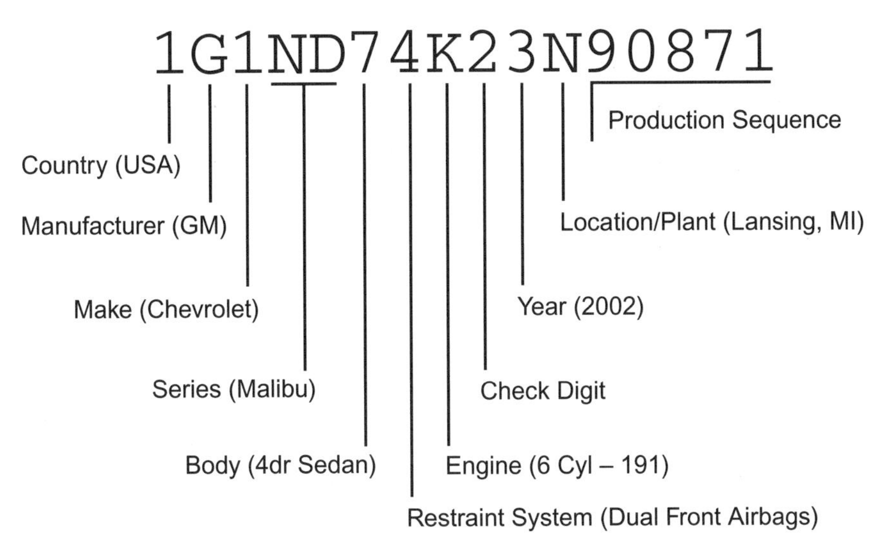

Figure 4.1 Vehicle Identification Number schematic detail

Aristotle's core disagreement with his teacher rests on the premise that truth and reality lie in what can be measured. As the father of modern science and rational thinking, Aristotle's realism view of nature leads to robust theory of causation, among other things.

> *Aristotle constructed his view of the Universe based on a intuitive feeling of holistic harmony. Central to this philosophy was the concept of teleology or final causation. He supposed that individual objects (e.g., a falling rock) and systems (e.g. the motion of the planets) subordinate their behavior to an overall plan or destiny. This was especially apparent in living systems where the component parts function in a cooperative way to achieve a final purpose or end product.*[2]

This duality between Plato and Aristotle has remained a central theme of knowledge and epistemology since their time. In fact, the duality of concrete and abstract truths can be found everywhere.

In computer science the duality is reflected in the relationship between schema and data, where the schema represents only the form of the data and tells us nothing of the data value itself. Likewise, the data only tells us value and fact, but without the schema we cannot know the data's form. For example, an automobile VIN number itself tells us nothing about model, make, or manufacturing date, but the schema for that VIN instructs us on how to dissect and understand the data.

In software programming the duality of "abstract" and "concrete" can be found in the relationship between classes and objects. When we program in OO languages such as Smalltalk and Java we program only the classes, the blueprint for objects,

[2] Miller, Barry, "Existence", *The Stanford Encyclopedia of Philosophy* (Summer 2002 Edition), Edward N. Zalta (ed.), URL = <http://plato.stanford.edu/archives/sum2002/entries/existence/>.

not the actual objects. The objects only exist when the program is instantiated and data is run through it. At run time, each class may have multiple unique instances, with unique object Ids. For example, in a product data management (PDM) system, the `PartsInvoice` class will define the blueprint for all invoices but represents no single parts invoice instance directly.

For information technology as a whole, the abstract/concrete duality can be found in the relationship between the software and the world around it. When we build systems in software we usually model some aspect of the real world in a business object layer. Then we operate the software to tell us something new about the world or assist us in managing our environment. Yet, often, there is no direct link between the real world and software. Software things like business rules, processes, work flow, and screen widgets have no direct physical representation in the world. But sometimes the things we model, like parts inventory, customers, and product assemblies, do have real-world representation. Today, the linkages between the software and the real world are becoming more and more explicit—especially with the advent of tools like radio frequency IDs (RFID).

In summary, Plato and Aristotle still have quite a bit to teach us about the world. Their relevance to the discipline of IT is a legacy that cannot be denied today. Every time we implement an IT system we are essentially reframing the classic debate in modern terms. Living with the duality of our world necessitates the codification of the idea, as with information technology, as an attempt to gain a better understanding of tangible and concrete business drivers. It follows that the better we get at representing both the concrete and the abstract separately, while improving the means to connect them in meaningful ways, the greater understanding we will have about the business, scientific, and technical worlds around us.

Kant Searches for What is Knowable

Perhaps the greatest German philosopher ever, Immanuel Kant is the intellectual giant of the Enlightenment period. As one of the most influential philosophers in the history of western philosophy, he made contributions to metaphysics, epistemology, ethics, and aesthetics that have had a profound impact on almost every philosophical movement that followed him. His most famous work, *The Critique of Pure Reason*, focused on the epistemology and metaphysics of reasoning. A significant portion of Kant's writing deals with the question "What can we know?"

Through his work, Kant became the father of modern information theory, drawing a distinction between data (sense) and semantics (understanding) and how their union is required for complete information and knowledge. His concepts of phenomena, things that can be directly known, and noumenon, things that cannot be directly known, are analogous to the concepts of classes and objects in computing. For Kant, objects take form and are bounded by space, time, and categories whereas a class blueprint, like an idea, is more about what could be. Kant determines that knowledge can be empirical or transcendental, but most importantly, Kant writes that information is context dependent and varies from person to person.

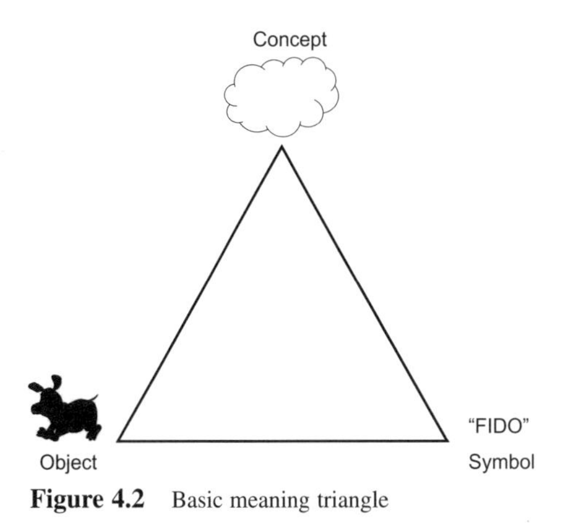

Figure 4.2 Basic meaning triangle

C.S. Peirce Redefines Logic and Meaning

As the founder of American pragmatism, Charles Sanders Peirce was a prolific writer on many subjects. Despite being a professional chemist, he considered philosophy and logic his primary vocation. Universally accepted as one of the greatest logicians who ever lived, he documented and studied a new kind of logic.

Abduction is *a method of reasoning by which one infers to the best explanation.* Most people consider abduction logically unsound but accept it as a very common reasoning model by which scientists make discoveries. An example of abductive logic is:

- All CPUs produced at this factory are 4-GHz chips
- All CPUs in my computers are 4-GHz chips
- Therefore all chips in my computers are produced at this factory

Peirce defined the scientific method as consisting of deduction, induction, and abduction—where abduction is the first step toward new information and discovery.

His greatest contributions were probably to the study of meaning. He pioneered several areas of the study of signs and semiotics. He evolved Aristotle's original conception of the triadic nature of meaning into a more comprehensive theory. For Peirce, meaning is understood by triangulating the object, its representation, and its referent.

According to John Sowa, a modern-day computational linguist discussed more fully later in this chapter, Peirce took the concept of the meaning triangle further than anybody else by recognizing that they can be linked together in arbitrary numbers of levels.[3]

Figure 4.3 shows how meaning triangles can be linked to show how the object is related to the concept, but also how the concept is related to the representation of

[3] Sowa, Chapter, 3.6 "Knowledge Representation".

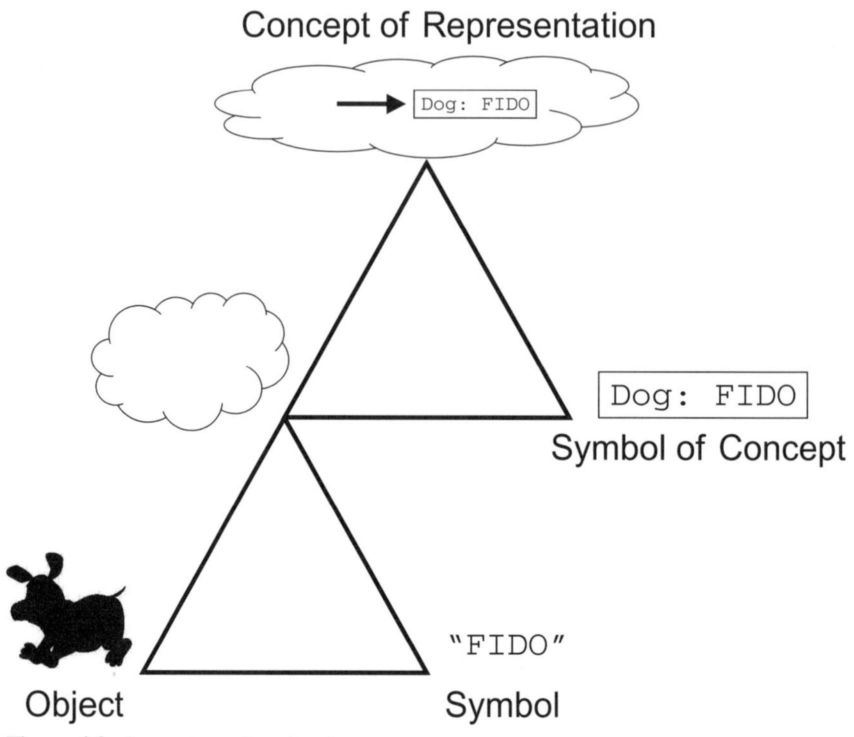

Figure 4.3 Layered meaning triangles

the concept in software. This is a useful technique for shearing the layers of meaning found in software systems.

Wittgenstein Finds Meaning and Nonsense in Language Games

Ludwig Wittgenstein is considered by many to be the greatest philosopher of the twentieth century. He played a central and controversial role in twentieth-century analytic linguistic philosophy. His works continue to influence current philosophical thought in topics as diverse as logic and language, perception and intention, ethics and religion, aesthetics, and culture. Two commonly recognized stages of Wittgenstein's thought, his early years and his later years, are pivotal in their respective periods. *Tractatus Logico-Philosophicus* characterizes early Wittgenstein. He provided new insights into the relations between world, thought, and language and the nature of philosophy by showing the application of modern logic to metaphysics, via natural language and semantics. *Philosophical Investigations* was the core of his later works. In his new approach he took a revolutionary step in critiquing all of traditional philosophy—including his own ideas presented in previous works.

For Wittgenstein, the meaning of a word is how it is used in language. In other words, dictionaries don't define words—they merely record their definitions. Dictionaries and thesauri are only snapshots in time. In practice, meaning is dynamic and changing based on usage and context. Think about the implications of this concept to data meanings in software.

Wittgenstein's language games are about the dynamism of language and meaning. They are rule driven but dynamically patterned. There is no set of universal rules for all games; each game begets its own structure. Wittgenstein suggests that people follow words through complicated networks of similarities across language games to get the best idea of what their true meanings are.

Eventually, he identified the ultimate paradox in the meaning of meaning:

> *These considerations lead to . . . [what is] often considered the climax of the issue: "This was our paradox: no course of action could be determined by a rule, because every course of action can be made out to accord with the rule. The answer was: If everything can be made out to accord with the rule, then it can also be made out to conflict with it. And so there would be neither accord nor conflict." Wittgenstein's formulation of the problem, now at the point of being a "paradox," has given rise to a wealth of interpretation and debate since it is clear to all that this is the crux of the general issue of meaning, and of understanding and using a language.*[4]

Bridging the Gap Between Philosophy, Linguistics, and Computer Science

John Sowa, a computer scientist, logician, and mathematician, is the leading expert on knowledge representation, model theory, and conceptual graphs (CG) in the world today. His work is of great importance because philosophers, psychologists, logicians, mathematicians, and computer scientists have been increasingly aware of the importance of multimodal reasoning. The natural offshoot of this emphasis has been in the area of nonsymbolic, especially diagrammatic, representation systems, of which Sowa's conceptual graphs[5] represent a significant step forward. For Sowa, CGs are a way of capturing aspects of natural language in a diagrammatic manner that facilitates its conversion to predicate calculus. Conceptual graphs consist of two kinds of nodes called concepts and conceptual relations.

Sowa is a modern-day bridge between the philosophy of meaning in natural language and the application of meaning in computational systems. Other leading experts in the field today include Nicola Guarino, Chris Partridge, and other scientists at the Universities of Manchester, Karlsruhe, and Stanford.

[4] Biletzki, Anat, Matar, Anat, "Ludwig Wittgenstein", *The Stanford Encyclopedia of Philosophy* (Winter 2002 Edition), Edward N. Zalta (ed.), URL = <http://plato.stanford.edu/archives/win2002/entries/wittgenstein/>.

[5] Sowa, John F., ed. (1998) *Conceptual Graphs*, draft proposed American National Standard, NCITS.T2/98–003.

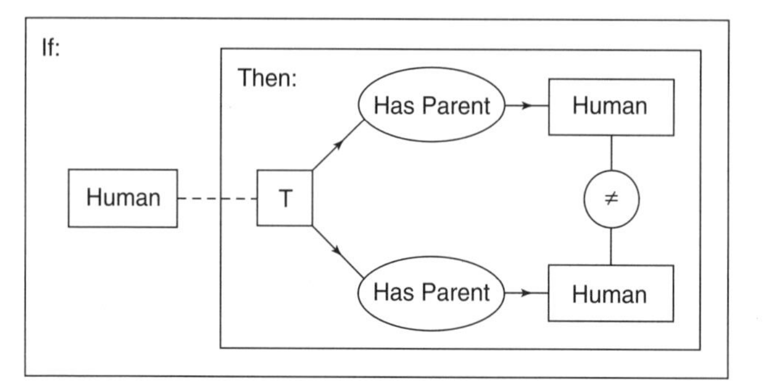

Figure 4.4 CG for "if there is a human then he or she has two distinct parents"

NATURAL LANGUAGE AND MEANING

The human mind does not function in neatly defined digital or first-order logic (FOL) paradigms similar to those that underlie computers. We draw connections and conclusions through means that are not yet fully understood by science. We know we compute, but we're not yet sure of how we compute. The activities of reading a book and engaging in conversation are fundamentally imprecise ways of sharing meaning and concepts. This is so because our perception of meaning is almost entirely influenced by our own personal contexts and background. However, digital communications does not have the luxury of neural or analog processing, the incredible adaptability of the human mind, and our mind's ability to handle information overload. The rule-based ways that software systems work with data are fairly rigid implementations of tightly coupled codified routines.

Definition: First-Order Logic (FOL)

Sometimes referred to as first-order predicate logic, this is a science of symbolic logic with a formalized system of primitive symbols, axioms, combinations therein, and rules of inference. In first-order logic only individual primitives may be used as arguments and variable quantifiers.

Most computational "semantic technologies" have their roots in FOL (or a subset of it). Even description logics, far more decidable than FOL, are regarded as fragments.

But many in the research community assert that there is a widespread lack of understanding and appreciation among IT practitioners regarding both the importance of formal logic and its application within IT systems.

Why Twentieth-Century Philosophy Stalled Out

The shift away from core issues in philosophy, and toward linguistics, that occurred in academia during the late nineteenth century is primarily due to the increased limitations that stem from the looseness and imprecision of natural language. As it became understood that a necessary precondition of rigor in language was required to even discuss basic concepts between individuals, new insight into language itself was required.

Since then, the dominant thread in philosophy has been in the area of linguistics. For example, many, if not most, of the great philosophers of the past 100 years have been linguists. As a consequence of this, our understanding of meaning and communication has come a long way. But there are still major hurdles to forming a complete theory of semantics in language.

Fuzzy Language

> *Like all other scientists, linguists wish they were physicists. They dream of performing classic feats like dropping grapefruits off the Leaning Tower of Pisa, of stunning the world with pithy truths like F = ma. . . . [But instead,] the modern linguist spends his or her time starring or unstaring terse unlikely sentences like "John, Bill, and Tom killed each other," which seethe with repressed frustration and are difficult to work into conversation.*[6]

Natural language is fundamentally imprecise and context dependent. The meaning of words can differ greatly between people and culture. Dictionaries document the various meanings as senses. Additionally, there are three kinds of lexical ambiguity:[7]

- **Polysemy**—several meanings that are related to another
  - Example: "open"—Open has many senses like unfolding, expanding, revealing, making openings, etc., but they are related.
- **Homonymy**—several meanings that are not related to another
  - Example: "bark"—the noise a dog makes, and the stuff on the outside of a tree
- **Categorical Ambiguity**—several meanings whose syntactic category varies
  - Example: "sink"—as noun as a plumbing fixture or as a verb as to become submerged

Language is truly dynamic; every day it changes and evolves as people and cultures evolve along with it. In some cultures, such as the British, this dynamism is celebrated, whereas in others, such as the French, it is abhorred and micromanaged. Because dictionaries (in all cultures) merely document usage, not define words, they must continually be revised and reedited to accommodate this fact.

Tried talking with a teen lately? The hip-hop lingo of this generation is changing the way our kids talk. In 2003 Merriam-Webster introduced the word "phat" into the dictionary and the *Oxford English Dictionary* legitimized the term "bling-bling"

[6] Joseph Becker: Becker 1975—"The Phrasal Lexicon." In *Proceedings, [Interdisciplinary workshop on] Theoretical issues in natural language processing*, Cambridge, MA June 1975. pp. 70–73.

[7] Graeme Hirst, "Semantic Interpretation and the Resolution of Ambiguity".

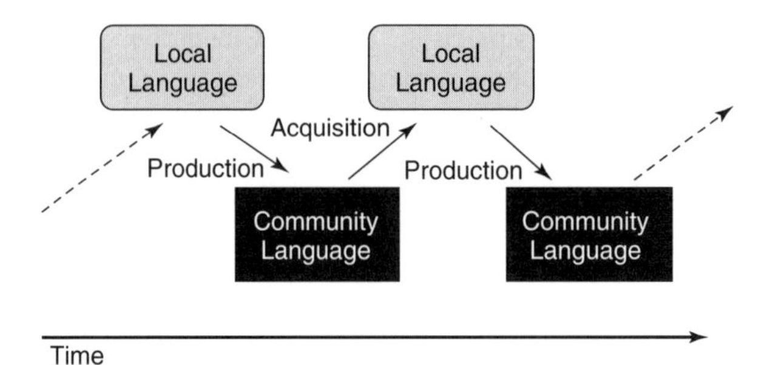

Figure 4.5 Evolution of language over time[8]

in its own records. Although not many people are likely to use these words in normal conversation, an entire generation uses these terms in everyday communications. Judging from the number of universities offering hip-hop language courses, and the number of hip-hop dictionaries on the market, it's easy to see that many are concerned about the breakdown of communication that is occurring across generations and cultures in America—largely due to the subtle differences in the way people are talking these days.

In the world of business the evolving nature of meaning and definitions is equally apparent. But rather than wholesale additions to language, we usually see differences in the way people use terms based on their individual viewpoints. For example, people from different organizations within a business may be discussing an order date for a particular product. However, "order date" might be interpreted as the date the order was placed by the procurement team, while the sales team might interpret it as the date the order was received.

Unfortunately, these same kinds of subtleties (in context, jargon, and meaning) lead to very expensive software communication problems as well. This is because the language barriers that exist between generations and cultures of people also exist between generations and instances of software.

CONTEXT AND MEANING

Just as context is critical for understanding natural language, it is critical for software systems to understand digital data. In natural language, the context of words is typically determined by the discourse context (domain cues) and local cues within the sentence. Likewise, context for digital systems is dependent on both domain cues and local cues.

Domain Context

- **Community Culture**—The people who implement information technology systems imbue the software and data with their own community knowledge, terminology, folklore, and jargon.

[8] Kirby, Simon, *Language Evolution Without Natural Selection: From Vocabulary to Syntax in a Population of Learners*. Edinburgh Occasional Papers in Linguistics, 1998.

- **Business Processes**—No software system operates in a vacuum; the work flow and processes that surround data processing greatly influence the interpretation of that data's meaning.
- **Business Rules**—As business rules are applied, at given points in a process, they can change the intent and meaning that underlie the data.
- **Data Usage Scenarios**—Answers to questions about where the data will be used, spreadsheets, E-mails, ERP applications, reports, and so forth, are crucial in understanding the contexts that influence meaning.
- **Application Functions**—When data moves from function point to function point, even within the software application, data can change meaning.
- **Reporting Formats**—When various user groups compile reports, either automated or manual, they can interpret and use data in different ways.
- **User Interface**—The appearance of data inside the user interface of portal and enterprise software can influence the meaning of data—from either the software or the user perspective.

Local Context

- **RDBMS Tables**—The metadata that describes type, column, length, and other schema information in relational database systems and other formats that use a tabular format is a crucial indicator of data's context.
- **Markup Tags**—metadata in markup languages like XML, SGML, and HTML all provide context in both their schema and the tag sets themselves.
- **Data-Layer Design Elements** (normalization and hierarchy)—The style of data representation, such as how deep the hierarchies are or the degree of normalization in relational sets, can provide insight into data meaning.
- **Application-Layer Design Elements** (objects and structure)—The modeling style employed by object-oriented designers, such as the use of inheritance and encapsulation, is an example of a local context cue.

Definition: Data Context

Data Context—anything that influences the interpretation of data; often a constraint that limits the applicability and meaning of data from particular viewpoints

By discussing these cues we are not discussing semantics directly or the definition of the data itself. The context cues are the scope in which the data is supposed to be understood, and they provide the primary constraints to the many possible meanings of data. Data can mean different things in different contexts. Understanding the role of context is crucial to understanding semantics in digital systems.

Context is the cornerstone—as well as a limiting factor—for all software applications that lay claim to semantic capabilities.

Counterpoint: Invariant Semantics in Context

Another way to think of context is to imagine a scenario where the sum of all context cues could provide a (boundary of interest) framework for specifying invariant semantics. Imagine the use of an ontology as a schema for specifying what concepts things are associated with, for example, that an `OrderDate` is a kind of `Date`. And that context model can be used to specify the invariant meaning of `Date` in certain contexts. For example, that `Date`, in the `Procurement` context, means the date on which the order was received and, in the `Sales` context, means the date on which the order was placed. In this way, the concept of date is consistent throughout the ontology, but can take on different specializations based on contextual constraints. In this way, ontology modelers may avoid the urge to make their ontology very implementation specific.

WHAT ARE SEMANTICS IN DIGITAL SYSTEMS?

For computers to fully process and understand information they must handle both syntax and semantics of data. Data comes in a wide variety of syntactical structures like Java, C++, XML, ASCII, DML, and more. But the semantics are much more subtle and may be found in a number of places. Some semantics are implicit in the structure of the syntax, and some semantics can be explicit in separate models that attempt to build relationships and draw associations between data. Furthermore, some semantics are explicitly defined in the structure of the data itself, as with various knowledge representation languages such as KIF and OWL.

Definition: Data Semantics

Data Semantics—the meaning of data. Meaning is subjective, and the interpretation of data semantics is always constrained by the perspective of the interpreter—context. Semantics are real time—they evolve and change as context evolves and changes—all the time. Data semantics are usually implicit but must be made explicit to be useful in data processing. Techniques for making semantics explicit include pattern analysis, dictionaries, thesauri, inference, semantic mapping, and conceptual graphing.

Semantics Are Real Time

As we pointed out for natural language systems, semantics are evolutionary in nature. Meaning is relative to context, and context changes with time. Software systems that work with semantics usually work with snapshots of data meaning. In many cases, the older the semantic metadata, the less accurate the data meanings (however, the metadata is still useful in determining "original intent"). In some approaches, thesaurus-based approaches, for example, it is quite possible to accommodate the real-time nature of semantics. In others, ontology-centric approaches, for example, it can be much more difficult to accomplish.

To differentiate real-time capabilities for semantic management is not to suggest that different knowledge representation languages impact the dynamism of semantics. However, the ease with which an infrastructure can accommodate the continuous evolution of semantics is very much a function of the chosen architecture and general infrastructure patterns. Solutions for handling this evolutionary aspect of semantics will be discussed in later chapters focused on metadata, ontology, and architecture.

Semantics Must Be Explicit to Be Useful

Interpreting data, for either computers or humans, requires some level of semantic processing to be successful. In fact, systems today already process semantics. But the semantics in most software systems today are implicit in the zeitgeist of the programmer and how he/she builds the algorithms in the system. Fundamentally, the problem of semantics lies not in how to insert them in software but in how to make them explicit so that the underlying and background knowledge applied by the programmer is made available to others who have not participated in the programming process. This accomplishment would lead to greater efficiency and dynamism for the system as a whole.

Consider the processing that occurs around a typical purchase order. An XML purchase order's components may include its date, part number, assembly, process information, delivery data, pick list attributes, and various logistics information. Some of the data semantics exists in the markup and schema information, but far more semantics exist in the processing that occurs around the actual purchase order document. The full extent of the relationships that exist in the elements contained within the purchase order and the relationship between the purchase order and other life cycle actors, such as invoices, bills of materials, applications, and business processes are almost never modeled or defined in a complete manner.

Most people assume that "everybody knows what a purchase order is," but in actuality no one really knows what a purchase order is—in the objective sense. For example, consider the vast number of standards bodies that have been defining purchase orders differently for decades, and no consensus exists today. In fact, a purchase order means altogether different things to different communities—yet each meaning is equally valid in its own context.

This should illustrate the point that purchase orders, along with every other collection of enterprise data, have their meanings encoded implicitly in the software that processes them. But that makes the semantics, and thus the data, entirely unusable by other processes that don't have the luxury of dedicated programmers to encode the semantics in programming logic. To make the meaning behind enterprise data more portable, the semantics needs to be made explicit outside of the programs. Today, almost nobody goes to this level of detail in their analysis because "everybody knows what a purchase order is."

This kind of rigidity in the processing of distributed data, and the interfaces among disparate systems, is what creates the myriad of problems relating to the management of change within connected software systems. Making semantics explicit is the first step in making software more resilient to change.

Continuum of Semantic Approaches

The most important thing about understanding semantics in digital systems is to realize that it exists at different architectural layers and can be made explicit by a number of different means.

There are many camps, comprised of academics, standards bodies, and vendors, who will each espouse different views of what data semantics are—and different approaches toward making data semantics in software explicit. We have attempted to provide an overview of the different approaches while describing their strengths and limitations. It is our belief that there are business and technical architectures that can benefit from each approach toward data semantics in software and that there is no single best way.

The following discussion of approaches to making semantics explicit in software progresses along a continuum representing the kind of data the approaches operate on.

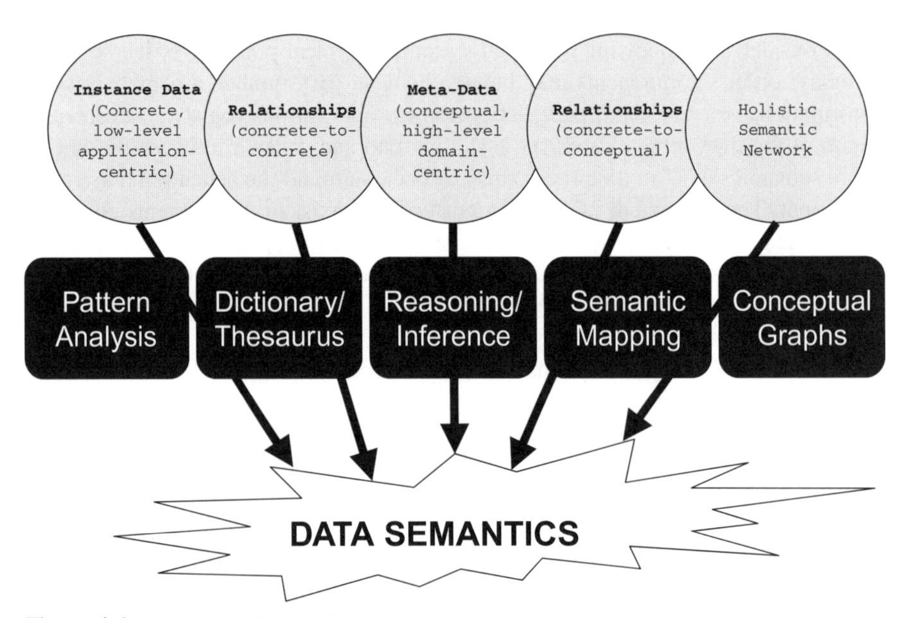

Figure 4.6 Continuum of semantic approaches

Semantics as Pattern Analysis

Pattern analysis has been used to analyze quantities of concrete data elements since the mid-1980s and is probably the most widely available application of semantic tools, data mining, and knowledge management software in use today.

Semantics as Definition and Synonym Relationships

Semantics in this category are implemented partially or wholly in a variety of commercially available knowledge systems, databases, and data automation middleware.

Semantics as Inference and Deductive Logic

Approaches utilizing this approach are available in only a very few middleware toolkits, but its maturity level is quite high—particularly for systems that make use of description logics.

Semantics as Context-Aware Schema Mappings

Software utilizing this technique has been implemented in fewer than a half-dozen commercially available middleware packages at varying levels of sophistication.

Semantics as Guesswork and Abductive Logic

This approach is mostly theoretical right now. Some AI programs have been written that highlight features of this approach, which involves a complicated mosaic of the previous four semantics plus the addition of abductive reasoning methods that overlay FOL and machine learning-type methods.

More details on each of these approaches toward data semantics are provided in the next five sections.

Semantics as Pattern Analysis

Collections of data can be analyzed to find patterns that are not easily discovered by humans and to produce taxonomy or other classification schemes in much more automated ways than humans could accomplish alone. These processes, and others that benefit from the pattern analysis of data, are ways to derive the implicit meanings of sets of data and make their semantics explicit.

These kinds of capabilities are often referred to as data mining techniques. Data mining techniques consist of three primary approaches:

- **Statistical Analysis**—Classic statistics embraces concepts such as regression analysis, standard distribution, standard deviation, standard variance, discriminate analysis, cluster analysis, and confidence intervals, all of which are used to study data and data relationships.

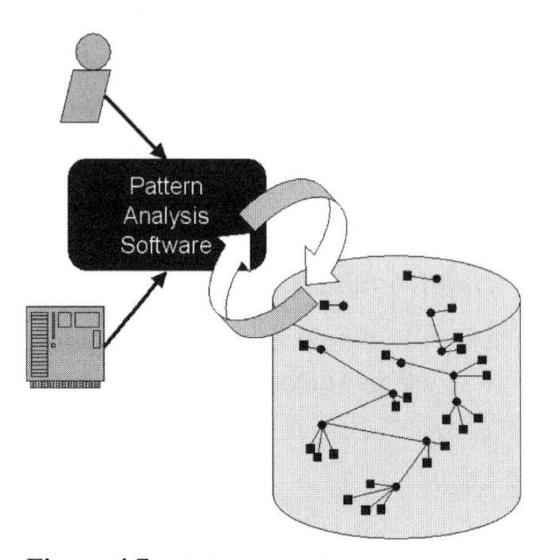

Figure 4.7 Pattern analysis on a data set

- **Artificial Intelligence (AI)**—This discipline, which is built upon heuristics as opposed to statistics, attempts to apply human thought-like processing to statistical problems. Because this approach requires vast computer processing power, it was not practical until the early 1980s, when computers began to offer useful power at reasonable prices. AI found a few applications at the very high-end scientific/government markets, but the required supercomputers of the era priced AI out of the reach of virtually everyone else. The notable exceptions were certain AI concepts that were adopted by some high-end commercial products, such as query optimization modules for Relational Database Management Systems (RDBMS).
- **Machine Learning**—This is the union of statistics and AI. Although AI is not considered a commercial success, its techniques were largely co-opted by machine learning. Machine learning, able to take advantage of the ever-improving price-performance ratios offered by computers of the 1980s and 1990s, found more applications because the entry price was lower than that of AI. Machine learning could be considered an evolution of AI, because it blends AI heuristics with advanced statistical analysis. Machine learning attempts to let computer programs learn about the data they study, such that programs make different decisions based on the qualities of the studied data, using statistics for fundamental concepts, and adding more advanced AI heuristics and algorithms to achieve their goals.

Each of these methods has achieved appreciable results in making previously unknown or implicit semantics explicitly known to both human and machine users. However, because they primarily operate on instance data and already codified metadata, they do not provide a comprehensive set of semantics for meanings that are not already represented (in some form) in computer systems. Other capabilities are required for this objective.

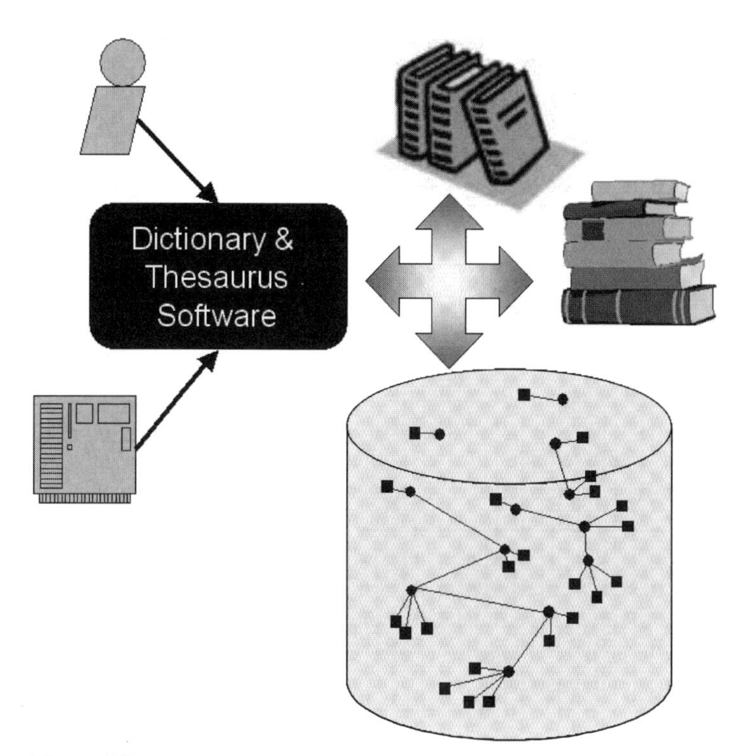

Figure 4.8 Use of external dictionary and thesauri for data semantics

Semantics as Definition and Synonym Relationships

One way that semantics can be partially derived is through the use of synonyms and antonyms. In this approach separate reference dictionaries and/or thesauri exist outside the realm of the software systems themselves. These external references can be queried to determine the definition of a given data element or to get a list of all synonyms and antonyms of that element. In this way, the software system can begin to "know" more about a new data element by comparing it with other elements that it already "knows" about.

The utility of this sort of semantic discovery can enable the software to do many interesting things and useful things, such as:

- Making informed guesses about the use and meaning of new data elements to generate transformation maps between different data sources
- Drawing associations between different data elements in discrete systems to generate an analytic report on the kinds of knowledge and information inside the enterprise
- Obfuscating the underlying structure and syntax of enterprise data in order to provide greater visibility into cross-system semantics and duplication of data

- Generating a taxonomy, with varying scope, of enterprise knowledge by going straight to the source, which would enable the ongoing classification of enterprise data in a fairly easy-to-operate infrastructure

Several tools on the marketplace make use of this approach to semantics. More details about specific tools and vendors can be found in Appendix A, Vendor Profiles, Tools, and Scorecards.

Semantics as Inference and Deductive Logic

This collection of approaches toward data semantics makes use of various techniques for finding relationships that are not already explicit. In some ways, these approaches can be thought of as techniques to automate what would otherwise be the difficult task of having humans make semantics fully explicit in models. Fundamentally, the use of first-order logic-based approaches relies on reasoning about data elements and the ways they are related in the enterprise data and application models. With these approaches, most notably the OWL-based approach that uses RDF and is a core component of the semantic Web activity, a level of strict formalism is required in the development of ontology. This formalism enables software to traverse data elements and discover implied relationships with 100% accuracy and decidability (when using non-monotonic reasoning).

Definition—Inference Engine

Inference Engine—A software program that is capable of inferring new, implicit, facts based off known, explicit facts

All computers today are Turing A machines, where the "A" stands for automatic. This kind of processing style is focused on computing with numbers and functions. Inference capabilities based on this method fundamentally rely on the foundations of predicate calculus. Proponents implicitly believe that these logic-based formalisms can capture and express unambiguous facts about the world. Developments in these logic-based approaches have led to greater capabilities by using networklike associations among data element structures while maintaining the decidability of the forms of reasoning in the logic-based system. These recent developments are embodied in a growing body of work around description logics.

Description logics are a small unit of predicate logic that create a formal, fact-based ontology to describe concepts and relationships for a given enterprise domain. Description logics are specified by three sets of features:

- **Concept-forming operators**—primitive concepts, conjunction and existential quantifiers

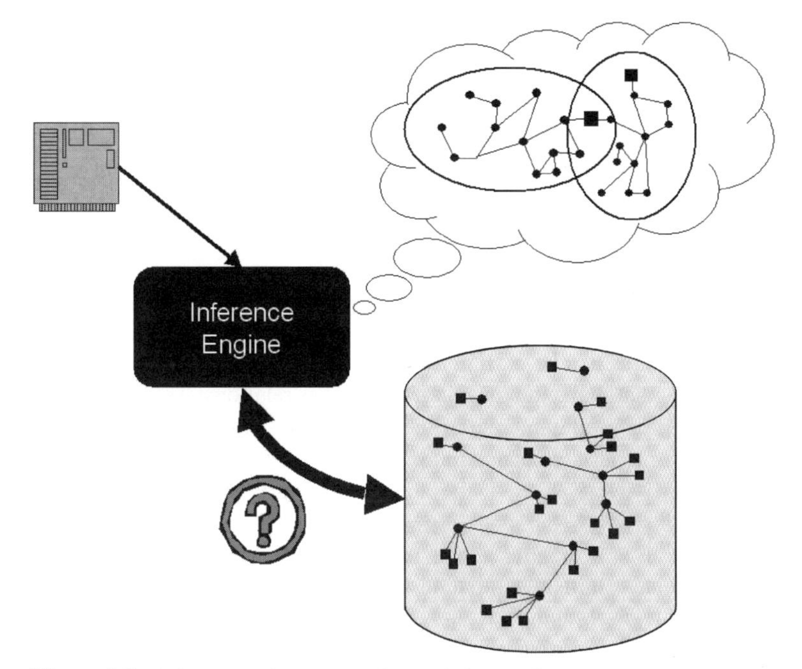

Figure 4.9 Inference engines use ontology to infer new facts

- **Role-forming operators**—primitive and inverse roles
- **Supported axioms**—concept implication, concept equality, and role implication

Inference engines are the run time components that make use of description logics to manipulate the data in interesting ways for client applications.

The basic use of an inference engine is to answer queries about the data contained inside the ontology that has been loaded into the inference engine as input. One of the essential benefits of this approach is its ability to overcome some of the issues implied in the schema-instance and class-object relationships, described above. Because the inference engine operates on data directly, it can be seen from a holistic, data instance level as well as being comprised of components that represent classes, individuals, relationships, etc., from a more schematic level. This capability is unlike any other run time-based approach and provides a great deal of flexibility to its users. It allows analysts to specify multiple levels of simple true/false queries arranged to answer complex questions that are highly relevant to government or business. For example:

- Have any travelers exhibited warning signs that may indicate terrorist activity?
- Has a new drug had an unanticipated effect on patients?

- Which customer orders are affected by the defect discovered in a production line?

The structures in the formalized ontology, which inference engines operate on, are also highly supportive of intelligent application interoperability scenarios. They enable the middleware to make decisions as to the applicability, structure, and form of the data being communicated across disparate systems.

Semantics as Context-Aware Schema Mappings

Another way that semantics can be partially derived is through the use of sophisticated mapping routines that make the meaningful business relationships among data elements explicit. One key problem in enterprise architecture today is the wholesale blurring of information models, exchange models, process models, data models, and object models. At nearly every layer of the architecture various models exist that contain some semantics about the data flowing through the system. Advanced mapping routines that align semantics across models, in a context-aware way, can offer dramatic operational returns when paired with a mediation platform that interprets data in these contexts. Mapping is an important capability because software system development cycles often sacrifice the conceptual integrity of the various information models for a number of good reasons:

- Performance
- Optimization
- Manageable scope
- Convention
- Implementation necessity (e.g., particular languages and data structures)

For example, a database under heavy load may be denormalized, tables may be merged, and descriptive data columns may be changed to obscure flags in order to improve performance. An XML schema may be simplified or constrained because an accurate model of the given information may result in arbitrarily deep nesting and complicated self-references throughout the schema's hierarchy. The net result of this loss of conceptual integrity is that the operational data and application models no longer reflect a great deal of the business reality in how the structures of information are actually related to each other. The process of defining relationships, in the form of schema-to-schema mappings, is the process of (re)defining the important relationships between data elements that have been denormalized because of other, business-as-usual, factors.

The utility of this sort of semantic discovery leads to several interesting applications:

- Making available the complete details about the data's true relationships, software can adapt to various data requests on the fly by dynamically building algorithms that transform the data from one view to another.

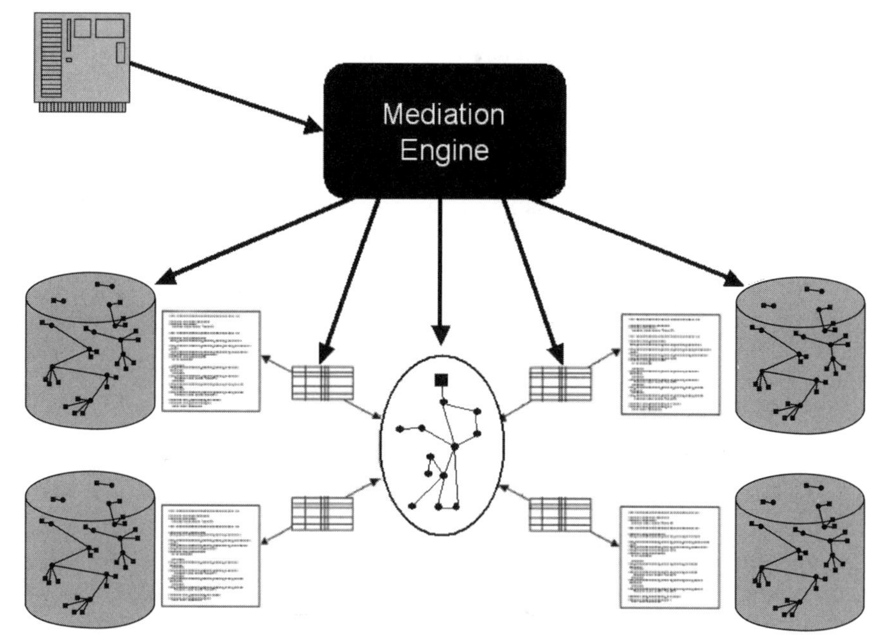

Figure 4.10 A mediation engine may use context-aware maps to derive semantics

- Optimizing mappings to capture the various contexts of data views enables the use of ontology as a pivot model for enterprise data, thereby simplifying the number of mappings and relationships to be modeled.
- Using a common approach for linking together various data sources, based on application schema, makes the underlying infrastructure irrelevant to the bigger picture of enterprise data as a whole.
- Renormalizing the enterprise information infrastructure is a great way to provide a foundation of visibility for many other tools that make use of data sourced from different applications.

Several schema-mapping tools on the marketplace make use of this approach to semantics. More details about specific tools and vendors can be found in Appendix A, Vendor Profiles, Tools, and Scorecards.

Semantics as Guesswork and Abductive Logic

I never guess.
—Sherlock Holmes

A final category of approaches for deriving data semantics is much more theoretical and largely dependent on capabilities achieved in the previously mentioned approaches. By first establishing a semantic network consisting of data semantics and contexts, algorithms and advanced reasoning methods can be applied to these collections, thereby deriving even more robust knowledge about enterprise information. This technique relies on collections of models, explicit semantic libraries, and model-based logic in the form of conceptual graphs. When taken in entirety, this holistic semantic net can be evaluated in light of higher-order reasoning methods like abductive logic.

As we learned in the beginning of this chapter, abductive logic is a quasiformal logic described by C. S. Peirce and used to facilitate the discovery of new connections without relying on induction or deduction. A way to familiarize yourself with this kind of reasoning is to think back to your understanding of Sherlock Holmes and the kind of reasoning that he used. Despite the fact that Sherlock Holmes is often understood to have been deducing the solutions to problems, he is most frequently using abduction.

Perspectives—The Science of Deduction or Abduction?

Holmes and Watson are discussing a French detective named Francois le Villard:

[Holmes]: "He possesses two out of three qualities necessary for the ideal detective. He has the power of observation and that of deduction. He is only wanting in knowledge. . . ."

[Watson]: ". . .But you spoke just now of observation and deduction. Surely the one to some extent implies the other."

[Holmes]: "Why, hardly. . . . For example, observation shows me that you have been to the Wigmore Street Post-Office this morning, but deduction lets me know that when you were there you dispatched a telegram."

[Watson]: "Right! . . . But I confess that I don't see how you arrived at it."

[Holmes]: "It is simplicity itself . . . so absurdly simple, that an explanation is superfluous; and yet it may serve to define the limits of observation and deduction. Observation tells me that you have a little reddish mould adhering to your instep. Just opposite the Wigmore Street Office they have taken up the pavement and thrown up some earth, which lies in such a way that it is difficult to avoid treading in it when entering. The earth is of this particular reddish tint which is found, as far as I know, nowhere else in the neighborhood. So much is observation. The rest is deduction."

[Watson]: "How, then, did you deduce the telegram?"

[Holmes]: Why, of course I knew that you had not written a letter, since I sat opposite to you all morning. I see also in your open desk there that you have a sheet of stamps and a thick bundle of postcards. What could you go to the post-office for, then, but to send a wire? Eliminate all other factors, and the one that remains must be the truth."

These leaps of logic, and educated guesswork, are examples of abduction—not deduction. This is an important distinction that takes nothing away from the value of this kind of logic. C. S. Peirce identified abduction as the first step of the scientific method and "the only kind of argument that starts a new idea."[9]

This kind of reasoning is highly relevant to the exchange of digital information because it offers a way to discover and use relationships among data sets that are not highly formalized. Work in various areas on alternatives to inference engines, sometime referred to as analogy engines, makes use of abduction to provide a broader range of capabilities. Analogy is a combination of induction and abduction,[10] which is why analogy engines can be described as a hybrid of neural net approaches to computing with more traditional rules-based approaches.

The Turing C machine, where the "C" stands for choice, was part of Alan Turing's vision of where the rote processing of rules stopped and a less deterministic form of computing begins.[11] Nobody has built a useful choice machine yet. Because of inherent problems with circular reasoning loops and nondeterministic outcomes, endeavors in this area have been less than productive. However, work is being done in some areas that blend classic ideas underlying the C-machine concepts with traditional rules-based processing that has the potential to reshape how software, particularly middleware, handles concepts, analogy, and information movement inside the enterprise. So, although this is the least developed approach for imbuing software with robust data semantics, it may well represent the finest opportunity to move beyond the limitations of rule-based systems—which data meaning does not adhere to well.

INFORMATION TECHNOLOGY

We are at the cusp of what has been called the "the third wave"—an information society. Ten thousand years ago the agricultural revolution changed a society no longer based on hunting and gathering into one based on farming and a less nomadic lifestyle. The second wave, based on manufacturing and production of capital, resulted in urban societies centered around the factory. The latest era, based on information, brainpower and mediacentric communications, is taking shape now. Only history will tell us the profound ways in which this shift will impact social, cultural, political, and institutional mores. However, the foundation of this new wave rests squarely on our capabilities to make information sharing ubiquitous and meaningful.

Information is Data with Meaning, in Context

Throughout the course of this chapter we have seen how the origins of information theory have taken shape for more than two thousand years. We've also seen how

[9] *Collected Papers of Charles Sanders Peirce*, volume 2, paragraph 97.

[10] *Collected Papers of Charles Sanders Peirce*, volume 7, paragraph 98.

[11] Turing A. M. (1936–7), On computable numbers with an application to the Entscheidungsproblem.

the various conceptions of semantics have given rise to a number of different technologies that handle aspects of semantic preservation in IT systems.

Unlike some writers on the subject, we will not attempt to give a definitive and narrow picture of what semantics is. Instead, data semantics in information technology seem to be comprised of at least five distinct characteristics:

- Patterns (as in statistical analysis and heuristics)
- References (as in dictionaries and thesauri)
- Relationships (as in data model associations)
- Core logic (as in deductive and inductive)
- Abductive logic (as in inference to the best explanation)

Each of these characteristics of data semantics is constrained by context. As with natural language systems, digital language systems have a limited scope of meaning that is highly dependent on context. These contextual factors shape and give meaning to semantics by providing a fixed reference point from which to understand the meaning being communicated via these data-intensive processes.

Information, on the other hand, is something that we take for granted as being well defined. However, the exact nature of what information is is still up for debate in the circles that wish to bring a mathematical clarity to concepts that can be inherently fuzzy.

For the purposes of this book, data is facts and information is data and meaning. Put differently, information = data + semantics. Remembering that semantics are constrained by context, a fuller equation would be information = data + (semantics/context).

With this as an understanding of the possible reach of information, we can begin to understand how much the industry has already grasped.

Where Is the Information in IT?

Modern IT systems can consist of hundreds of millions of lines of code, millions of objects, hundreds of components, and dozens of external interfaces. Nobody is saying that these massive systems are free of data semantics. As was pointed out

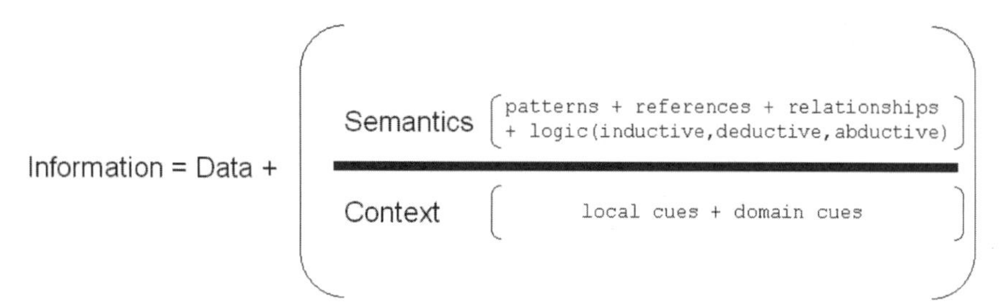

Figure 4.11 Information is data- and context-constrained semantics

```
<!-- IT in modern business -->
<font size="tiny"> Information </font>
<font size="huge"> Technology </font>
```

Figure 4.12 Very little information in IT

above in this chapter, the semantics are already in these systems. However, they are *implicit* in code-base, interpretable only by human programmers, and *unavailable* to external software routines for the purposes of further automation.

Making semantics explicit is crucial for significant innovation to occur in the IT world.

Introduction to Knowledge Representation and Ontology

Knowledge representation (KR) is schema (explicit or implicit) describing that which something knows about. This definition is intentionally broad because the study of knowledge representation encompasses many disciplines and takes many forms. One way to more fully understand knowledge representation is to observe the roles it can play[12]:

- **KR is a surrogate**—To think about something external to ourselves we need to establish a surrogate of that external thing internally.
- **KR is a set of ontological commitments**—To think about things internally we need to make decisions about how and what we see in the world.
- **KR is a fragmentary theory of intelligent reasoning**—Often driven by the desire to know more via inference, KR can require logic to be useful.
- **KR is a medium for efficient computing**—To compute we require a representation of that which we are computing.
- **KR is a medium of human expression**—Human communication is indelibly affected by individual weltanschauung, or personal worldview.

Ontology, like knowledge representation, is a broad and fuzzy subject. Historically, the concept of ontology is tied to metaphysical philosophy and the structure of discussing what is. From a software perspective, ontology has been a study of how to represent knowledge for computing. The working definition for ontology in the computer science field is along the lines of *an explicit specification of a conceptualization*, meaning that ontology is a technically constrained and processable set of data about some collection of concepts describing the world within a given context (in some capacity, for some purpose, within a given scope, and from a particular viewpoint).

[12] R. Davis, H. Shrobe, and P. Szolovits. What is a Knowledge Representation? *AI Magazine* 14(1):17–33, 1993.

One definition provided by Esther Dyson is a particularly nice way of understanding how ontology is important to managing change in business[13]:

> *An ontology, then, is an active model that contains a variety of data structures and some way of propagating changes through itself. It can comprise a host of things: taxonomies of data objects; taxonomies of relationships or typed links (often expressed as verb phrases), from "is associated with" to "is a kind of" to "contains" or "produces" or "consumes" or even "enjoys" or "prefers" or "burns." Those relationships can usually be modeled or represented by combining other more elemental components, or through applications that implement (for example) all the things that a customer can do or can have done to her and her account. (Another example: Burning is a specific kind of destruction; it is also a chemical process. Which representation you use depends on the context.)*

Ontologies themselves can come in a wide variety of implementations, each varying in content as well as the degree of complexity and capabilities. For the purposes of this book, the following taxonomy of ontology classification has been used:

- **Interface Ontology**—as in service and API interface descriptions
- **Process Ontology**—both fine-grained and coarse-grained procedural descriptions
- **Policy Ontology**—access, privilege, security and constraint rule descriptions
- **Information Ontology**—all things about business contents
  - Industry Ontology—domain concept descriptions
  - Social/Organizational Ontology—organizational and social networks
  - Metadata Ontology—published content descriptions
  - Common Sense Ontology—general facts about life
  - Representational Ontology—metainformation about classifications
  - Task Ontology—term and task matching descriptions

The pragmatic use of ontology is limited only by our imagination. In the broadest sense, we have been utilizing ontology since the inception of computing. In a more narrow sense, we are only now beginning to understand how ontology can reshape our notions of information technology. More on the uses and specifications of ontology is found in Chapter 7, Ontology Design Patterns.

Information Is at the Eyeball, but Also at the Interface

One alternate view of data semantics, held by people in the camp who argue that meaning cannot exist inside computers, that meaning is an inherently human experience that occurs when data is processed by one of our five senses. In this conception, information and knowledge are restricted to the human mind because meaning is subjective and subjects capable of deriving meaning must have consciousness.

[13] Esther Dyson, Release 1.0.

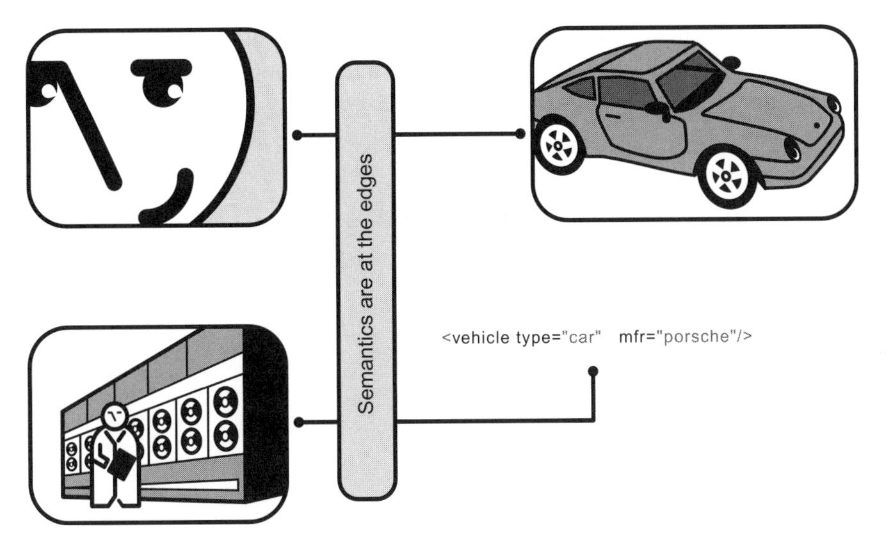

Figure 4.13 Semantics are at the edges, on the interface

Any proof as to the true nature of semantics, information, and knowledge must be a metaphysical proof well beyond the scope of this book. The thesis here is not to negate that semantics is at the eyeball (a human experience) but to show how semantics is, more broadly, at the interface of any given communication—including software interfaces. Middleware is the software that facilitates communication between discrete software systems by managing data flow between interfaces. It is in this framework that the nature of data's meaning is highly relevant to the efficient interoperability of different software systems. The approximate position of semantics in a machine-to-machine data exchange is at the point where data moves from one interface to another.

Knowledge representation, ontology, and mechanisms for information exchange are critical for the advancement of frictionless information sharing because without them semantics will always be locked away in uninterpretable software algorithms.

CONCLUSIONS

The crucial importance of semantics in successful communication should be clear by this point. In human communications the questions of semantics have preoccupied the greatest minds for thousands of years. In technical communications, the processing of data's meanings simultaneously provides the greatest value and the most significant roadblock to progress in today's information-intensive technology systems. Common practices for manipulating data semantics (e.g., hard-coding and programming interfaces) are the largest sinkholes of time and money when responding to dynamic changes in the organizational environment. New technologies that work with data semantics, at various levels, hold much promise in driving down IT costs and creating new capabilities for dynamic, frictionless, and automatic enterprise networks.

Chapter 5

Semantic Conflict Solution Patterns

KEY TAKEAWAYS

- To understand something new, it must connect with something known.
- Semantic conflict, among data and schema, is the most significant barrier to fully realizing automated and adaptive system-to-system interoperability.
- Semantic conflicts are an inherent aspect of system interoperability—they cannot be prevented, only handled in a more efficient manner.
- Four key patterns of semantic conflict solutions exist: machine learning, third-party reference, model-based mapping, and inference.

Chapter 4 was a look at the foundation of semantics from its history, its multiple definitions, and its role in information technology. In this chapter, we continue by examining the ways that semantics can clash (semantic conflicts), how semantics can be resolved without custom code (solution patterns), and how semantics are typically constrained (context).

FINDING SEMANTICS IN ENTERPRISE SOFTWARE

Sometimes it can be easy to drown in all the terminology, concepts, and technicalities of a discussion about semantics. Once a simple understanding of data semantics has been reached, everything starts to look like a nail ready to be pounded down by the semantic interoperability hammer. Uses for semantic computing are indeed everywhere. Usually the trouble is just figuring out where to start.

One reason that this book is focused narrowly on the interoperability challenge is because of the vastness of reach that semantics can have—a single book on the subject would hardly do it justice. But as we start to consider some of the applied technologies swirling around the semantics technology space, such as metadata, schema, and other context representation tools, it will be useful to step back and reflect on how these concepts impact content across media.

Adaptive Information, by Jeffrey T. Pollock and Ralph Hodgson
ISBN 0-471-48854-2

Explicit semantic metadata provides both software and people with the ability to produce, aggregate, interoperate, catalog, index, integrate, syndicate, personalize, and market all sorts of digital content like audio, video, photo, data, documents, and forms. Semantics can be both infrastructure and tools that run over multiple channels, including broadcast, wired, wireless, and satellite. The scope and application of semantics-based systems is unlimited.

Because of the distributed nature of digital content, much of its native context, and thus semantics, is lost. Consider the situation of somebody giving you a gift of new digital content, but the gift is just a bunch of jargon and you can't figure out what to do with it. You might ask some of the following questions:

- Where is the content?
- Whose content is this? Who created this content?
- What is this content about?
- What does this content mean?
- How does this new content relate to content that I already know about?
- What is the right content for me?
- How can I monetize this content?

These are just some of the questions that metadata attempts to answer about atomic units of digital content so that you (and your machines) can use the content wisely.

One way in which people have managed to get around the process of specifying explicit context and semantics as metadata has been to repurpose digital content into new formats that other systems can easily understand. This process is pretty straightforward; it involves the authoring of a software component or program to decompose and recompose the digital content—a digital adapter of sorts.

But this common solution has many hidden problems, which tend to lie in details and become apparent at large scales. Conflicts of semantics, schema, and syntax create the need to hardwire adapters in very inflexible ways—resulting in large quantities of brittle code that cannot adapt with change.

The remainder of this chapter will examine the details of various kinds of data conflicts that can occur between two or more silos of information. Bear in mind that these conflicts are typically overcome with code, but the purpose here is to examine

Figure 5.1 Using code to process data and semantics

```
...
<serial_number>RF235</serial_number>
<model>Saab 900</model>
...

  ...
  <serial_number id="RF235" name="Saab 900" from="EU" to="US"/>
  ...
```

Figure 5.2 Simple style difference in XML usage

ways in which they may be overcome through semantic solution patterns—without the use of custom-written code.

SEMANTIC CONFLICTS IN COMMUNICATION

Conflicts that arise in the structure and meaning of information in digital systems are implicitly known by most experienced programmers and sometimes explicitly known to researchers in the area of ontology creation and transformation. These conflicts represent the core of what needs to be more effectively handled in application middleware to eliminate brittleness, tight coupling, and lack of adaptability to business change.

The following discussion about the categories and descriptions of schematic and semantic conflicts has been significantly influenced by Cheng Hian Goh's doctoral work at MIT.[1] As a pioneer of the understanding of semantics in computer science Goh brought much form to a very chaotic field of study.

A Short Note About Structure and Style

Just because different pieces of data share the same syntax does not mean that they are automatically interoperable. Take, for example, the following value representation conflict:

The example shown in Figure 5.2 demonstrates how the same vehicle instance can be represented by two completely different, but also completely valid, XML structures. Mismatches in schema form usually result from the style of a given developer being different than the style of another developer. This style problem in XML is exacerbated by the rapid proliferation of XML schemas from 1998 through the present. These same kinds of style differences can also be found in other data representation languages—databases, object oriented, semistructured—they arise throughout enterprise systems because designers design artifacts differently. Design style may be recommended and best practices can be espoused, but style cannot be reliably mandated or enforced. Thus semantic data conflicts are a fact of life in IT that will not go away.

[1] Goh, Cheng Hian, *Representing and Reasoning about Semantic Conflicts in Heterogeneous Information Systems*, 1997.

Summary of Data Conflicts

Table 5.1 Data Table—Summary of Data Conflicts

Conflict	Description	Type	Solutions
Data Type	Different primitive or abstract types are used across schema with similar or same information.	Schema	Model-Based Mapping
Labeling	Synonyms and antonyms have different text labels used to describe same objects or hierarchies.	Schema	Model-Based Mapping Third-Party Reference Inference
Aggregation	Different conceptions about the relationships among concepts in similar data sets.	Schema	Model-Based Mapping Inference
Structure	Collections and associations are modeled differently in similar data sets.	Schema	Model-Based Mapping Inference
Cardinality	Systems have been modeled with different constraints on the association of component parts.	Schema	Model-Based Mapping Inference
Generalization	Different abstractions are used to model the same environment.	Schema	Model-Based Mapping
Value Representation	Different choices are made about what values are made explicit in a schema (vs. inferred values).	Schema	Inference Model-Based Mapping
Impedance Mismatch	Data moves between fundamentally different representations systems.	Schema	Model-Based Mapping Inference
Naming	Synonyms and antonyms exist among data values where there may or may not be a common concept identifier.	Semantic	Inference Machine Learning
Scaling and Unit	Different units of measures are used with fundamentally incompatible scales.	Semantic	Model-Based Mapping Third-Party Reference Inference
Confounding	Similar concepts with different definitions are exchanges among systems.	Semantic	Inference Model-Based Mapping
Domain	Fundamental incompatibilities exist in the underlying domains of what are being modeled.	Semantic	Model-Based Mapping Third-Party Reference Inference
Integrity	Disparity exists among the integrity constraints asserted in different systems.	Semantic	Machine Learning Model-Based Mapping Inference

Data Type Conflicts

A kind of schema conflict, data type conflicts occur when different primitive system types are used in the representation of data values for a given programming language. For example, actual dates may be represented as strings in a custom XML tag, but the XML Schema type `dateTime` can also represent them. Sometimes these data type problems are subtler, but just as real. For instance, when a Java program uses the object `Integer` (which has an upper limit value of 2147483647) and another program uses the `Short` object (which has an upper limit value of 32767), a data type conflict can cause a significant incompatibility between the systems.

The data type conflict can be solved pretty easily on a case-by-case basis with application code. Solving data type conflicts in a model-driven semantic interoperability environment can be more problematic. In most systems there is not enough explicit metadata for automated routines to make decisions about the conversion of data from one type to another. For example, changing from a `Date` object to a `String` object requires very specific rules about how to structure the data in string format. This kind of metadata is context dependent, because different systems using strings to hold dates structure them differently. Although we have used dates as an example, this holds true for integers, Booleans, doubles, complex objects, and other data types.

Conflict Solution:

- See Model-Based Mapping Pattern

Labeling Conflicts

Also a schema conflict, labeling conflicts occur when synonyms and antonyms appear with different labels inside the schema specification. In other words, the same data element could be referred to by different naming conventions inside two discrete information systems or, conversely, the same naming convention may be used in identifying data elements that are not equivalent. Examples of these kinds of conflicts can be found in XML tag names, Java object names, and RDBMS table/column names. Typically, the cause of labeling conflicts can be found in style mismatches, which are further complicated by the complete discretion of the programmer in naming objects, tags, and data elements.

This is perhaps the most commonly understood semantic conflict because it is so intuitive—people name same things differently. As such, the solution for label-

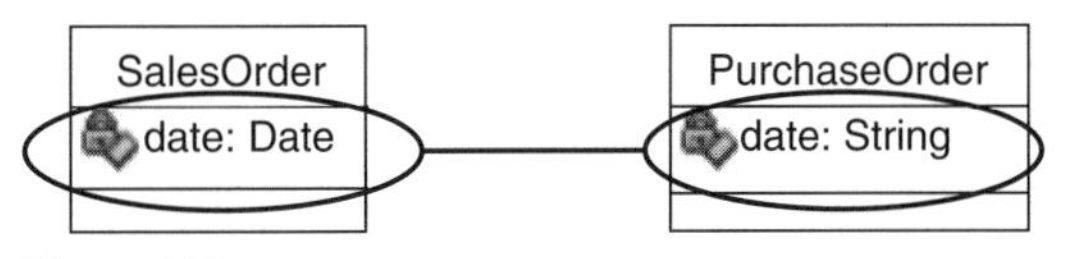

Figure 5.3 Data type conflict

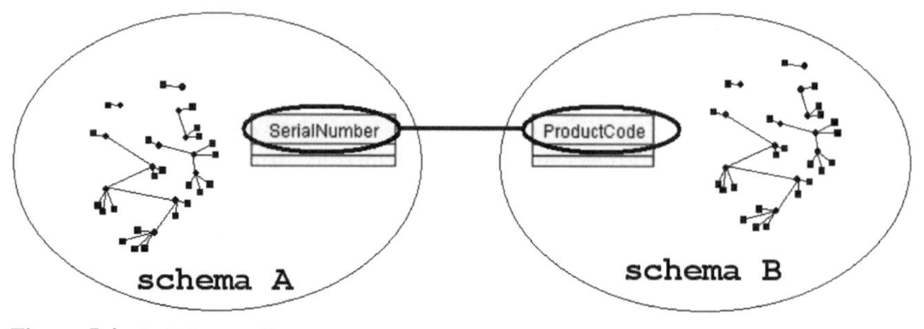

Figure 5.4 Labeling conflict

ing conflicts is found in a number of patterns. The Third-Party Reference Pattern is quite sufficient for most labeling conflicts; by associating data labels with synonyms and antonyms a software program may interpret the semantics of a label without having to understand its specific label. Likewise, the Inference Pattern relies on the development of a rigorous ontological model, often relying on modelers using a shared meaning for classes in that model, which can act as a placeholder for a label's meaning. The most flexible solution approach for this conflict lies in the Model-Based Mapping Pattern—it enables modelers to associate data classes to a neutral ontology without agreeing on a shared meaning.

Conflict Solution:

- See Model-Based Mapping Pattern
- See Third-Party Reference Pattern
- See Inference Pattern

Aggregation Conflicts

Aggregation conflicts, also schema level conflicts, arise from different conceptions of relationships among a given data set. Decisions about the type of association, nesting, or object scope contribute to these structural problems with aggregation. For example, modelers may design a supply chain system with the products being associated with processes or, conversely, with the processes being associated with parts—where each focuses on a different "center of the universe" and results in different aggregations inside the model.

Differences in Structure

When similar concepts are expressed using different structures in different schemas, there is a need to navigate the structures when moving data from one to the other during the transformation process. For example in the first schema, there exists an author entity with an attribute called name and a book entity, which is related to the

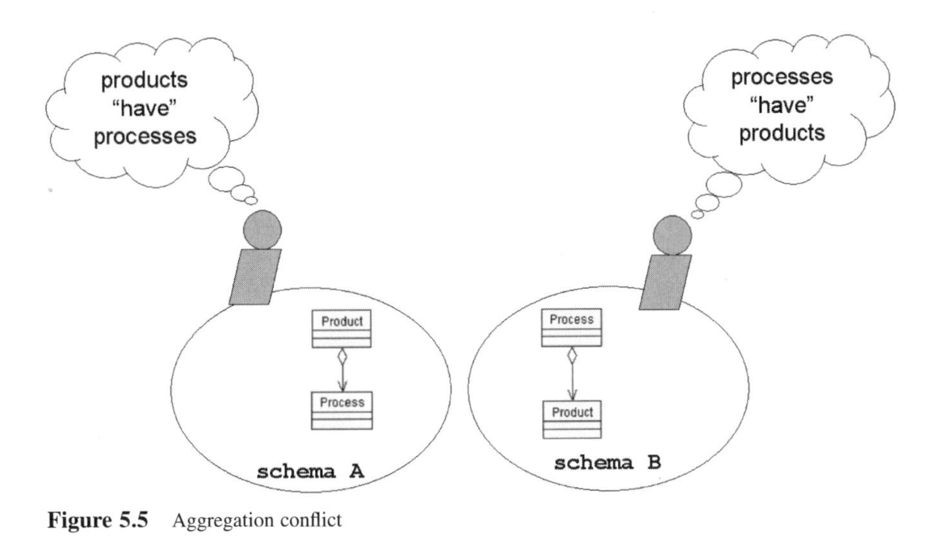

Figure 5.5 Aggregation conflict

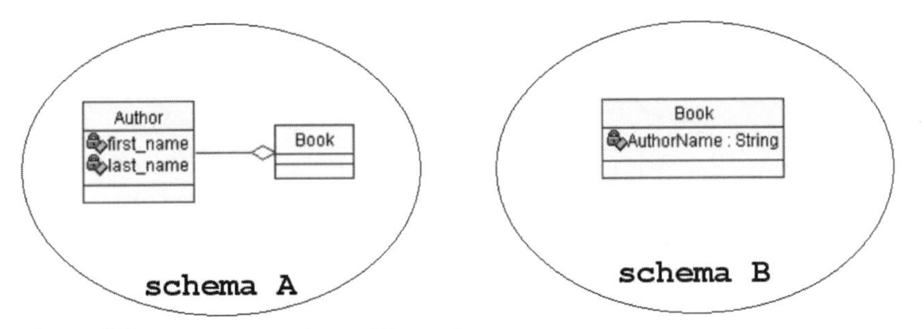

Figure 5.6 Aggregation conflict—difference in structure

author entity. In the second schema, there is no author entity and the author name is simply an attribute of the book entity.[2]

Differences in Cardinality

Differences in cardinality occur when the systems have been modeled with different constraints on the association of their component parts. For example, if a product data management system has been modeled with a `Car` consisting of exactly one `Engine`, it may preclude the system from accepting data from another system modeled for newer chassis that contain four electric engines, one for each wheel.

Solutions for the aggregation conflict problem are twofold. The Model-Based Mapping Pattern is the most flexible approach for the range of aggregation

[2] Goh, Cheng Hian, *Representing and Reasoning about Semantic Conflicts in Heterogeneous Information Systems*, 1997.

problems because the rules for structure and cardinality can be modeled in the context-sensitive mappings that unite the disparate schema. However, the Inference Pattern may also be used, especially for structural differences, but the Inference Pattern may have to rely on program code to resolve certain cases in which cardinality problems are severe.

Conflict Solution:

- See Model-Based Mapping Pattern
- See Inference Pattern

Generalization Conflicts

Generalization conflicts occur when different abstractions are used to model the same environment. For example, one system may have a supertype called `Vehicle` with an attribute called `Type` that can contain types like "car," "truck," or "tractor."

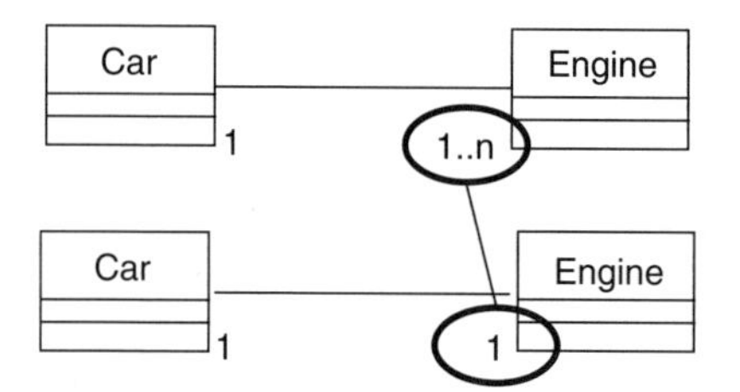

Figure 5.7 Aggregation conflict—difference in cardinality

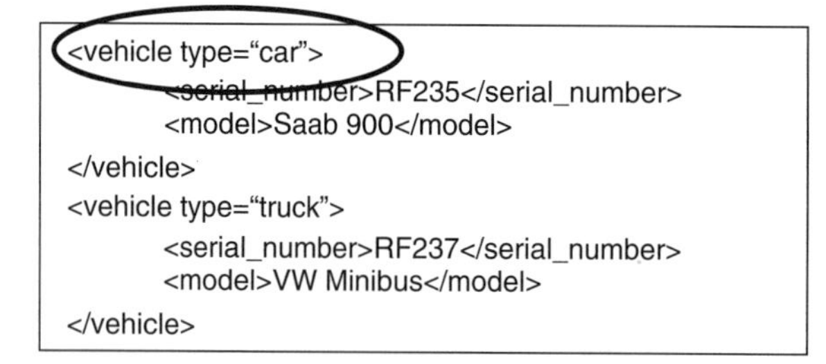

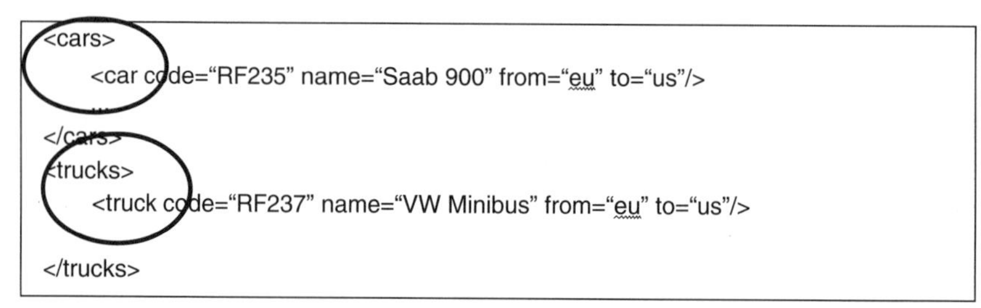

Figure 5.8 Generalization conflict

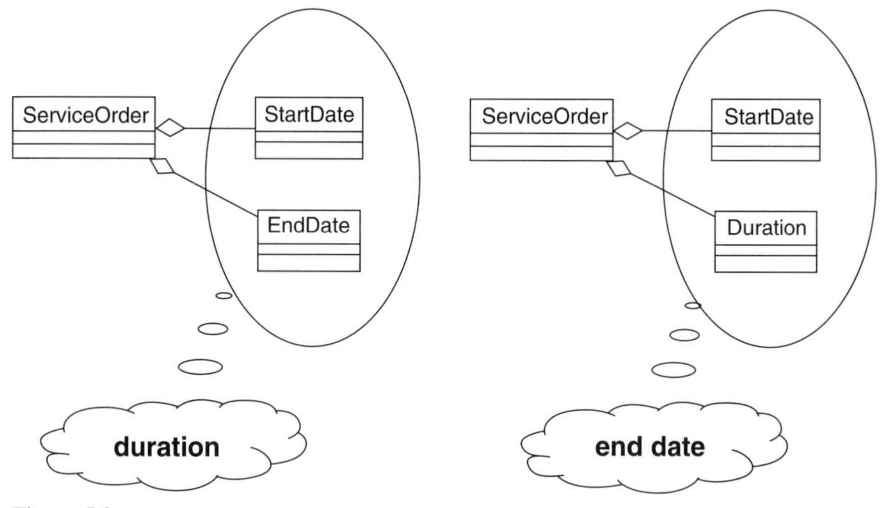

Figure 5.9 Value representation conflict

In another system, the supertypes may be modeled separately and called `Automobile`, `SUV`, and `Tractor`. The resulting logic required to share information among systems with different generalizations can cause issues with data correlation and many-to-many table conflicts with RDBMS environments.

Generalization conflicts are most efficiently resolved by using the Model-Based Mapping Pattern. A neutral ontology may specify a highly abstract entity like `Product`, which can behave as a slot for various vehicle types. Then the contexts of the two diverse schemas can be modeled into the ontology capturing the scope of where the divergent generalizations are appropriate—thus enabling a run time routine to reassemble the structures appropriately for each system on demand.

Conflict Solution:

- See Model-Based Mapping Pattern

Value Representation Conflicts

The choices data modelers make about the explicit representation of their data can result in another schematic conflict known as value representation. The classic example[3] of the value representation problem is when a one schema models a `StartDate` and an `EndDate` while another stores a `StartDate` and `Duration`. The software program can derive either the `EndDate` or the `Duration` from either schema, but it must perform a calculation to get one of them. In a middleware scenario, calculations would have to be done on the fly to get the right data into the correct fields.

[3] Goh, Cheng Hian, *Representing and Reasoning about Semantic Conflicts in Heterogeneous Information Systems*, 1997.

Value representation conflicts are classic case studies for inference capabilities. With little or no additional effort, beyond modeling a description logic-based ontology, an inference engine can effectively infer new values based on the contents of the ontology at run time. Alternatively, the Model-Based Mapping Pattern can be used to determine implicit values; however, it has the added disadvantage of requiring case-by-case algorithms to instruct the run time components during the data movement.

Conflict Solution:

- See Inference Pattern
- See Model-Based Mapping Pattern

Impedance Mismatch Conflicts

Impedance mismatch is a generalized conflict that refers to the problems of moving data between fundamentally different knowledge representations, for example, moving data between relationally structured RDBMS systems to an objectcentric Java system and then to a hierarchically structured XML document or a tokenized flat file. The foundations and theory on which these data structures rely are different in their core conception of relations. This results in a very real structural misalignment of how data is modeled and persisted. Further complicating this scenario are the culture and best practices that motivate modelers from each discipline to

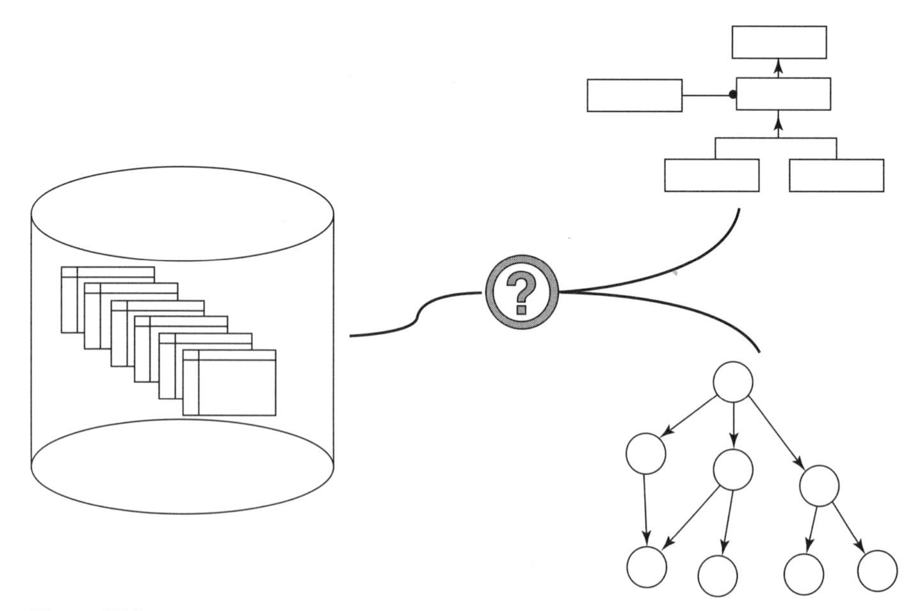

Figure 5.10 Impedance mismatch conflict

follow different patterns of representation.[4] In actuality, a thorough decomposition of the technical consequences of impedance mismatch will result in the more granular conflicts mentioned elsewhere, but as a whole, the nature and scope of conflicts that arise from mismatched representation structures justify a separate discussion.

The only plausible, non-code-based solution to the impedance mismatch problem is to use a neutral conceptual model that is more expressive than any of the implementation models causing the mismatch. Either the Model-Based Mapping Pattern or the Inference Pattern may be used because they rely on a neutral third schema that is typically a very expressive ontological model. The Third-Party Reference Pattern is generally not acceptable because thesaurus/dictionary structures tend to be hierarchical in nature, which makes them impractical for negotiating differences between complex object-based models.

Conflict Solution:

- See Model-Based Mapping Pattern
- See Inference Pattern

Naming Conflicts

A purely semantic conflict, naming issues arise when there are synonyms and antonyms among data values, for example, when the name of a business entity is represented differently:

- Daimler-Benz AG
- Daimler Benz Corp
- Daimler Chrysler
- Chrysler
- Daimler-Benz

Each of these names can be synonymous with each other, or in some cases homonymous as well. In the case where historical searches need to be run, say for historic financial performance, the name "Daimler-Benz" could represent a functionally and legally different company.

Naming conflicts are considered some of the more difficult semantic challenges. This is because there is very little machine-interpretable metadata to provide clues to software about the "sameness" of clear-text patterns. One solution is to use inference capabilities to operate on the instance data as if it were an ontology and then to assert the "sameness" of the classifications that are generated from the instances. This is a two-step process, and it requires human intervention to assert the "sameness." Another approach is to use the Machine Learning Pattern and apply statistical analysis algorithms to associate records with one another by evaluating the whole

[4] The Object-Relational Impedance Mismatch, Scott Ambler (URL: http://www.agiledata.org/essays/impedanceMismatch.html).

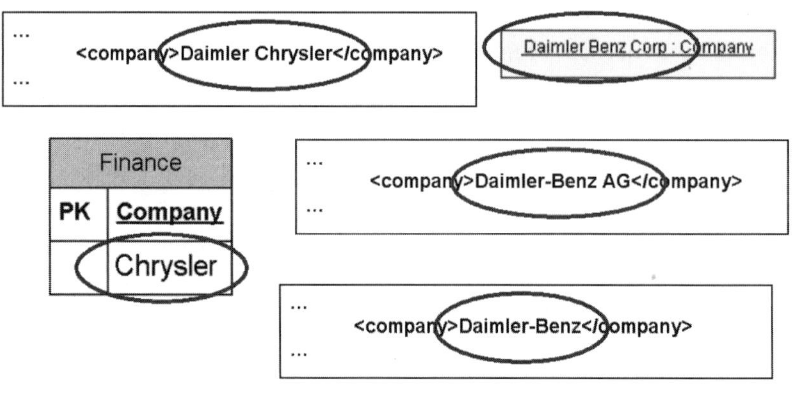

Figure 5.11 Naming conflict

record data and finding related patterns. This method is employed by a number of data cleansing tools in an effort to clean up quantities of human-keyed data—which is error prone.

Conflict Solution:

- See Inference Pattern
- See Machine Learning Pattern

Scaling and Unit Conflicts

Scaling and Units Conflicts refers to the adoption of different units of measures or scales in reporting. For example, financial data are routinely reported using different currencies and scale-factors. Also, academic grades may be reported on several different scales having different granularities (e.g., a five-point scale with letter grades A, B, C, D and F, or a four-point scale comprising excellent, good, pass and fail).[5]

Solutions to the scaling and unit conflicts are inherently lossy—in other words, a certain amount of expressiveness is lost when moving between scales. They can be solved in fairly straightforward ways by using Model-Based Mapping, Inference, and even Third-Party Reference Patterns—but the modeler must decide where the loss in meaning will be absorbed. For example, it is possible to convert a five-point scale to a four-point scale with great precision and minimal loss if the grade point average (GPA) is available. But if the GPA is not available, the modeler must make the decision to equate a "C" with "good" or "pass," thereby losing distinctions present in one scale that are not available in the other.

[5] Goh, Cheng Hian, *Representing and Reasoning about Semantic Conflicts in Heterogeneous Information Systems*, 1997.

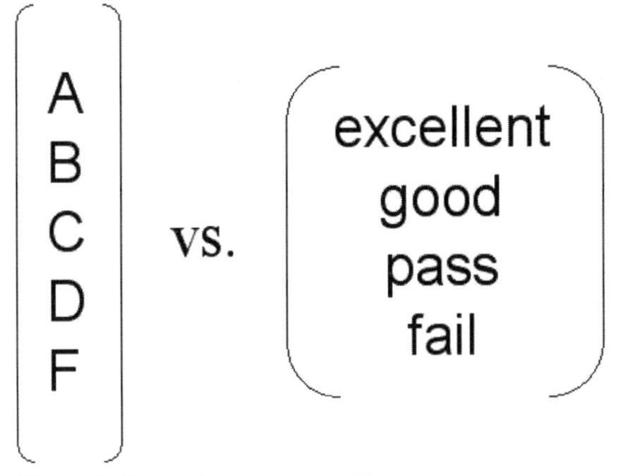

Figure 5.12 Scaling and unit conflict

Conflict Solution:

- See Model-Based Mapping Pattern
- See Third-Party Reference Pattern
- See Inference Pattern

Confounding Conflicts

Also a purely semantic conflict, confounding conflicts arise when concepts are exchanged that have different definitions in the source systems. A common example of these kinds of conflicts is illustrated by the financial indicator P/E. *The Wall Street Journal* divides the combined market capitalization of the 500 companies in the index by their most recently reported four quarters of earnings. However, Standard and Poor's updates earnings statistics for the index just once a quarter and doesn't revise earnings from previously reported quarters to account for additions and deletions to the index.[6] These kinds of conflicts are frequent and require a deep domain expertise by the programmers and designers to prevent integration errors.

Confounding conflicts are classic semantic conflicts. The underlying background meanings of same-named data create confusion in cases where that data needs to be blended. The solution for many confounding conflicts, as in the example stated above, are most efficiently solved with the Inference Pattern. By decomposing the fundamental units of complex ideas, such as a P/E, into an ontology, the inference engine makes 100% accurate decisions about how to reconstruct the complex idea from its fundamental parts and enable interoperability among systems that implement different semantics. It should be noted that for this method to fully

[6] *Financial Information Integration In the Presence of Equational Ontological Conflicts*, Aykut Firat, Stuart Madnick, Benjamin Grosof, MIT Sloan School of Man agement, Cambridge, MA.

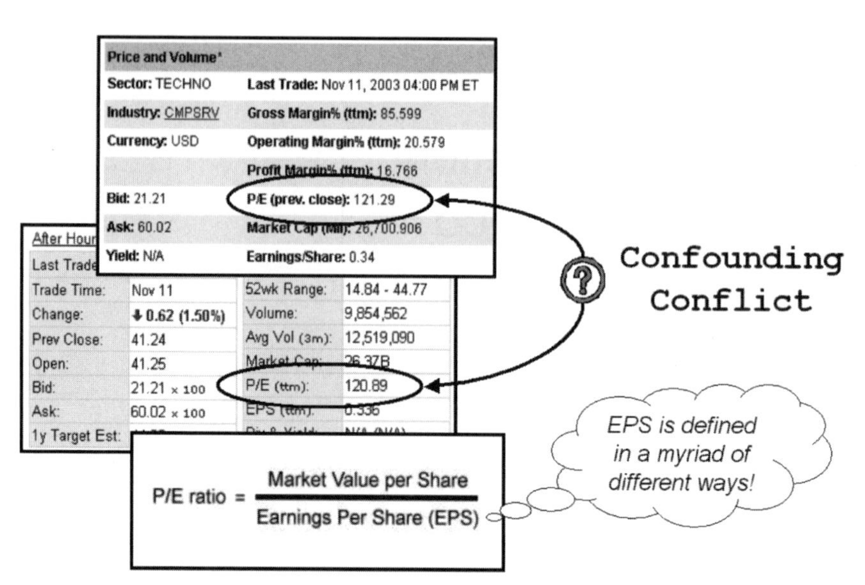

Figure 5.13 Confounding conflict

operate access to the fundamental data units (such as the values used for EPS calculation) must be present. If the interoperability system is only working with the mismatched data units, and not their parts, then an automated process for conflict resolution is not possible. In this case, the best approach would likely be a Model-Based Mapping approach, whereby the context factors can be captured and the rules can be modeled in relation to a neutral ontology.

Conflict Solution:

- See Inference Pattern
- See Model-Based Mapping Pattern

Domain Conflicts

Domain conflicts are sometimes classified as intensional conflicts. They refer to problems with the culture and underlying domain of what has been modeled inside the software system. For instance, an automaker might model its system around the notion of a main assembly (whether it is labeled as such or not)—just as its suppliers might. However, the main assemblies in an automaker's system are vastly different from the main assemblies in the supplier's system. These differences in domain can result in a number of problems with overlap, subsets, or disjointed concepts rippling throughout a model and its parts. The assumptions that underlie the model's development are critical in overcoming this far-reaching semantic conflict.

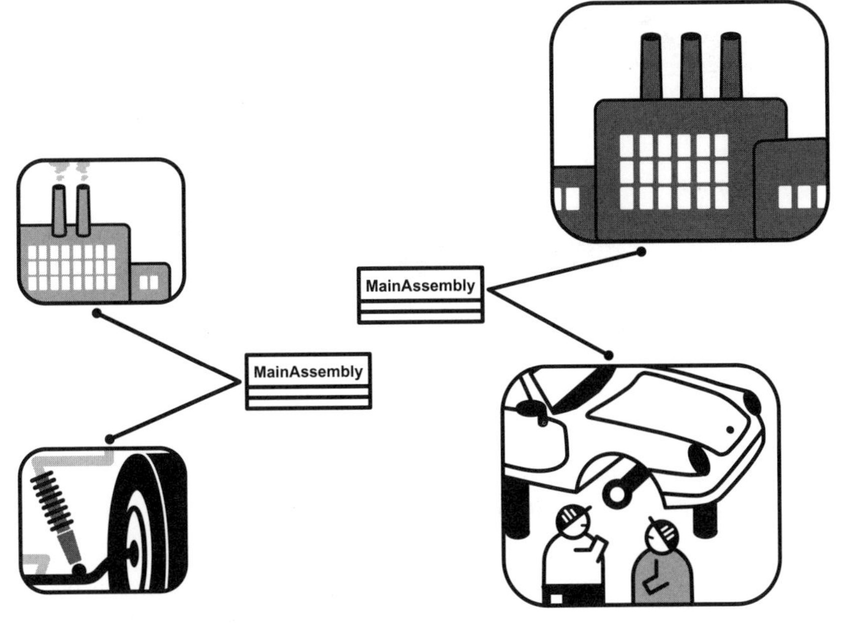

Figure 5.14 Domain conflict

Domain conflicts are among the most common semantic conflicts. The background cultures that influence structural concepts and classification schemes greatly influence the minor nuances in modeling that result in severe interoperability problems. Model-Based Mapping is the best pattern for dealing with this conflict because it enables analysts to capture context and metadata in mappings—without having to agree on a shared set of semantics in the neutral ontology. At run time the software can disassemble and reassemble the data parts based on the context factors used in the mappings. Both the Inference Pattern and the Third-Party Reference Pattern may also be used to resolve domain conflicts, but they typically require an ontology or taxonomy with a shared meaning throughout a broad community of interests—which is sometimes impractical or impossible.

Conflict Solution:

- See Model-Based Mapping Pattern
- See Third-Party Reference Pattern
- See Inference Pattern

Integrity Conflicts

Integrity constraint conflicts refer to disparity among the integrity constraints asserted in different systems. The simplest (but potentially most troublesome) form of these is conflicts over key constraints. The name of an individual may be unique in one com-

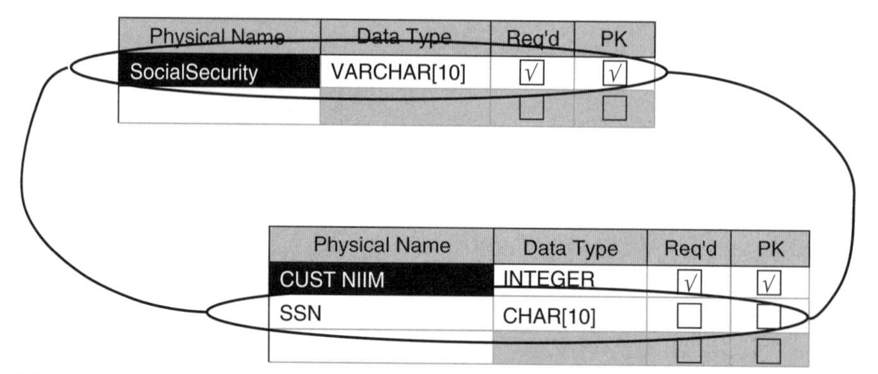

Figure 5.15 Integrity conflict

ponent system (hence allowing it to be used as a key), but not in another. In general, many different possibilities exist given that integrity constraints can take on many different forms.[7]

Moving quantities of data across schema that have integrity conflicts is one of the most problematic semantic conflicts. Realistically, there is no reliable one-step automated solution for this kind of conflict. This is so because the integrity conflict can be decomposed into two discrete kinds of semantic conflicts: naming conflict and aggregation conflict. The solution for this conflict is a two-step process. First the data in the non key fields (which are keys in the other schema) need to be scrubbed and cleansed to make that data acceptable for entry into a keyed field. Second, the structural differences in the schema element aggregations need to be resolved so that the data instances can be populated in the variant schema. A conflict resolution pattern for the first step would be the Machine Learning Pattern, whereas the appropriate patterns for the second step would include the Model-Based Mapping Pattern and the Inference Pattern.

Conflict Solution:

- See Machine Learning Pattern
- See Model-Based Mapping Pattern
- See Inference Pattern

SEMANTIC SOLUTION PATTERNS

The following presentation of semantic solution patterns closely follows the structure of digital semantics presented in Chapter 4 because each solution pattern leverages a different approach toward discovering and utilizing semantics in digital

[7] Goh, Cheng Hian, *Representing and Reasoning about Semantic Conflicts in Heterogeneous Information Systems*, 1997.

content. These patterns are generalized solutions to a number of the semantic conflicts presented above and are not intended to represent granular implementation solutions for specific conflicts or data interoperability problems.

Machine Learning Pattern

The Machine Learning Pattern relies on statistical analysis and complicated algorithms to find implicit semantics in quantities of data instances. This pattern is typically used as a first stage of a multistage conflict resolution scenario. Key outputs may include reports on instance data that is likely semantically equivalent (as with the value representation and integrity conflicts) or pattern analysis trends that indicate implicit semantics, which may have been previously unknown. The key advantage that machine learning approaches have over other semantic discovery/processing mechanisms is that they can operate on vast quantities of instance data directly, with little or no reliance on imperfectly stated schema, metadata, or rigorous ontology.

Applicability

Use the Machine Learning Pattern when:

- Discovering semantics within instance data (vs. schemas).
- Large quantities of batched data require conflict resolution.

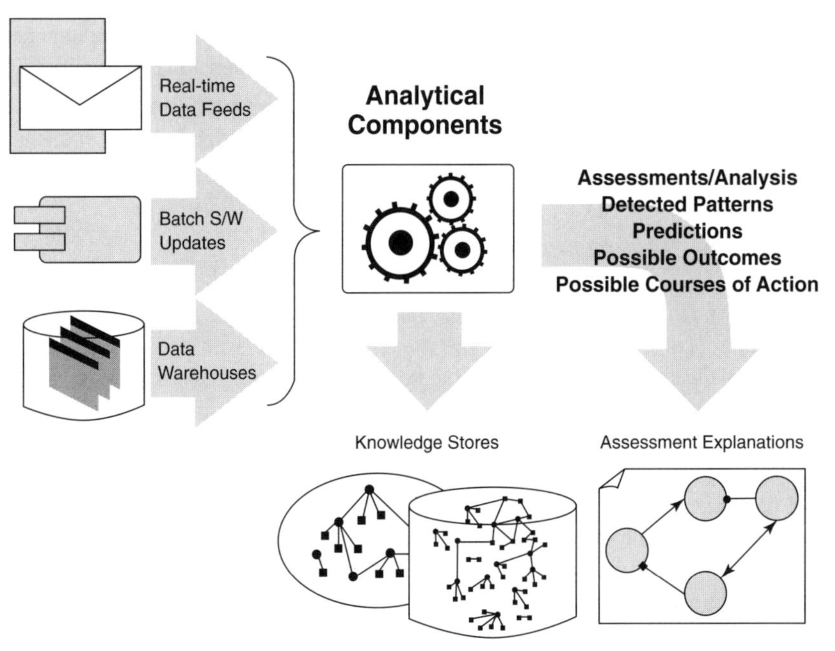

Figure 5.16 Machine learning inputs and outputs

- Data quality is poor.
- Seeking patterns that are not known in advance.

Structure

The structure of machine learning algorithms can take many forms. Algorithms specializing in different aspects of learning and pattern discovery can be used in parallel or in unison to achieve desired results.

- Pattern recognition
- Neural networks
- Bayesian learning
- Gaussian processes

Known Implementations

Generally speaking, most software vendors with tool sets specializing in knowledge management, data cleansing, classification/taxonomy generation, and data mining will implement some form of the Machine Learning Pattern. A representative set includes:

- **Verity**—A market-leading search/retrieval for enterprise portals, Verity has developed proprietary algorithms for automatic classification of unstructured data.
- **Cognos**—A market-leading business analytics tool suite, Cognos implements a variety of algorithms to build and analyze three-dimensional data cubes.
- **Ascential**—A market-leading data integration platform, Ascential's QualityStage product family implements probabilistic and fuzzy matching algorithms.

Third-Party Reference Pattern

This pattern is deceptively simple. The Third-Party Reference Pattern is when an external reference is used to define the meaning of data. This external reference usually takes the form of a dictionary or thesaurus, but may also be a more robust ontology. Dictionaries are well-understood elements of semantic capture, but, as anyone who has ever created or maintained a data dictionary knows, they are notoriously impossible to keep up to date. In addition, data dictionaries tend to rely on natural language descriptions of meaning, which are completely uninterpretable by software. However, the thesaurus aspect of this pattern is much more tractable. The explicit definition of synonyms, antonyms, and other linguistic characteristics to data gives software adequate rules on which to base processing decisions. This strength can facilitate the automatic generation of maps and/or programming code that drives the transformation of disparate data structures.

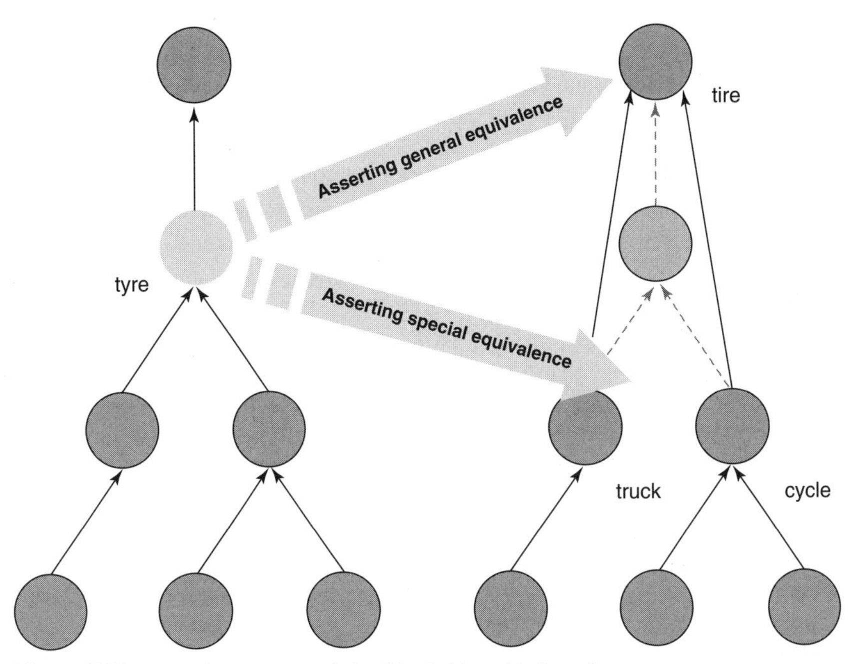

Figure 5.17 Asserting synonym relationships in hierarchical graphs

An exemplary feature of thesaurus-based patterns is the ease with which they may adopt an evolutionary approach toward semantics. Simply increasing the number of synonym mappings to a thesaurus increases the richness of semantics in the overall system. As the mappings change, or are used differently, in real time, the semantics of the overall system changes along with them. Problems with thesaurus-based approaches include their reliance on hierarchical representation structures for modeling information, their dependence on shared meaning in the thesaurus itself, and their lack of capabilities in the arena of managing complex contexts among constituent sources. Likewise, thesaurus-based approaches often output code or scripts that, as much as the pattern assists the design process, do not fix the high degree of brittleness embodied in popular run time architectures.

The key differences between this pattern and the Model-Based Mapping Pattern are:

- This pattern relies on the thesaurus or ontology having a shared meaning across sources and targets (Model-Based Mapping does not require that an ontology contain a shared set of meanings).
- This pattern's primary output is in the form of scripts and code, which can be deployed in a separate run time environment (Model-Based Mapping requires a dedicated run time component to mediate among sources and targets).

Applicability

Use the Third-Party Reference Pattern when:

- Time to market is a primary business driver.
- Data sets are uncomplicated (compared with complex part structures).
- A new run time engine in the architecture is not feasible.

Structure

The Third-Party Reference Pattern may utilize either a thesaurus or an ontology as the external reference.

- **Thesaurus**—A hierarchically structured thesaurus is used as the target for all mappings, thereby providing a common reference point for disparate schema.
- **Ontology**—An object-oriented schema is used as a target for all mappings; it contains a shared vocabulary for mapping sources into it.

Known Implementations

- **Contivo**—leverages a patented external thesaurus for mapping schemas; generates XSL/T for run time execution
- **Unicorn**—leverages an ontology as a target for mapping schemas; generates XSL/T and SQL for run time execution

Model-Based Mapping Pattern

Unlike simple mappings between data structures, the model-driven approach utilizes well-defined metadata about context to constrain the scope of its semantic assertions. The main benefit of this kind of approach is that a neutral ontology does not have to represent a shared meaning among participants—making it highly attractive for diverse communities. Mappings, such as Tabular Topic Maps or proprietary models available from several vendors, capture both the relationships and the scope for which that those relationships are valid. This pattern relies directly on the use of a neutral ontology to capture both the relationships and context of those mappings. Typically, the ontology should be more abstract than any of the schemas being mapped to it, to avoid problems capturing the conceptual semantics of the more concrete implementation schema.

This pattern is very powerful for schema-level and some purely semantic conflicts but falls short in providing mechanisms for resolving semantic conflicts with instance data. Additionally, it is incapable of providing inference-like capabilities, which enable derived or inferred semantics about the models it is operating on.

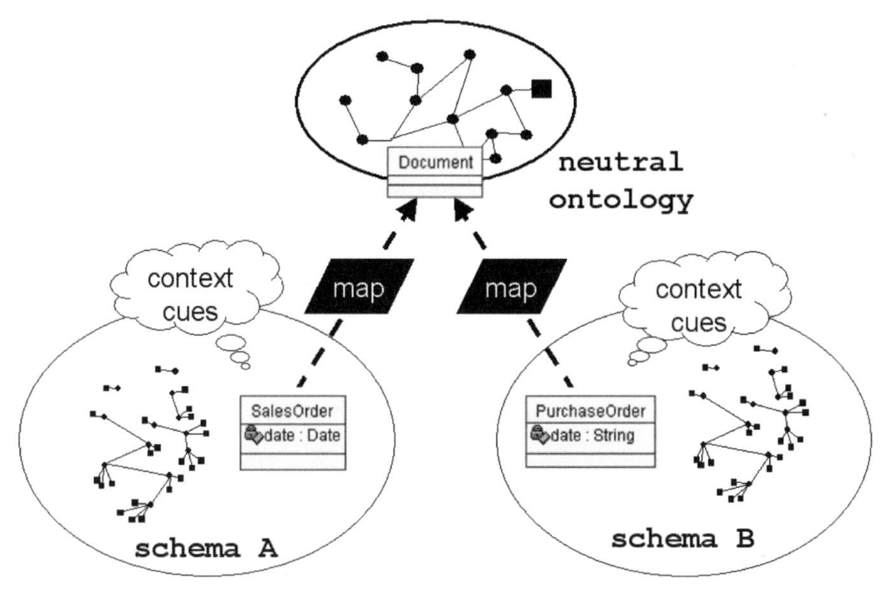

Figure 5.18 Relating different concepts and contexts in a conceptual model

Applicability

Use the Model-Based Mapping Pattern when:

- Creating an ontology with a shared meaning is impossible or impractical.
- Schema-level differences are a primary source of conflict.
- Combinations of data sources cannot be reliably predicted before run time.
- A priori script generation and deployment is impractical.

Structure

Although the overall pattern is consistent across various implementations, the actual mapping approach varies by vendor and mapping language. In general, the structure of this pattern requires that a model be built that explicitly defines the meaning of one schema in terms of another—using context cues as a mechanism to bound assertions about the possible meanings of a single data unit at a time. With Topic Maps, the context cues are identified by "scope" constraints. Proprietary vendor approaches typically require a fairly deep understanding of a given methodology. For instance, Modulant's Context Mapping approach implements techniques documented in their Context-based Information Interoperability Methodology (CIIM).

Known Implementations

- **Modulant**—provides a tool set and methodology for context-based information maps among disparate schema; the mapping target is an Express-based ISO 10303 derivative focused on product data.

- **Topic Maps**—multiple vendors support topic map specifications, which do not specify that an ontology target be used.

Inference Pattern

A crucial aspect of semantic conflict resolution is the degree of automation in the process. The Inference Pattern strikes a balance between the machine learning pattern, which requires very little metadata preparation, and the model-based mapping pattern, which requires a significant degree of model, map, and metadata to function well. An inference engine does require a model, with some type of built-in formalism (description logics, for example) to describe the structure of relationships in the system. Unlike some other semantic patterns, the inference engine can operate on very low-level, granular ontology as well as very conceptual, high-level ontological models.

Inference is a very powerful pattern for working on well-formed, robust schema with explicit logics, but it falls short on working with the kinds of schema that are already widely deployed in the enterprise (such as XML Schema, UML, Express, DTD). Likewise, typical distributed Inference Pattern deployments require that a shared meaning exist for the ontology being used as the unified model—increasing the probability that widely divergent communities will be bogged down in the process of aligning data semantics for the central model.

Applicability

Use the Inference Pattern when:

- A mixture of instance- and schema-based data conflicts exists.
- Creating a robust RDF or OWL model is feasible.

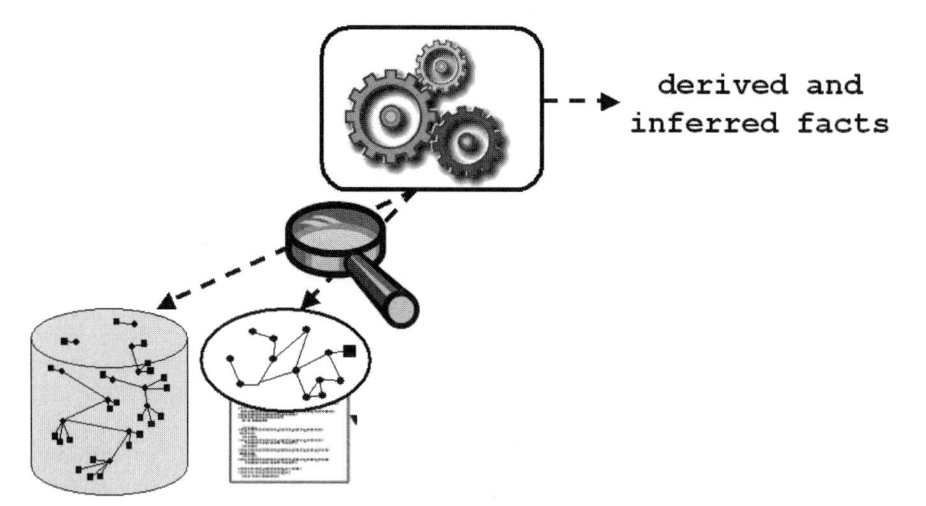

Figure 5.19 Inferring new facts from schema and instance data

- Ontology can represent a shared meaning among sources.
- A significant amount of semantics are implicit and inferable.

Structure

Typically an inference engine is used as a component of an overall system. The inference engine provides true/false answers to queries regarding the structure of data and schema, which in turn allows other software components to make better decisions with the new data available. For example, interfaces such as CORBA, SOAP, or Java can be used to ask questions of the ontology about the concepts, taxonomy, and roles it contains—without specific a priori knowledge of the implementation details behind that ontology. This is a very powerful pattern for accomplishing adaptive behavior in system-to-system data exchanges.

Known Implementations

- **Network Inference**—provides a design time and run time tool set that allows developers to create OWL-compliant ontology and then use that ontology to reduce friction in ultra high-performance data exchange scenarios.
- **OpenCyc**—Based on the CyCorp commercial inference engine, the open source version shares many of the features and retains access to the worlds' largest and most complete repository of commonsense assertions about the world in general.

CONTEXT BOUNDARIES FOR SEMANTIC INTERPRETATION

Most semantic conflicts are resolved by understanding the contexts of the data sets. As previously mentioned, context provides the cues that we, and machines, need to assess data and make decisions about how it should be reformatted.

In lieu of a lengthy discussion of context, it is most useful to think of context as *anything that influences the interpretation of data.* Unlike metadata, context cannot always be succinctly embodied in application models or software programs; it is often the intangible elements of events and space that are completely unique to a given individual or software program.

To Computers, All Context Is String

A group of colleagues would reflect back on STEP data standards work they did several years past, where lengthy discussions about business and process contexts would nearly always end with one of German participants observing, "Yes, but to the computer, all context is string." Invariably, his observation is crucial—because, in life and business, context remains a particularly subjective and dynamic experience. Discrete subjective and dynamic experiences do not lend themselves to predefinition in either standards or software. The net result of this fundamental fact is that we are left sending signs to the software about the context in the form of data—ultimately reducible to strings.

Context Models and Taxonomy

Although context will likely still be strings in future semantic interoperability architectures, this does not mean that context need be chaotic. A significant opportunity exists for communities of practice to come together and define a context model—defining a structure through which individual discrete contexts can be understood. This is not to say that a data model of context can be created, but that a conceptual model of context should be created. If a comprehensive framework for information context were to be introduced, much could be done to align the various patterns of semantic conflict resolution into a broad semantic net, whereby the patterns could all work in unison with a common grounding of context categories.

Insider Insight—Julian Fowler on How Context is Leveraged for Interoperability

In terms of utility for processing and sharing information, as it is represented by data, context can be regarded as having two interrelated aspects:

- the circumstances within which data is valid (i.e., is relevant and accurate with respect to some purpose or set of purposes)
- the circumstances within which the meaning of that data is invariant (i.e., every actor within the context will interpret the data in the same way)

In order to support functions such as sharing of information among people, organizations, and applications—the context in which data is created and used must itself be captured and represented as data. There are several different levels at which such capture takes place: Each enables more complex and sophisticated semantic technologies, but also demands more complex and sophisticated methods and formalisms for context capture.

Collecting Context Cues

Information about context can be captured in human-readable documents that are attached to the data and/or schemas to which it relates. Although such an approach actually fails the criterion stated above: *"the context in which data is created and used must itself be captured and represented as data,"* this can be of benefit simply because it recognizes the existence of context and gathers information about it in one place. Because context is at least documented, this approach represents a major advance on the common situation where context is not formally recognized and information about context is chaotically distributed across schemas, application code, specifications, and in the heads of programmers and domain experts.

Inserting Context Cues as Code Comments

A more formal technique that extends the first approach makes use of a framework within which context description—still in human readable form—is embedded within a data definition or schema. In some respects this can be seen as a relatively minor extension to good practices for schema documentation: Every element, entity, table, or class, and every column or attribute, is described in terms of its range of validity and its range of applicability. Such a framework and approach supports the need of programmers and analysts who need to understand and make decisions about the nature of data—but, since this would still largely take the form of comments in code, it does little for run time operations or the automation of semantically aware tasks and processes.

The first two approaches summarized above are useful in aiding programmers or analysts to understand the data that they are working with. However, unless universally applied they do little to aid true semantic interoperability, since the representation of the context is effectively hidden from *other* creators and users of data. In order to address this, we need to look at approaches that place information about context *in* the data, so that people and/or software, not involved in the processes of data or application creation, can achieve a level of understanding of the context that bounds the validity and applicability of that data.

Context as Schema Accessibility

The first—and very limited—approach that fits into this category is that of exposing the data's governing schema as metadata (table/column names, markup tags, etc.). This is so common—especially in today's environment where XML is becoming a near-universal format for data exchange between applications—that we may not even recognize this as a contribution to understanding and sharing of context.

Context as Standard Data Vocabulary

A more potentially effective approach is to add metadata that reflects the usage (or the intended usage) of the data in question. This is a common thread in the development and adoption of "Standard Data Vocabularies" (usually expressed as XML schemas with shared meaning), the use of Topic Maps, and many of the initial efforts under the "Semantic Web" banner. The basis here is that data creators provide sufficient additional information about their data—information about the circumstances that we refer to as "context"—that other users can arrive at equivalent interpretations of the data + metadata combination as it moves across boundaries between applications, organizations, or user communities. This approach typically takes one of two flavors:

- Every user, or every user community, applies its own metadata / markup scheme—meaning that every receiver has to anticipate translation or mapping of the supplied metadata into their own context
- Every user and every user community is required to apply a common metadata / markup scheme—meaning that every creator has to apply a translation or mapping from their local context and vocabulary—in effect creating a single common context for the community.

Such an approach *can* be effective, *if* all the users and organizations within a community agree to "follow the rules" and adopt a shared context for exchange. Practical expe-

Continued

rience shows us that *agreeing* on the rules is difficult, and following those rules is consistently an order of magnitude more difficult.

Context as Definitional Metadata

The next alternative that we consider takes the benefits of a metadata approach—i.e., that information about context is embedded in and associated with the data to which it relates—but decouples it from the requirement for users to either develop localized translation capabilities, or to commit to wide-ranging but often inaccurate or inappropriate "standard" vocabularies.

In this approach, one or more interoperability schemas—which should correctly be considered a first-class ontology—are defined for a specific, well-understood and bounded, environment or community. Such interoperability schemas are typically an ontological model defined by systems integrators, and used as "pivot models" to mediate communication between the different parties in the environment.

The keys to this approach are threefold:

- The interoperability schema allows contextually significant information to be represented as metadata inside the model.
- The data created and used by each participant in the environment may be adequately described using the interoperability schema, via model-based maps or programming scripts.
- Run time tools can use the combination of model-based maps, scripts, and the interoperability schema to automate the processes of transforming data between different applications and users.

Although this approach provides significant advantages compared to those summarized above, it has two important limitations: first, its effectiveness depends on the definition of a fixed interoperability community; second, it typically has to be supported by a single systems integrator defining the interoperability schema(s) and model-based maps, often using proprietary software tools. Further, the metadata that is defined to represent contextual information is still in the form of simple textual labels, processable only at the level of relatively simple equality and inequality operations.

Context as Model-Based Ontology Bridges

Finally, and at the bleeding edge of today's applications of semantic technology, we can envision solutions in which context is itself represented—not as human readable text, or machine-parsable metadata, but as ontologies.

Now, rather than attempting to process sets of context labels ultimately reducible as strings to arrive at interoperability solutions, software engineers can build systems that represent the environmental circumstances people recognize as "context" as real-world concepts, and can therefore formalize the relationships between sets of context factors and the data whose meaning they affect and constrain.

This approach can be supported by the use of languages such as RDF/S and OWL, and by inferencing tools that operate on these ontologies. This enables an overall methodology that is simultaneously small scale (some might even say piecemeal) and yet potentially globally applicable.

- Any data can be characterized by and tied to the contextual factors that bound its validity and applicability.
- Contextual factors should be interrelated within and across ontologies defined at industry, enterprise, or domain level.
- Links or bridges can be defined between different contexts and user communities.

Coupled with inferencing technologies, Web Services, and software agents, this opens up the possibility of a true "Semantic Web" in which commonality and differences among diverse data sets can be *discovered in context* from their declared relationships to local, federated, and global ontologies.

Julian Fowler
Partner
PDT Solutions

FINAL THOUGHTS ON CONFLICT AND CONTEXT

In the realm of semantic technologies conflict and context is the crux of everything. The reason why semantic technologies exist at all is because of the unavoidable fact of semantic conflict. Software engineers cannot escape it; business leaders pay a hefty price because of it. The only way semantic conflict can be dealt with successfully is by fully understanding the context in which the conflict exists. In fact, context is crucial regardless of whether software engineers program a solution in a popular programming language or model the solution in a scaleable data representation modeling language. So, whether Java is used or OWL is used, the engineer and analyst will always have to understand the business context that created a given data conflict.

The remainder of this book is dedicated to exploring alternative solutions for solving semantic conflicts in enterprise environments. The role of metadata in providing cues for resolving conflict is the topic of Chapter 6, *Metadata Archetypes*. The role of ontology in designing enterprise solutions that dynamically resolve semantic conflicts is the topic of Chapter 7, *Ontology Design Patterns*. The various service modes of semantic architectures for semantic conflict resolution and model-driven integration are the topic of Chapter 8, *Multimodal Interoperability Architecture*. Finally, the architectures are examined in context of infrastructures and E-business models in Chapter 9, *Infrastructure and E-business patterns*.

Chapter 6

Metadata Archetypes

KEY TAKEAWAYS

- As a term, metadata has multiple meanings and can refer to different things.
- All IT systems are metadata systems.
- For semantic interoperability, metadata is necessary but not sufficient.
- Of the five core metadata layers, referent metadata is the most valuable for semantic interoperability—but it is also the least developed.
- The value of metadata in communication should be canonized by an expanded conception of the OSI communications stack (that includes metadata roles).

This chapter is a somewhat unconventional look at metadata. Many books on this subject offer deep insights in the specific uses of metadata for data warehouse implementations and even for XML management issues. In fact, a number of books on the subject of metadata specifically address the creation and management of metadata repositories. Unlike these books, our focus in this concise chapter is to show how metadata exists at a number of architectural layers and to describe how metadata at each of these layers contributes to the interpretation of data meaning—semantics.

WHAT IS METADATA?

It all seemed so simple in the beginning. Metadata was simply data about data. But then we started seeing the term pop up in so many disconnected places. Metadata was in the database, it was in the mainframe, in our cell phones, in a J2EE application, and also in the XML documents. Metadata is just about everywhere—at least in name.

The Ultimate Overloaded Term

When we use the term metadata, we usually mean some particular kind of metadata. It is impossible to talk about metadata in general because it has so many manifes-

Adaptive Information, by Jeffrey T. Pollock and Ralph Hodgson
ISBN 0-471-48854-2

tations. This makes it difficult to really understand the strengths and capabilities of vendor tools classified as "metadata management platforms"—which tend to "manage" only a small part of the metadata inside the enterprise.

Insider Insight—Mike Evanoff on Why Metadata Matters in Design and Execution

The problem of information interoperability has largely gone unrecognized by senior stakeholders, like CIOs, within large enterprises. To truly appreciate the information interoperability problem requires one to think "outside of the integration box" and recognize the importance of using metadata as part of the design time and run time solutions. It is now possible to utilize commercial-off-the-shelf tools to model this metadata and then use the metadata at run time mode to perform a higher-quality and more efficient form of information integration. Many in the IT industry have called this new form of information integration "information interoperability", in part because of the decoupling of physical, logical, and virtual models of the information assets. ManTech's e-IC team strongly believes that information interoperability is the key capability that will drive integration value for years to come. Our commitment to this vision is demonstrated by our deep interoperability experience, metadata expertise, customer roll-outs, and our innovative MEIIM methodology designed to add value to every stage of your interoperability development and support needs.

Mike Evanoff, Technical Director for
ManTech's Enterprise Integration Center

Since the rise in popularity of XML-based exchanges, the metadata management platform has tended to focus in this area. Before XML, it made little sense to call out metadata as a discrete component of the overall enterprise architecture because mainframe and database tools are mostly self-contained. They also do a pretty good job at managing the metadata generated by their own systems: Why bother with another tool?

However, XML and the increasing importance of distributed information sets (documents and unstructured content) have changed all of that. Information is starting to have a life of its own, outside of previous enterprise application boundaries. This, of course, gives rise to a wide range of tools intended to manage all of this information separately from the applications, mainframes, and databases that gave rise to much of that information in the first place.

Today, all software systems are metadata systems and metadata is pervasive in all things digital.

All Systems Are Metadata Systems

Whether one is considering a legacy product data management (PDM) system or a newly released off-the-shelf small business financial package, their internal data stores and external interfaces seethe with metadata. Data about product data, data about interface APIs, data domain models, and data about XML structures are just some of the ways that metadata live in our software applications.

Counterintuitively, metadata is also present when it is not formalized. For instance, an EDI document—manifested as a flat text file with a series of string values—may not contain any explicit metadata in the form of schema or tags or column names or data type information, but metadata is still present. Metadata can exist as the informal context that surrounds a message. The EDI document, despite its total lack of any explicit metadata, still conforms to an implicit set of rules about the structure, position, delimiters, and contents of that message. It is the context surrounding that particular EDI message that gives rise to its metadata. Therefore, metadata need not be explicit for it to impact the interpretation of data semantics.

Metadata Is Necessary, but Not Sufficient

Today metadata is indeed everywhere. But that does not necessarily mean that it is useful. Metadata, like data itself, does not strictly conform and express its own scope in ways that software can easily understand without human involvement. For instance, the metadata contained within an Oracle database is vastly different than the metadata used to describe a set of XML documents. Worse, the metadata contained within databases from different vendors (e.g., Oracle, IBM, Microsoft) can all be different as well. In the XML world, individual human modelers can design XML schemas that model precisely the same domain—with completely different metadata content. Understanding newly discovered metadata is still largely a human endeavor.

HIERARCHY OF METADATA

Decomposing the different kinds of metadata is a first step toward improving the ubiquitous processability of metadata regardless of its origin. Overcoming issues with the diversity of metadata in enterprise systems is not a simple task. There are many layers of metadata,[1] which do not necessarily relate to one another or easily bend to standardization. Let's take a look at some of the different kinds of metadata.

The hierarchy of metadata shows how layers of descriptiveness (syntax, structure, references, domain, and rules) build upon one another to augment the semantics of raw data. Note that none of the metadata layers is called a "semantic layer." This is because *all* metadata contributes to the meaning and understanding of raw data.

[1] Dr. Amit Sheth is an early pioneer of metadata. His work has greatly influenced this section on metadata layers—in particular, his work presented at the IEEE Meta-Data conference in Maryland, April 1999, on the subject of Semantic Interoperability of Global Information Systems.

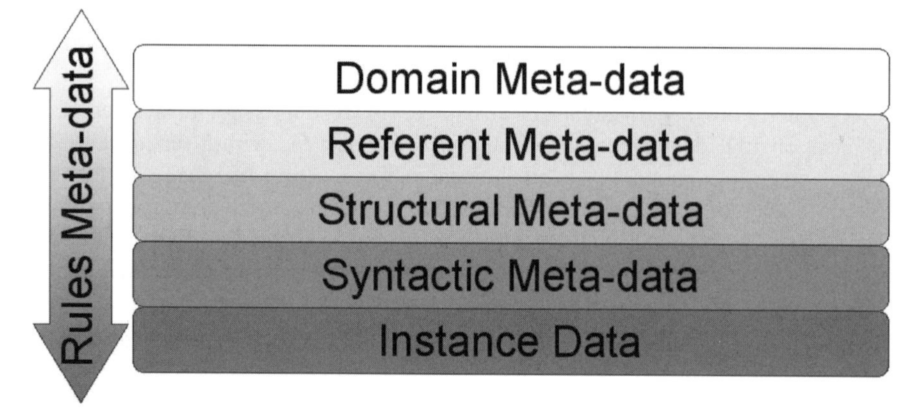

Figure 6.1 Layers of metadata

Layer 1: Instance Data

This layer in the metadata hierarchy is included for completeness. In actuality, there is nothing "meta" about "data" by definition. One way to think about raw instance data is to think about data value as being distinct from data itself. For example, the value "21" alone doesn't really say anything about whether the data is ASCII, binary, generalized, aggregated, a gambling game, or the drinking age. We as humans might infer that, because it is composed of numbers, it is a numeric value, but without additional metadata we could be making a mistake if that particular data value is a mechanical part code, which could just as easily be "ACW19." Raw data is value-oriented and tells us nothing about anything else. However, *collections* of raw instance data do contain metadata.

Collections of instance data contain implicit semantics that are frequently indiscernible by the schemas that govern that instance data. Pattern analysis tools and machine learning approaches to semantics work at this level. By assessing the patterns of volume, scale, and position in both numeric and textual data sets, the software can recognize and learn facts about the data that humans may have never even thought to ask for. Utilities like this are quite valuable in discerning semantics in places where a schemacentric approach cannot venture.

Layer 2: Syntactic Metadata

The first explicitly defined metadata a machine needs to "know about" in order to process data is syntactic metadata. This kind of metadata is inherent in all types of digital information but can vary widely based on data syntax. Syntactic metadata can include data type, language format, message and/or document length, date stamps, source, bit rate, permissions, encryption levels, and more. The extent of syntactic metadata is constrained by the expressiveness of the language syntax. ASCII

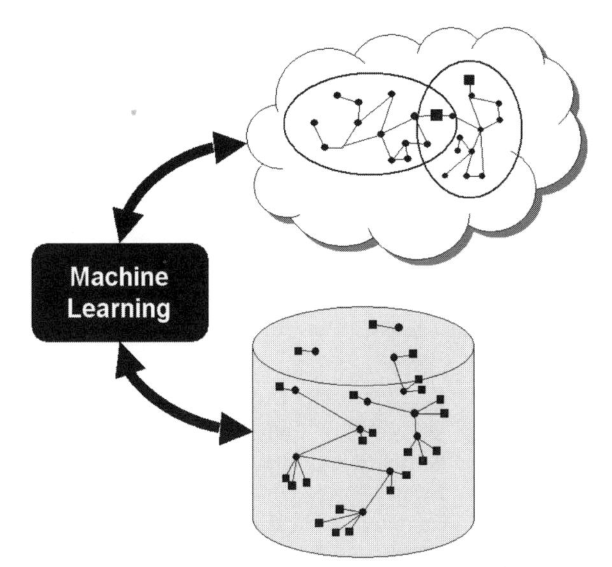

Figure 6.2 Patterns of implicit metadata contained in collections of instance data

Figure 6.3 Examples of various XML model syntax

data has a vastly different syntactic structure than unmarshalled Java objects or binary audio streams. Also included in this category of metadata would be syntax intended as display cues for human users—HTML, for example.

Layer 3: Structural Metadata

Building upon syntactic metadata is structural metadata, which gives form and structure to "units" of data. This is far and away the most common conception of metadata used in both tools and document standards. Structural metadata is simply a way

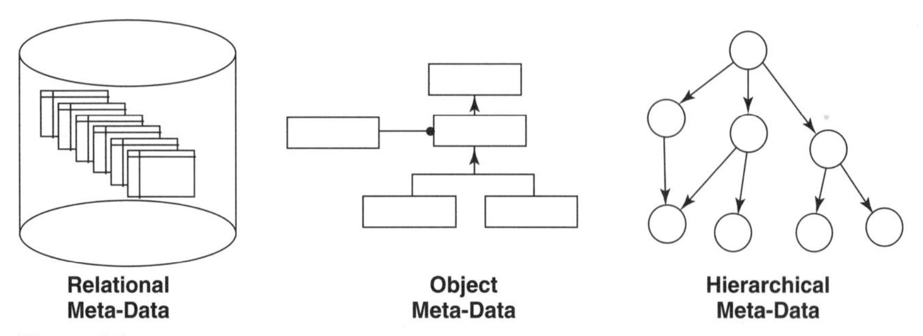

Figure 6.4 Metadata about structure varies based on the given structure and syntax

for machines to understand varying levels of scope within a single data stream, document, or message. The three main kinds of structure that we can create within digital data are hierarchical, relational, and object-oriented. The structure of data defines how it can be searched, modeled, visualized, and expressed. Unfortunately, the structure of data is usually arbitrary—highly dependent on the skills and norms of the humans who create the structures.

Within enterprise systems we see structural metadata as the choice of document structure given to an XML document (Is it two-dimensional or deeply nested? Is it hierarchical or objectlike?), the degree of normalization in a database (first, second, third normal form), or the abstraction levels used in OO modeling (Is a square a rectangle or not?). Structure has all sorts of metadata, describing relationships among data elements, but the forces that influence structure derive from even higher levels of metadata: data references and domain context.

Layer 4: Referent Metadata

Whereas structural metadata is usually most relevant within the sphere of a single data model, the sole purpose of referent metadata is to provide linkages between different data models. The form of referent metadata can vary widely. In most applications where data is exchanged between different data models, the referent metadata exists inside the program's application code. For example, a product life cycle management (PLM) application may host several application adapters that receive data from outside sources. These PLM adapters may implement Java, C++, SQL, or some other application program logic to receive the data stream, disassemble the data structure from its source, and reassemble the data in a form that the PLM system uses internally. In this example, the referent metadata is locked away in the program logic of the application—as opposed to being specified separately from the application logic.

Conversely, many new systems implement architectures that separate the referent metadata from the application layer by leveraging other mapping or scripting technologies. Technologies like XSL/T, Topic Maps, and data-driven mapping tables allow architects to separate the concerns of referent metadata from the programming logic of individual applications. By enabling integration architects to separate these concerns, this crucial functionality (providing model-to-model conversion instructions) can be leveraged by different components within the overall architecture.

When object-oriented systems began to rise in popularity, the notion of converting data structures from object systems to relational (RDBMS) systems (and vice versa) became an important topic of systems management. The term "impedance mismatch" succinctly described the problems encountered when moving similar data between different schematic representations. Today, the problem is much larger. Today's most popular data representations (XML, OO, RDBMS, EDI) all contain fundamentally different schematic representations. The problems of converting similar data across these formats can sometimes be nontrivial.

Referent metadata is the metadata that describes how data should be converted from (a) one schema to another in same languages, (b) one schema to another in different languages, and (c) one schema to another at very different abstraction levels. The authors strongly believe that this area of metadata specification is ripe for standardization; more will be presented on this subject in Chapter 12.

Topic Maps, Tabular Topic Maps, and semantic "context-aware mapping" technologies offered by vendors such as Modulant, Unicorn, and Contivo all take great strides toward strengthening the value of referent metadata in the interoperability architecture. Their capabilities are far beyond the typical "basic mapping table" approach taken by Microsoft in Biztalk and most other EAI vendors in their tool suites. But much more work needs to be done—particularly in the area of spanning abstraction levels and unambiguously capturing context in shared context ontologies. These capabilities, or shortcomings as they may be, will become ever more important as the next level of metadata matures and begins to take center stage in the enterprise interoperability architecture: contextual metadata.

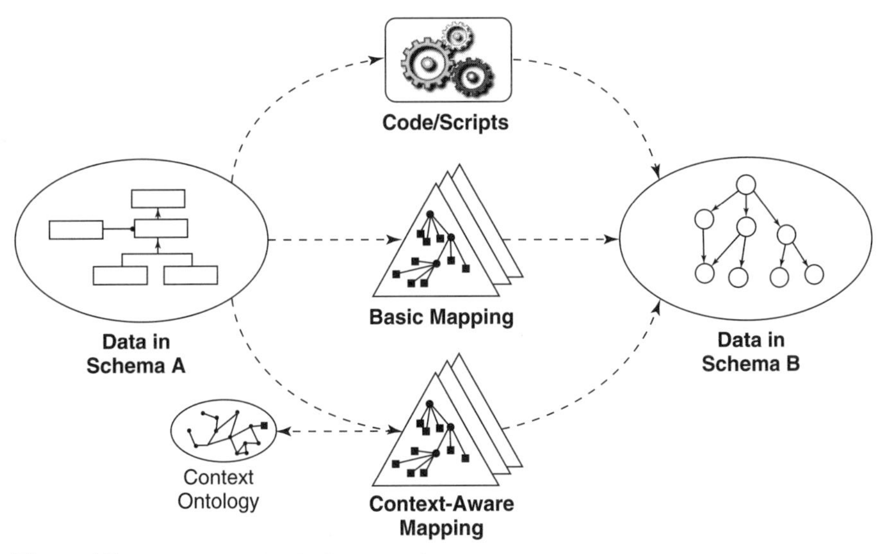

Figure 6.5 Three core kinds of referent metadata

Layer 5: Domain Metadata

As described in Chapter 4, context is what you get when express something local in terms of something global. In the metadata stack this translates as providing an overall conceptual domain model from which a wide variety of local application models can be associated (via referent metadata mappings). Practically speaking, contextual metadata typically takes the form of ontology. It is through ontology that the information your individual application "knows about" can be understood within the larger sphere of "all things known about" within a given community—thus providing context to your local data and information.

This capability is a crucial requirement for information sharing architectures to accurately convert data meaning across multiple contexts. As data moves from one system environment to another it is quite literally changing context—and thereby needs to change form, structure, and syntax. The use of ontology as contextual metadata gives applications an external reference point to base the transformation and interpretation of data from a wide array of different and seemingly incompatible contexts.

Although it can be said that all metadata provides context, only the metadata in this layer (e.g., conceptual domain ontology) provides a sufficiently abstract reference point on which the rest of the metadata can be understood within appropriate context. Much more will be covered on this subject in Chapter 7, Ontology Design Patterns.

Layer 6: Rules

This final layer cuts across all the previously discussed layers. Rules, in all shapes and varieties, can be used to constrain and crystallize the semantics of metadata

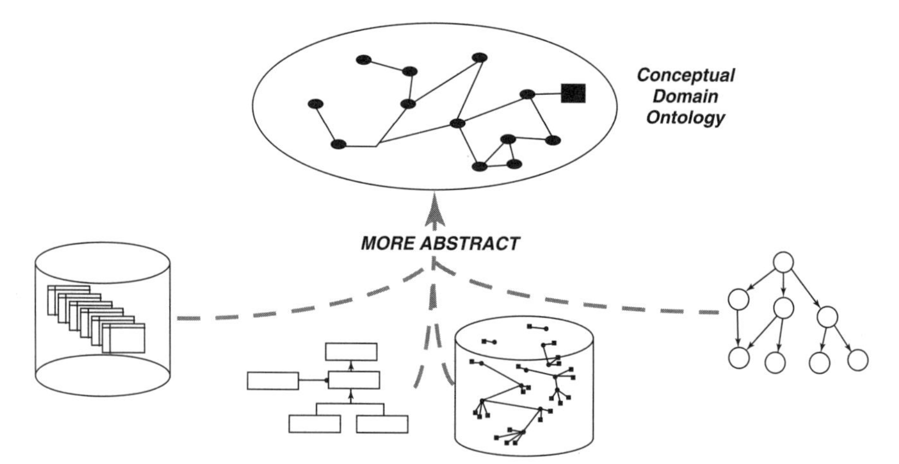

Figure 6.6 Domain metadata is more abstract than implementation schema

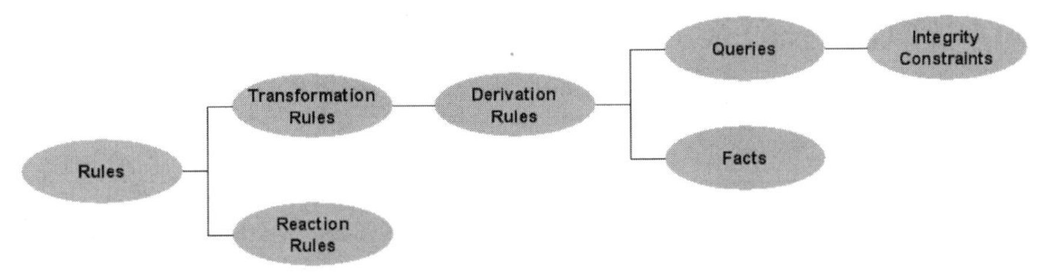

Figure 6.7 RuleML rule hierarchy

specifications and models at any abstraction level. Often in software applications business rules are not treated as separately managed metadata. Like structural and syntactic metadata, they are frequently bound up in processing logic and tightly coupled with application code. This makes the management and refinement of business rules a difficult process. Recent efforts, Rule Markup Initiative (RuleML) and the Web Ontology Language Rule (OWL-R) activity; do much to alleviate this problem, but a more systematic treatment of business rules across all relevant layers of metadata is required.

Business rules can represent definitions of business terms, data integrity constraints, mathematical derivations, logical inferences, processing sequences, and relationships among facts about the business. Some modeling languages, such as OWL, have some built-in mechanisms to account for rules in the models. Others, like UML and ERD, rely on external formalisms like the Object Constraint Language (OCL) to assert rules on top of models. The key capability in any of these applications is to make the rules actionable by the processes that read the models and constraints.

ENVISIONING A MODERN OSI COMMUNICATIONS STACK

Given the importance of data semantics in the role of network communications and the pervasiveness of conflicts in information exchange, a new formulation of the communications technology stack should be examined.

The OSI, or Open System Interconnection, model defines a networking framework for implementing network communications protocols in seven layers. Control is passed from one layer to the next, starting at the application layer in one context, proceeding to the bottom layer, over the channel to the next context, and back up the hierarchy.

Network Communications Stack

It is important to start with a basic understanding of the protocol stack (OSI layers) in order to explain where barriers to communication exist among information systems. Cross-system communications begin with the plumbing concerns repre-

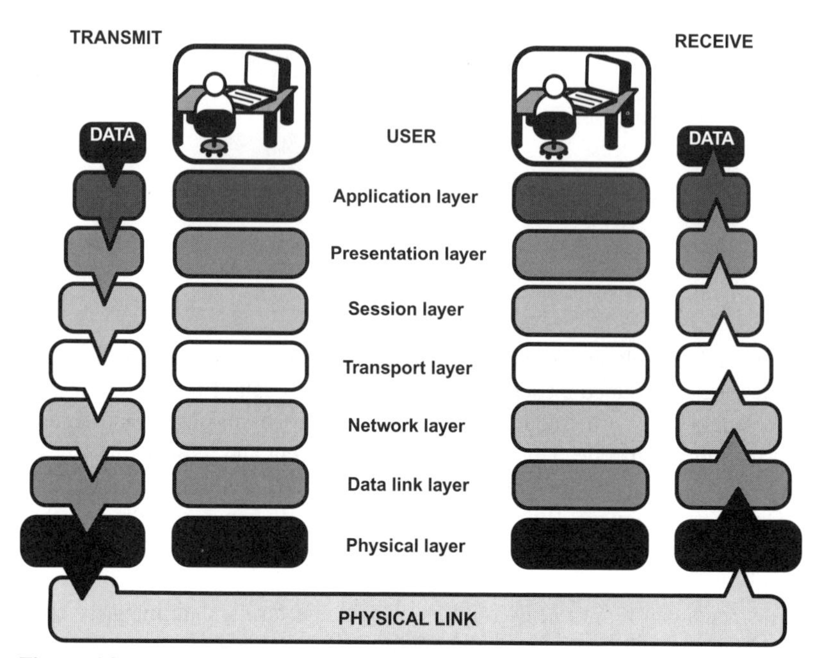

Figure 6.8 OSI communications stack

sented in the OSI model but continue further into how information models, which are required to exchange meaningful data between heterogeneous software systems, are implemented and used.

Layers of the OSI model can be understood as shown in Table 6.1.

Payload, Not Just Packets

The OSI model was designed with the physical nature of communications in mind. Allusions to application models and syntactic structures are as far as the model goes to acknowledge the importance of the data within the communications framework. The utility of this emphasis is based in the historical significance of the OSI model. Developed by the International Organization for Standardization (ISO) in 1984, the OSI was originally intended to provide the reference architectural model for system-to-system interconnectivity.

Today, the demands placed on middleware are far greater than they were in 1984. The explosion of digital data has reached astronomical levels and continues to grow exponentially. The OSI framework is necessary yet insufficient in providing an end-to-end communications architecture.

A more complete conception of a communications model would account for the metadata required for software applications to accomplish successful communication. These metadata layers would include:

Table 6.1 Data Table—OSI Communication Stack

Application (Layer 7)	This layer supports application and end user processes. Communication partners are identified, quality of service is identified, user authentication and privacy are considered, and any constraints on data syntax are identified. Everything at this layer is application specific. This layer provides application services for file transfers, E-mail, and other network software services. Telnet and FTP are applications that exist entirely in the application level. Tiered application architectures are also part of this layer.
Presentation (Layer 6)	This layer provides independence from differences in data representation (e.g., encryption) by translating from application to network format and vice versa. The presentation layer works to transform data into the form that the application layer can accept. This layer formats and encrypts data to be sent across a network, providing freedom from compatibility problems. It is sometimes called the *syntax layer*.
Session (Layer 5)	This layer establishes, manages and terminates connections between applications. The session layer sets up, coordinates, and terminates conversations, exchanges, and dialogues between the applications at each end. It deals with session and connection coordination.
Transport (Layer 4)	This layer provides transfer of data between end systems, or hosts, and is responsible for end-to-end error recovery and flow control. It ensures complete data transfer.
Network (Layer 3)	This layer provides switching and routing technologies, creating logical paths, known as virtual circuits, for transmitting data from node to node. Routing and forwarding are functions of this layer, as well as addressing, internetworking, error handling, congestion control and packet sequencing.
Data Link (Layer 2)	At this layer, data packets are encoded and decoded into bits. It furnishes transmission protocol management and handles errors in the physical layer, flow control, and frame synchronization. The data link layer is divided into two sublayers: the Media Access Control (MAC) layer and the Logical Link Control (LLC) layer. The MAC sublayer controls how a computer on the network gains access to the data and permission to transmit it. The LLC layer controls frame synchronization, flow control, and error checking.
Physical (Layer 1)	This layer conveys the bit stream—electrical impulse, light, or radio signal—through the network at the electrical and mechanical level. It provides the hardware means of sending and receiving data on a carrier, including defining cables, cards, and physical aspects. Fast Ethernet, RS232, and ATM are protocols with physical-layer components.

- **Syntax Layer**—binary format of the application layer message
- **Schema Format Layer**—physical structure of information classifications
- **Referent Layer**—relationships among conceptual and implementation schema
- **Domain Context Layer**—abstract model used to align context

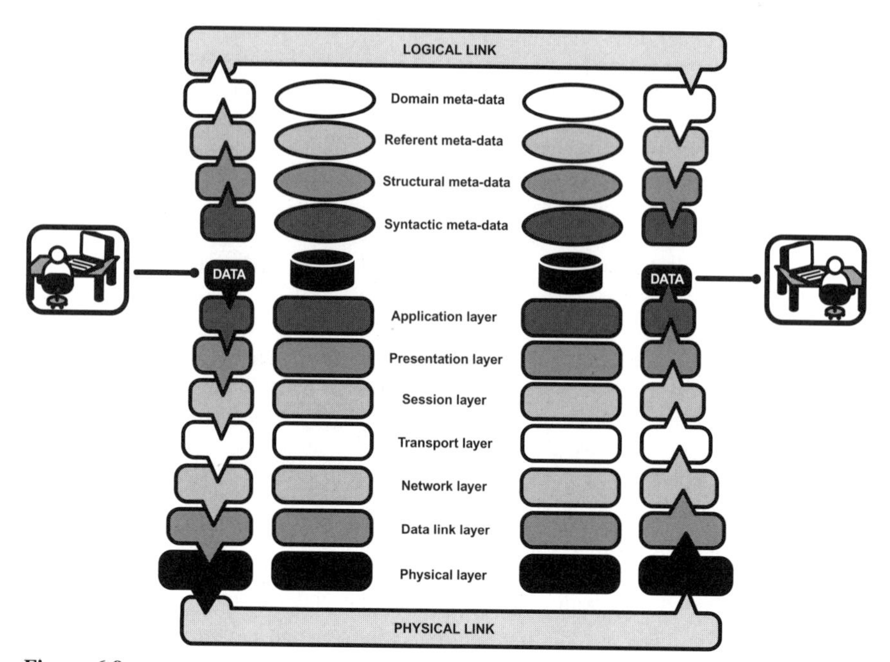

Figure 6.9 A more complete communications stack

The resulting model might look something like Figure 6.9.

Trends in middleware continually point to an era where information's value is inexorably tied to the network, and that network's ability to share that information. Expanding the scope of an OSI-like model to include elements of an information infrastructure—a reference model for a schema architecture—would serve the greater community.

In this expanded view, the emphasis and focal point would be on the duality of the payload and the packets—not just the packets. Payload is the contents of the message packet, whereas the packet is the container that enables delivery. Including a representation of the information models and metadata in the message container would shift the paradigm from a physical, switching and routing, connection-centric view of system-to-system communication to one that accurately reflects the true value of communications—in its *meaning and understanding*.

METADATA ARCHITECTURES FROM THE OBJECT MANAGEMENT GROUP

The OMG has taken center stage in the proliferation of metadata frameworks for software application development. Both their Common Warehouse Metamodel (CWM) and Model Driven Architecture (MDA) frameworks provide substantial assistance in specification of robust metadata architectures. However, the primary focus of the OMG efforts remains fairly limited to the development of individual software applications. Both the CWM and the MDA are substantially geared toward the implementation of single software applications—as opposed to system-to-system interoperability.

Although the focus of this book is on the interoperability of multiple applications, the utility of CWM and MDA is very high. In the remainder of this section we examine briefly the key architectural features of the CWM and MDA and evaluate their relevance in an interoperability scenario.

Common Warehouse Metamodel

The CWM was ostensibly created to facilitate the exchange of metadata among business intelligence systems and data warehouses. By virtue of its design, the implementation of the CWM relies on technology vendors to implement the CWM specifications inside their highly proprietary tools. Some vendors will choose to adopt the CWM specifications as the core of their metadata repository, whereas others will choose to simply offer data export features to extract the metadata in the CWM specifications.

The net result of this effort is a standard that was created to provide an on-the-wire format for metadata—which, when implemented by leading vendors, will allow for the exchange of metadata in a common format. The design parameters for the CWM are consistent with other OMG efforts and other popular standards. CWM relies on the XMI model syntax to implement UML models and XML (using CWM DTDs) as the core exchange format for the metadata. Likewise, the CWM effort proposed the use of CORBA-like IDL interfaces for physical access to warehouse metadata in the CWM framework.

As previously mentioned, the CWM was primarily intended to solve problems encountered by a fairly narrow range of applications—business intelligence and warehouse systems. It does offer as a premise the idea that a data warehouse is a useful way to store and disseminate the vast amounts of corporate data contained within the enterprise. However, as we have explored earlier in the book, data warehouses are not at all ideal for widespread information-sharing schemes for many reasons. Where CWM takes great strides is in the specification of models for metadata and in the way it grounds them in a self-describing Meta-Object Facility (MOF) to contain model creep. These modeling advances are significant well beyond the scope of typical data warehouse and business intelligence applications.

The CWM breaks down the classification of metadata into four main packages: Core, Behavioral, Relationships, and Instance.

However, it is the classes provided for in the Core package that are the foundation upon which the rest of the CWM rests. It is also in this model that we can begin to see how the classification of core metadata concepts in a model can be highly important to a broader community.

One of the most important aspects of the CWM effort is that it provides a baseline understanding of how to approach the metadata specification problem. Whether a software architect is dealing with data warehouse issues, package application

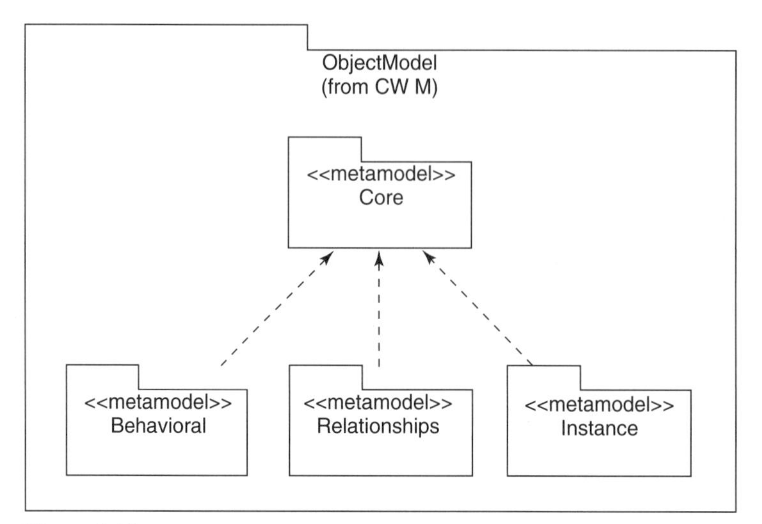

Figure 6.10 CWM Core package model

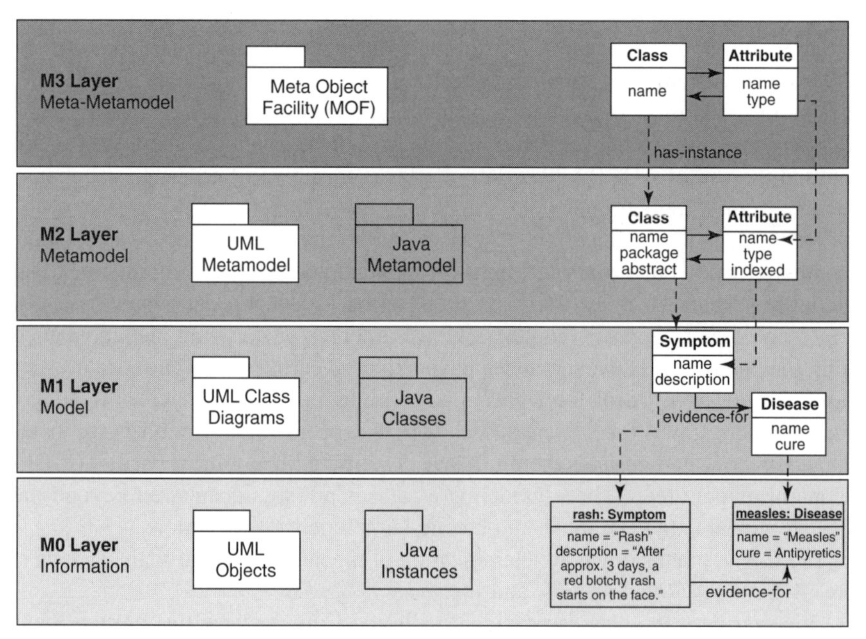

Figure 6.11 OMG Metamodel hierarchy[2]

[2] Image originally published in "Ontology Design and Software Technology" presentation to Stanford Medical Informatics Colloquium on June 12, 2003 by Holger Knublauch.

issues, or information interoperability issues, the CWM offers an approach for modeling and specifying the metadata that is bound to move alongside data in any possible exchange.

Although the authors believe that the CWM can be substantially improved to account for the wide range of metadata that can flow in non-warehouse-type exchanges, they also believe that it provides the best possible foundation upon which a comprehensive metadata interoperability framework can be built—in other words, it can provide an effective ontology for metadata.

Model-Driven Architecture

The scope of possible applications of the MDA is vast—and the literature on the subject reflects this vastness. MDA was initially conceived as a way to separate the underlying technology of applications from the specification of business rules, models, and metadata about those applications. This follows a core architectural pattern of separating the concerns of an implementation by recognizing that software behavior is fundamentally different from the specific technology that implements that behavior. MDA promises a future where models and metadata enable software developers to (re)generate any software on any technology platform.

Today much of the focus of the MDA has shifted toward the interoperability problem in distributed environments. However, the core specifications are still focused on the implementation of specific applications. Additional detail on the use of MDA in federated interoperability environments can be found below in this section.

The core business model in the MDA architecture is the computation-independent model (CIM) that shares many common characteristics of ontology—as we have described in this book. The CIM is typically a model of the environment in which a system will operate and acts as a source of shared understanding and shared vocabulary for the given domain. The platform-independent model (PIM) represents a finer-grained look at the application in question and describes the details of its operation—without specifying the technology in which it is implemented. Finally, the platform-specific model (PSM) specifies the details of a given technology or platform implementation of that PIM. Usually the PSM will reflect some technology design choice such as a J2EE or .NET platform decision.

Recent efforts in the OMG have focused on the difficult area of providing mappings between the various model specifications. Importantly, the mapping specifications between the PIM and the PSM are crucial for the ultimate vision of the MDA to operate as promised. Mappings at this architecture layer will provide developers the ability to generate functional code in multiple formats from common models, thus reducing the development time for new technology deployments. Much additional information can be found on OMG's website, which discusses the various technical considerations for model instance mappings, metamodel mappings, and the use of model marks and templates to facilitate those mappings.

The long-term goal of many of these MDA efforts is to move from application-specific frameworks for programming automation to federated interoperability environments. Conceptually an entire distributed framework can be generated from a series of models that describe the interactions among a diverse population of applications. This could result in a future solution whereby Web Services, middleware, and an application's data import and export capabilities may be generated directly from the models.

This is precisely where the efforts of the MDA begin to overlap substantially with the scope of this book and the central ideas of semantic interoperability. To create a truly dynamic middleware capability that removes significant portions of the human element from the development and deployment of technical solutions, a significant level of model-driven development is required.

OMG leadership recognizes this and is actively pursuing development of MDA in this direction. Their focus is primarily in the area of using models to specify the behavior reflected in the underlying technical framework of middleware—which, of course, leaves much to be desired in doing the same for the information framework. However, the combination of advances in the MDA approach for application and middleware deployment will dovetail with approaches for augmenting the flexibility of information frameworks and produce widespread benefits to companies that adopt these approaches.

Criticism of OMG and Metadata Architectures

Starting this section with sincere praise and appreciation for the OMG is only appropriate. Without the OMG, the global information technology sector would be much less advanced and would likely still be struggling to efficiently adopt and implement object technologies. However, the OMG's past does not guarantee its future performance—and many would argue that the OMG is losing relevance in today's World Wide Web Consortium-dominated Internet standards. This is at least partly due to some very glaring problems with OMG's basic offerings:

- **Focus on single applications**—Whereas the many other standards groups have focused on integration and exchange standards, the OMG (until recently) steadfastly remained committed to standards and approaches for developing single applications.
- **Premature marketing**—OMG consistently promotes and markets its new visions very early in the hype cycle—resulting in long cycles while tools and vendors play catch-up.
    - MDA tools support is still in its infancy
    - MDA and CWM are still partly vaporware
- **UML inconsistencies**—Persistent inconsistencies with the specification of UML continue to result in incompatible CASE tool implementations.
- **Behind the curve**—much of the industry hype continues to focus on Web Services and Semantic Web activity, both areas that the OMG is woefully lacking in. However, the OMG has a tremendous opportunity to add substantial value in these areas—its legacy with CORBA and its dominance in

Table 6.2 Data Table—Key differences between OWL and UML[a]

	OWL	UML
Orientation	Knowledge Representation	Information Technology
Modeling	Conceptual	System
Usage	Inferencing/Reasoning	Software/Hardware Design
Primary Construct	Property (Relationships)	Class (Entities)
Secondary Construct	Class (Entity)	Association (Relationships)
Naming	URI Based	Package Based
Development	Distributed and Open	Centralized and Closed
Primary Format	Syntactic	Visual

[a]IBM submission to OMG Ontology Working Group, Dan Chang, Yeming Ye, 08/2003.

metadata architectures could substantially drive new W3C standards development.

Future Developments of UML—A Visual Ontology Language?

Given the worldwide adoption of the Unified Modeling Language, it is no wonder that many people in the semantic technologies space wish to use it for specifying ontology. Today, nonstandard tools (such as Sandpiper Software's ontology modeler) provide substantial capabilities in this area, but widespread development of UML ontology modeling capabilities is hindered by the lack of OMG-supported standards.

Starting in 2003, the OMG opened the Ontology Definition Metamodel (ODM) Technical Committee. This working group has many objectives related to the sharing of expertise among OMG members and augmenting the MDA mapping specifications. However, the most important goal of the Ontology Working Group is extend the UML for effective ontology development and support of the W3C's web ontology language (OWL). Although this ontology committee got a late start—UML revisions were already well under way when it was formed—there is still much hope for substantial improvements to UML in order to support ontology modeling.

Depending on the definition of ontology in use, any and all UML models may appropriately be considered ontology. However, in the context of the Semantic Web activity there is only one true benchmark for ontology languages—OWL. Because the overall objective is to enable UML to visually represent OWL, it is crucial to understand some of the key differences between these languages.

Although some of these differences may seem superficial, others are quite substantial, resulting in significant modifications to traditional UML modeling approaches when attempting to model UML in OWL-compliant ways. Some of the hurdles that the ODM working group has to face include:[3]

[3] IBM submission to OMG Ontology Working Group, Dan Chang, Yeming Ye, 08/2003.

- The metalevels of the MOF (metaobject facility) framework don't align well with the core abstraction levels defined in OWL.
- Role naming in UML is very loose, whereas in OWL tighter constructs are required.
- Naming and referencing are accomplished in different ways—in OWL, they are URI based, whereas in UML they are package based.
- Properties and associations are abstracted differently between OWL and UML—in OWL a property is a relationship that is defined independently of any class, whereas in UML a relationship is a binary association whose instance is precisely between two classes.
- Visibility and scope are more granular in UML—in OWL everything is "public," whereas in UML the designer may choose "public," "protected," or "private."

At the time of writing, the ODM working group has made substantial progress in many of these areas and continues to work on use cases for ontology-aware applications, metamodel requirements, and layering of the ODM specification. It is clear that by 2005 an initial draft of the ODM specification will likely be approved by the OMG. Although this first draft will make substantial progress in defining the use of UML for OWL-based ontology development, it is still unclear as to what degree the UML specification itself will actually change to support advanced OWL Full and OWL DL features.

FINAL THOUGHTS ON METADATA ARCHETYPES

Metadata is a confusing subject. However, for its diversity and complexity there are surprisingly few major archetypes. These simple archetypes—instance, syntax, structure, referent, domain, and rules—can account for all types of metadata in the enterprise regardless of implementation details. Although the industry is still developing robust solutions for some of these layers (specifically the referent and rules layers), overall maturity is occurring quite rapidly. OWL-Rule and Topic Map standards look very promising and will likely provide sound alternatives to proprietary solutions for the rules and referent metadata layers, respectively.

In Chapter 7 a special kind of metadata will be covered in depth. All ontologies are metadata. In fact, ontologies can capture metadata at every archetype layer with very sound formalisms. This is in stark contrast to other metadata persistence mechanisms that do not have sound ontological foundations. Ontology types, classifications, and usage patterns are the central focus of Chapter 7, Ontology Design Patterns.

Chapter 7

Ontology Design Patterns

KEY TAKEAWAYS

- Ontology and ontology patterns are the applied use of long-time, fundamental engineering patterns of indirection and abstraction.
- Ontology can come in a wide variety of types, usage, and formalism.
- Classic ontology is the kind of ontology that deals with domain concepts, metadata, representation, and commonsense logic.
- Ontology architecture patterns help identify places in systems architecture where ontology can be useful and guide the creation process.
- Ontology transformation remains a crucial topic of research and development.

As a term, ontology has been with us for several hundred years. As an important emerging discipline of computer science, ontology has had a much shorter lifetime. Today, ontology is a tool used by system architects that can enable greater efficiency and quality for information technology. Indeed, the use of ontology is inexorably tied to the processing and understanding of semantics in digital systems.

There is some irony that ontology, a tool to assist with the processing of data semantics, has so many definitions. Apart from being a tool for software, ontology is also a tool for philosophers thinking about the nature of being. But even for pragmatic software jockeys, ontology can still be a label for a score of different model types with varying degrees of expressiveness.

In this chapter, we will examine various system and software architectures that ontology can augment within the enterprise. Essential patterns of ontology deployment and creation will be discussed in terms of software architecture. Finally, a technical overview of some of the most common languages for ontology specifications and transformations will be offered.

Throughout the course of this chapter it should become clear that ontology is not just one kind of thing; nor is it used in one kind of way. In fact, ontology is simply an enabler for software engineers and architects to apply core problem solving patterns in new and innovative ways.

Adaptive Information, by Jeffrey T. Pollock and Ralph Hodgson
ISBN 0-471-48854-2

CORE ENGINEERING PATTERNS

A useful way to start this section on ontology patterns is to consider the fundamental software engineering patterns that influence the development and utility of ontology in practice. Both indirection and abstraction are core principles of object-oriented development and software architecture planning. They are the kinds of tools that allow software engineers to overcome the increasing complexity they are faced with in the day-to-day operational environment of business.

> *All software design problems can be simplified by introducing an extra level of conceptual indirection. This fundamental rule of software engineering is the basis of abstraction, the primary feature of object-oriented programming. Get in the habit of using indirection to solve design problems; this skill will serve you well for many, many years to come.*
>
> —Bruce Eckel, *Thinking in Java*[1]

Indirection

A basic tenet of systems design and architecture is indirection. Indirection is a concept that is used to plan for future uncertainty. Simply put, indirection is when two things need to be coupled, but instead of coupling them directly, a third thing is used to mediate direct, brittle connections between them. A simple example of indirection is one that you probably use every day. When you type a web address, such as `www.yahoo.com`, into your web browser it will not go to Yahoo directly. First it is directed toward a DNS (domain name server) that tells your computer what IP (internet protocol) address the simple name `www.yahoo.com` is mapped to. These numeric addresses, like phone numbers, identify the correct location of the Yahoo website.

Indirection is used to insulate separate concerns from being too tightly connected with each other. For instance, when Yahoo changes its IP address, which may happen all the time, your computer system doesn't break when you try and connect to it—a simple solution enabled by indirection.

In object-oriented software systems, indirection is used all the time in patterns such as proxy, facade, and factory, among others. In a collaborative interoperability framework indirection is a fundamental component of both the physical and information infrastructures.

[1] *Thinking in Java*, 2nd ed 2000, Bruce Eckel.

Indirection in Web Services—UDDI and Dynamic Discovery

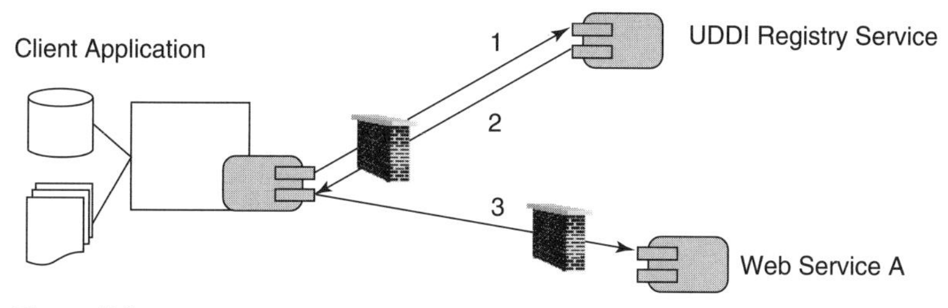

Figure 7.1 Indirection in Web Services

Indirection in Semantic Computing—Ontology and Semantic Mediation

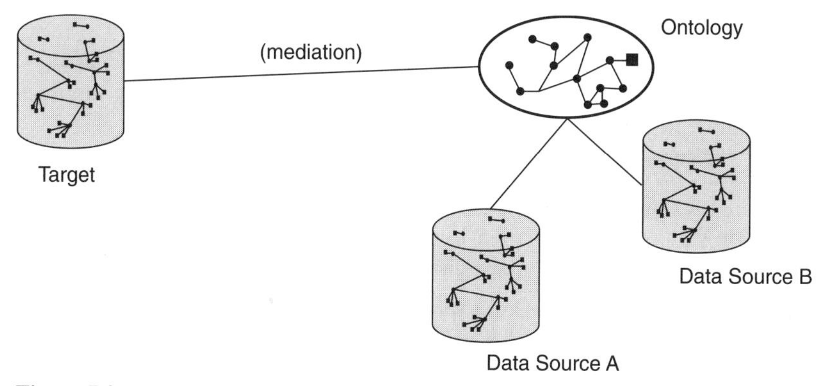

Figure 7.2 Indirection in semantic interoperability (mediation)

Using ontology as indirection is only one possible way, among many, to leverage ontology in semantic technologies, but it is a common pattern because of the value it creates by loosely coupling the information in disparate systems.

By leveraging indirection in the fundamental aspects of the technology, semantic interoperability is built for change. It's this built-in capability to manage change, adapt, and to be flexible that differentiates semantic technologies from other information-driven approaches.

Abstraction

One of the most difficult skills to describe and teach, abstraction remains a crucial technique for engineers of all disciplines. A commitment to abstraction can break down problems and provide simplicity when faced with high degrees of complexity in a system. Architects of both software and physical structures routinely use the principle of abstraction to isolate complex components and reduce the scope of a problem to be solved. Used in this way, abstraction is a mechanism for providing a view of a system that is free of important, but noncritical, technical details. It is a way to see the forest for the trees.

Ontology is the ultimate abstraction tool for information. By focusing on a conceptual representation of data, implementation details such as data format, types, performance optimizations such as binary flags, and meaningless object associations often found in object-oriented software systems can be isolated to a discrete portion of the overall architecture. The uses for this kind of abstraction through ontology are nearly limitless. Ontology can be a useful tool to reduce system complexity at all phases of a development life cycle and in several aspects of a run time environment.

By definition, ontology is abstraction. However, the pragmatic usages of ontology as an abstraction layer can vary greatly depending on how it is being leveraged.

WHAT IS ONTOLOGY?

Although we covered the basic definition and utility of ontology in Chapter 4, Foundations in Data Semantics, it will be useful to review here what an ontology is. Fundamentally, this book is consistent with at least three definitions:

Layperson's, More Narrow Definition: Ontology is a conceptual data model used to link together other, more granular data models in order to reach agreement on data meanings—either by committing to shared vocabularies or by explicitly modeling vocabulary differences, context, and meaning.

Technical, More Inclusive Definition: Ontology is a model of any bounded region of interest expressed as concepts, relationships, and rules about constructs and associations of interest to a modeler.

Most Popular, but Vague, Definition: Ontology is a specification of a conceptualization.[2]

Insider Insight—Dr. Jürgen Angele on Semantic Interoperability with Ontologies

Better, faster, cheaper—The so-called information-based approach promises to reduce time to market by 30% and costs by 40% and to help companies to produce with higher quality. By focusing on better information management rather than processes, companies can dramatically boost their [. . .] performance (Source: *McKinsey Quarterly* 2003). To be able to organize processes around information flows the supporting systems have to become more integrative, more flexible and more powerful. Semantic technologies will allow companies to set up this approach by delivering information on-demand, in-context, and real-time.

[2] This is adapted from Tom Gruber's widespread definition.

Integrative and flexible: Ontologies enable companies to accomplish semantic interoperability to reduce integration costs by more than 50%, which already has been proven in industrial applications. Within companies and organizations information is distributed over a variety of heterogeneous sources. Companies must therefore manage a lot of different data systems, each with its own vocabulary. This heterogeneity results in poor information quality, an inflexible IT infrastructure, and high IT costs. Thus the way knowledge is accessed and shared in organizations has to overcome these barriers to interoperability. Ontologies provide a common logical information model for heterogeneous distributed data and are thus an ideal means to realize information integration. They represent the agreed-upon business meaning of distributed data sources and enable higher-quality information. Through the separation of data, application, and business logic the management and setup of applications is more flexible and can thus enable companies to better manage the ever-changing demand for information.

More powerful: The key to setting up a successful information support system is to be able to capture the meaning and context of data. Describing information using ontologies can make information interpretable and understandable by computers and define and incorporate information context to a wider community. They allow for the representation of domain knowledge, complex dependencies and expert know-how, as well as the dynamic inferencing of implicit knowledge through inference engines like OntoBroker™. Thus ontologies enable the automation of information delivery and the reuse of knowledge at the right time, in the right place, and with the right context.

The use and implementation of ontologies gives companies the flexibility to set up processes around information flows and not vice versa. This will help them to reach the next level of productivity and improve their readiness to compete. Sounds like a vision? It's a compelling vision that has already arrived today.

Dr. Jürgen Angele
Chief Technology Officer
Ontoprise GmbH

Regardless of the precise definition employed, or the technologies used, the reality is that people are using ontology in a wide variety of capacities and the term itself is being attached to models that others would say are not ontology at all. In recognition of that chaos, the remainder of this chapter attempts to present a coherent view—in the context of semantic interoperability—of how ontologies can be classified and leveraged in an enterprise environment.

ONTOLOGY TYPES AND USAGE

The number of possible ontology types is limited only by the imagination of the human mind. Therein lies its greatest strength as well as its greatest weakness. As a tool for modeling concepts, the many forms ontology can take give us windows into the true nature of what we are attempting to model in the first place.

The idea of creating an ontology of ontology types is almost absurd; any such classification scheme could never be wholly correct. But for the purposes of this text, and for the benefit of the reader, it is valuable to consider at least four main categories of ontology: interface, process, information, and policy. These distinctions are loosely related to typical tiers found in an *n*-tier enterprise architecture, thus providing a basis for understanding how ontologies are used within typical systems architectures for the purposes of interoperability.

Interface Ontology

The interface ontology is a straightforward concept: an ontology that models an application programming interface (API) of some kind. As easy as the concept is, implementations are quite difficult, particularly in relationship to specific technology choices. Generally speaking, these interface ontologies are used to describe expected API behavior and results so that machine agents may "discover" how a given interface should be used and what to expect from it. In theory these ontologies are used to create loosely coupled services and enforce "design by contract" software engineering principles. In practice it has proven quite difficult to implement generic interface ontologies. Instead, language-specific uses dominate the landscape today.

Some influential interface ontologies include:

- **WSDL** (Web Services Description Language)—One of the more basic interface ontologies, WSDL was initially conceived of as a way to describe web service interfaces in a machine-readable way. Newer developments may fully or partially supplant WSDL with some derivative of OWL-S.
- **OWL-S**—The Web Ontology Language Service descriptions go beyond what WSDL offers by linking primitive concepts—such as a data type—to outside ontologies. In this way, an OWL-S user can infer that a `<price>` tag with a `float` data type is really a currency and denomination value because OWL-S enables the service interface to link that tag to an outside currency ontology. Semantic capabilities for service descriptions will greatly enhance an application's ability to dynamically discover and use new services.
- **FIPA QoS Ontology Specification**—The Foundation for Intelligent Physical Agent Quality of Service Ontology Specification is an ontology that describes the interface and protocol APIs that intelligent agents would use to determine the QoS of a given resource. Its specification enables the use of predicate ontology to describe the "domain of discourse" for real-time agent-based inquiries.

Process Ontology

Closely tied to content ontology because of the role process plays in providing context, the category of process ontology is distinct because of the requirements to accurately model concepts related to time, cause, effect, events, and other nonphysical aspects of the real world. Process ontology has innumerable applications in software that span from providing event services for single applications, to work flow utilities for human users, to orchestrating complex machine-to-machine interactions among multiple businesses in various industry segments.

To date much of the progress in the work flow and business process sectors of software providers has been due to proprietary approaches to process ontologies. Recently, the momentum for standard metamodels and ontology describing process-related activities have been increasing.

Industry consortia like OASIS and UCC have sponsored large efforts to standardize key industry processes. In the process they have created unique process ontologies that are similar but largely incompatible. Likewise, efforts with the business process modeling language (BPML) have conflicted with parallel efforts in the Web Services community. Process experts from each of these major initiatives recognize the overlap and thus the cause for concern regarding future inoperability of discrete process models. However, it remains likely that some sort of common process ontology will unify at least a portion of these competing groups.

Policy Ontology

In the context of this book, "policy" is used loosely to mean anything that specifies rules of usage—which could include security and authorization, network management and software components, digital rights assessments, or simply logic rules and constraints. In contrast to the other types of ontology previously identified (interface, process, and content) the policy ontology is purposefully limited to a description of constraints and logic-based assertions. Typically a policy ontology would be used in conjunction with interface, process, or content ontologies, but it may exist outside of them as well.

One use of a policy ontology is with IT networks: The ontology is intended to represent the systems, functions, and components deployed inside a large enterprise. The purpose of this type of ontology is not for interoperability, but for visibility and monitoring. Several organizations have developed proprietary tools, based on custom-implemented ontology, to monitor system resources such as J2EE applications, distributed IT installations, and operating IT systems within a given enterprise. These models define concepts in the enterprise IT domain and classify the systems, software, components, services, and network protocols that operate within a defined context.

Another use of a policy ontology is with rights management, which may be used to model complex agreements that cover intellectual property rights throughout a variety of domains (media, law, science, technology, etc.). These complex agreements can be overlapping, nesting, and overriding with each other. An ontology would help tame some of this complexity and aggregate metadata content from a variety of different sources. Furthermore, services for rights could be offered (via the ontology) to users, creators, and controllers of intellectual properties—

simplifying the management of these frameworks. A number of academic and commercial applications are under way that will leverage ontology for policy management; see Chapter 10, Capability Case Studies, for more information.

Still another way to imagine a policy ontology is a generic rules layer that provides constraints over other models—a set of ontology policies. In this way policies can exist independently of service, process, and content ontologies and provide additional metadata about transformations, queries, mappings, work flow, and more. These ontology policies may be stated in a rules language such as RuleML.[3]

Information Ontology

Sometimes referred to as "classic ontology," the use of this kind of ontology is discussed in the bulk of this book. Information ontology is a specification of a set of concepts for a given scope—usually some sort of business or technical domain. Although all the types of ontology listed here conceptualize content, the classic information ontology is typically restricted in scope to modeling objects in the real world that have been abstracted as data elements inside computer systems.

Because the basic working definition of ontology, a specification of a conceptualization, is so broad, it can be said that nearly every modelable aspect of software is an ontology. But, in practice, ontology is used to identify concepts free of implementation details that could obscure important truths about a specific domain or area of interest.

Within the sphere of information ontology, there are at least six distinct categories of ontology[4]:

- **Industry Ontology**—used to capture concepts and knowledge useful to a particular industry domain such as manufacturing, life sciences, electronics, logistics
  - *Example*: HL7 Health Care Ontology
- **Social Organizational Ontology**—used to represent humans and the human organizations they establish. This kind of model captures the structure and goals of an organization in terms of companies, divisions, groups, roles, locations, facilities, objectives, strategies, and metrics of the enterprise alongside a model of human communications, human skills and competencies, cultural dispositions, and other social concerns.
  - *Example*: SKwyRL Social Architectures Project
- **Metadata Ontology**—used to describe unstructured published content, which resides either on-line or in print.
  - *Example*: Dublin Core RDF Schemas

[3] See the Rule Markup Initiative at www.dfki.uni-kl.de/ruleml.

[4] *Ontologies: A Silver Bullet for Knowledge Management and E-Commerce*, Dieter Fensel.

- **Common Sense Ontology**—used to capture general facts about the world like "water is wet." These ontologies often have a unique notation and are considered valid across multiple domain spaces.
  - *Example*: OpenCyc Upper Ontology
- **Representational Ontology**—used to define information about information; sort of a metaontology to describe the relationships, associations, and containment among data elements.
  - *Example*: OMG's UML Meta-Object Facility
- **Task Ontologies**—used to associate terms with tasks within fairly narrow scopes, for instance, task description for computer-based training programs. The notion of a task ontology can be closely associated with both process and interface ontologies but comes at the problem from a datacentric angle.
  - *Example*: Computer-Based Training task ontology

SPECTRUM OF ONTOLOGY REPRESENTATION FIDELITY

Content ontologies can be represented in a variety of ways, ranging from informal to very formal; this results in a spectrum of data representation sophistication.

Controlled Vocabulary

A very informal, usually nonschematic way of creating an ontology is via a controlled vocabulary set. By defining valid terms, as in a data dictionary for instance,

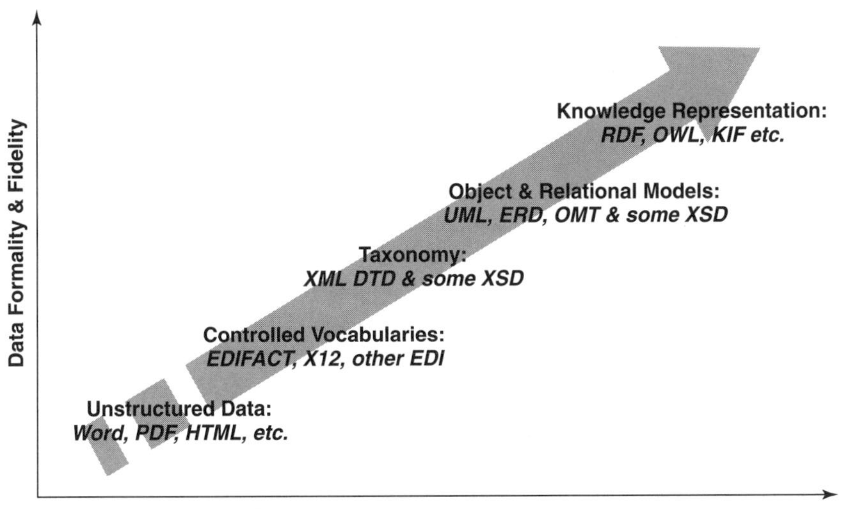

Figure 7.3 Spectrum of ontology representation formats

a user population may narrowly scope a set of concepts and their lexical relationships with each other. Likewise, a rudimentary thesaurus, with identifiable synonyms and antonyms, can comprise a controlled vocabulary function as ontology.

Taxonomy

More formal, a hierarchical data association like a taxonomy can provide a useful ontology that specifies implicit parent-child relationships between data elements. Commonplace examples of this kind of ontology are found anywhere an XML document type definition (DTD) is specified or in more classic systems like the Dewey decimal taxonomy for book classification.

Relational Schema

The relational schema is a formal way of capturing a variety of data relationships through the use of classes, attributes, and some relations (such as foreign key relationships). Although the relational approach is quite flexible, the use of denormalized structures for an implementation-specific rationale limits the applicability of most existing relational schemas as a candidate for true information ontology in the classic sense. Additionally, relational modeling rarely captures the complete underlying semantics of a relationship. For instance, software might know that the customer number from the customer table is a foreign key in the invoice table, but the software would not know what *kind* of relationship exists between a customer and an invoice.

Object-Oriented Models

Also a very formalized approach toward modeling, the OO techniques specified through the unified modeling language (UML) utilize classes, behavior, and attributes as core components of the model. UML also encompasses a wide array of binary relationships that can be defined among classes in addition to the recently approved constraint language (OCL) that provides further qualifiers for class relationships. Although some may argue that UML (plus OCL) is as capable as OWL DL, the fact remains that OWL and UML exist for completely different reasons—UML is intended to visually model data *and* behavior whereas OWL was intended to syntactically model knowledge.

Knowledge Representation Languages

The most formalized of the ontology representations, languages such as KIF, RDF, and OWL provide a high degree of expressiveness and precision for modeling the world. KR languages tend to subsume the modeling capabilities in the previous ontology representations while adding additional layers of logic, relations, constraints, disjunction, and negation.

ONTOLOGY ARCHITECTURE PATTERNS

With all the variety of ontology uses, it is probably no surprise that they are being implemented and architected in a wide range of ways. This section provides insight into the possible architectural uses of ontology for the purposes of interoperability and large-scale application development. This is not intended to be a comprehensive pattern library for ontology because it does not directly address important application areas such as:

- Ontologies for unstructured data (e.g., Internet and word processing)
- Ontologies for networks (e.g., systems visibility and monitoring)
- Ontologies for processes (e.g., work flow and BPR functions)
- Ontologies for human interface (e.g., GUI and industrial design)

However, the patterns identified here should be of value for architects and engineers seeking solutions to common problems with large application development and systems integration projects. Special care has be taken to validate the applicability of each pattern and to tie them to Chapter 10, which includes capability case studies and implementation details.

Pattern Summaries

The patterns are classified in a slightly modified Gang of Four system:

- **Structural pattern**—The value of this pattern is embodied in the role the ontology plays inside the structure of the overall system or process architecture.
- **Behavioral pattern**—The value of this pattern is embodied by the resulting behavior of the component or ontology within the overall system or process architecture.
- **Creational pattern**—The value of this pattern is manifested during the process of defining a set of concepts, relations, and constraints to comprise a given ontology.

Core Three Schema

The core three-schema pattern is the well-established ANSI-SPARC database pattern that calls for the separation of the client view schema from the physical view schema via a third conceptual view schema—an ontology.

Although this pattern has been classically associated with relational database design, it embodies the principle of abstraction apparent in all other ontology pat-

Table 7.1 Data Table—Ontology pattern summary table

Ontology Pattern Name	Description	Pattern Class
Core Three Schema	This classic generalized pattern describes the utility of using three schemas to separately embody user views, conceptual views, and physical views of data.	Structural
Conceptual Identity	This pattern is the use of an explicit ontology as the design centerpiece for the development of a single software application.	Structural
Design Moderator	Design Moderator is the design time use of a cross-application ontology to align vocabularies among multiple disparate applications or software components.	Structural
Ontology Broker	Ontology Broker is the central distribution of ontology, either component parts or in whole, to third-party ontology consumers.	Behavioral
Decoupled Blueprint	This pattern is the manual use of ontology by architects and designers as a way to derive new insight into an applications domain and scope.	Behavioral
Federated Governance	This pattern is the automated use of ontology to implement and control various constructs and behaviors in multiple applications and data artifacts in a holistic IT ecosystem.	Behavioral
Logical Hub and Spoke	Logical Hub and Spoke is the behavioral use of ontology to synthesize, aggregate, and augment data from multiple disparate applications or software components for the purpose of re-presenting that data in a new context that is meaningful to the other application(s) and their context.	Behavioral
Smart Query	This pattern involves the use of ontology as a conceptual query interface for diverse sources of information.	Behavioral
Herding Cats	Herding Cats occurs, when a large community process, characterized by many classes of heterogeneous data experts from related industries, attempts to reach agreement on a homogenized set of industry terminology.	Creational
Structure Reuse	The Structure Reuse pattern of ontology creation consists of the careful selection and assembly of concepts in previously available ontology.	Creational

terns, and is therefore well worth reexamining here. The Core Three Schema pattern specifies the use of three distinct schemas in a database application:

- **External Schema**—different based on context and users
- **Conceptual Schema**—a high-level representation of the entire database domain
- **Internal Schema**—physical storage structures in the RDBMS

Class

Structural pattern—The value of this pattern is embodied in the role the ontology plays inside the structure of the overall system or process architecture.

Intent

- To provide a decoupling of different users' view of data
- To decouple the user view from the physical implementation
- To enable structural changes to an RDBMS without impacting user views

Motivation

- The same physical database may require access to different sets of data from multiple clients.
- The consequences for changing internal structures may have profound ripple effects on many user views and other system schemas.

Applicability

Use the Core Three Schema pattern when

- A database represents a larger scope of data generally required by various clients who access it.
- The affects of change over time are likely to create problems with the initial physical properties of the database.
- Data independence is a business requirement.

Structure

- For a given RDBMS instance, the Core Three Schema pattern consists of a physical view, a conceptual view, and one or more user views.

Participants

- Data Modeler/Data Architect
- RDBM system

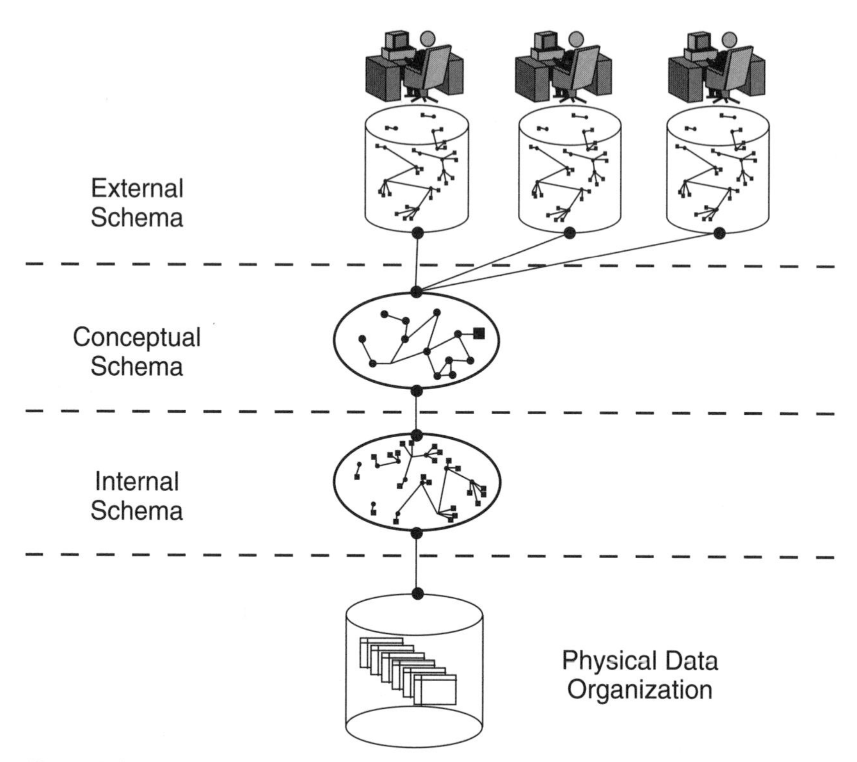

Figure 7.4 Core three schema pattern

Consequences

- A user's view of application data should be immune to changes made in other views.
- RDBMS clients should not need to know about physical database storage details.
- The database administrator should be able to change underlying structures without impacting user views.
- The conceptual structure of the database should be able to morph over time without impacting all the client views.

Implementation

- Typically the Core Three Schema pattern is implemented internally to a given RDBMS and administrator tools are provided to make use of its capabilities.
- In the broader concept of ontology usage for enterprise information architecture, the Core Three Schema pattern can be implemented in a variety of ways—see:

- Design Moderator
- Decoupled Blueprint
- System of Systems Controller
- Logical Hub and Spoke

Conceptual Identity

Conceptual Identity is the use of an explicit ontology as the foundational element for the development of a single functional application. It is a common practice in the object-oriented community to begin the modeling process with a conceptual model of the domain space relevant for the application being designed.

Typically used in the analysis and design phase of the object-oriented project, the conceptual domain model of a single system is produced to understand the system scope and to provide useful direction for the further design of the implementation-specific application model. This class of ontology is useful in a fairly narrow scope because these kinds of ontology are intended only to model a single application, its functions, and its data.

In most object-oriented design processes, this kind of ontology is useful to get a firm grasp on the kinds of objects and behaviors for the final application model. However, it is usually viewed as an intermediary model to be discarded and forgotten once the design process starts to move beyond the conceptual abstraction level.

Class

Structural pattern—The value of this pattern is embodied in the role the ontology plays inside the structure of the overall system or process architecture.

Intent

- To provide a foundational basis for the application design model
- To identify crucial object classifications and relationships
- To educate project participants about the scope and terminology of the system

Motivation

- Object-oriented software projects should provide a systematic methodology: The movement from conceptual models to implementation models provides a crucial component of this methodology
- Modeling software applications is extremely complicated: Ontology provides a way to get past this complexity and conceptualize an essential view of the model

Applicability

Use the Conceptual Identity pattern when

- Organizations wish to improve the overall quality of their object-oriented development project.
- Software engineering teams seek a structured approach for system decomposition.
- Project teams are in need of an educational tool for communicating the scope and relationship among critical business domain objects.

Structure

- In a software development life cycle, the ontology serves as a project conceptual domain model.

Participants

- Object-oriented analysts
- Domain experts
- Requirements analysts

Consequences

- Additional time required in the early stages of the life cycle
- More thorough coverage of the application scope
- Consistent, distributable model of the application domain
- Early identification of possible problems and application complexity

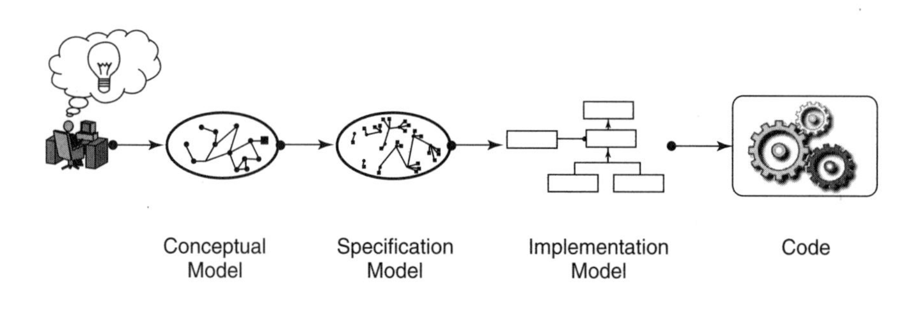

Figure 7.5 Conceptual identity pattern

- Logical intermediary step between informal requirements and formal application model

Implementation

- The Model-Driven Architecture (MDA) espoused by the Object Management Group (OMG) operationalizes this pattern by advocating automated routines that derive implementation models and code from the initial conceptual model—see also the System of Systems pattern.
- Most large object-oriented software projects would implement some derivative of this pattern. However, with the recent rise in popularity of lightweight development methodologies, this kind of rigor in software design is becoming less frequent.
- Implementation of this pattern would usually occur inside some sort of application modeling tool such as Visio, Rational Rose, Together J, or any number of other options.

Design Moderator

Design Moderator is the design time use of a cross-application ontology to align vocabularies among multiple disparate applications or software components. This static use of ontology is frequently used to generate code for later execution at run time.

Tools that implement this pattern are considered design tools and usually do not make use of the ontology at run time. Instead, the tool supports an ontology during the design process by enabling modelers to relate disparate schema to a common set of vocabulary terms found in the neutral ontology. The process of relating the terms may be accomplished through a variety of means including:

- Thesaurus (synonym and antonym) mapping
- Structural mapping (hierarchies and associations)
- Pseudocode and logic-based functions

The end result of this type of pattern is for the design tool to automatically generate a set of transformation routines in a particular programming language. Because this type of pattern separates the actual run time transformation routines from the design time activity, it is easy to enable a variety of output program languages such as XSL/T, SQL, and Java. Although a powerful pattern, this use of ontology still requires human intervention to make changes and generate new transformation routines before deployment.

Class

Structural pattern—The value of this pattern is embodied in the role the ontology plays inside the structure of the overall system or process architecture.

Intent

- To unify disparate vocabularies into a common set of ontological semantics
- To provide an application independent view of a domain
- To provide automation in the code generation process
- To create an application-independent framework for managing disparate vocabularies

Motivation

- It is very expensive and resource intensive to build, but especially to maintain, data vocabularies within enterprise computing environments with a great need for cross-system communication.
- It is difficult to manage distributed, remotely deployed program components inside various run time environments that all perform the same kinds of logical functions—vocabulary alignment.
- Problems with supporting multiple programming languages to resolve semantic and schematic differences among application schemas create chaos in an internal enterprise development team.
- Often a laborious part of development, the process of writing code should be automated to the greatest degree possible.

Applicability

Use the Design Moderator pattern when

- Your organization's expenses for maintaining systems-to-systems interoperability are a large portion of annual budget.
- Data vocabularies change frequently, new vocabularies are often adopted, or ERP systems upgrades are frequent.
- The chaos of distributed execution code, throughout the enterprise, for vocabulary transformations is becoming unmanageable.

Structure

- During the design process, the use of a common ontological model for aligning disparate schema and data vocabularies provides the foundation of this pattern.

Participants

- Data modelers
- Domain experts

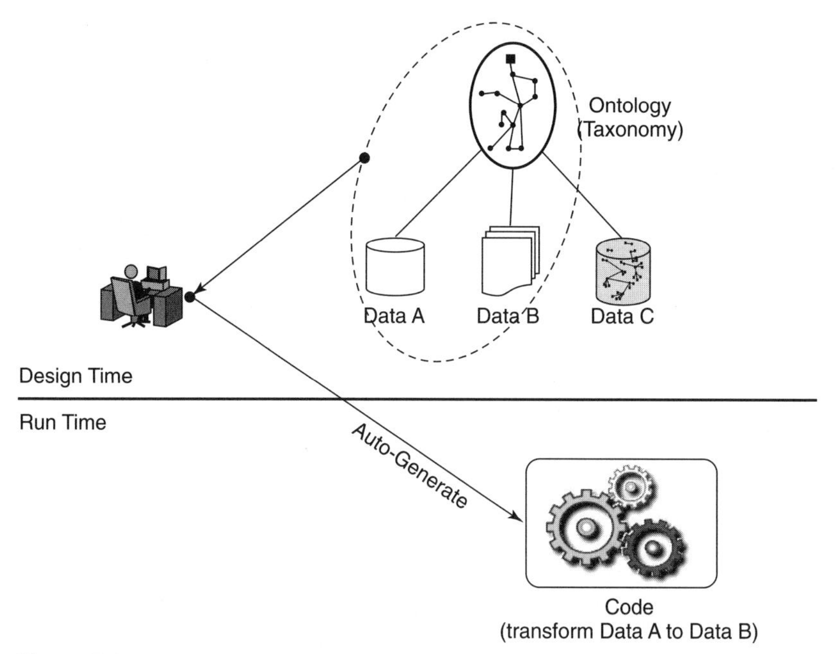

Figure 7.6 Design Moderator

- Requirements analysts
- System architects

Consequences

- More time focused on modeling vocabularies and domain
- Less time spent writing transformation routines to align vocabularies
- Unified view of vocabulary concepts and terminology
- Application-independent management center for distributed transformation code modules
- Faster response to changing vocabularies and shifting business requirements

Implementation

- Used during the design process, independent of the run time architecture. The basic architecture chosen by the organization remains unaltered—enabling typical and known uses of application middleware, application servers, and other enterprise components that enable interoperability.
- Contivo provides a known implementation of this pattern in a commercially available tool set.[5]

[5] For more information, see: www.contivo.com.

- The Institute of Computer Science in Greece has provided an implementation framework for thesaurus-based mappings that addresses many semantic issues.[6]

Ontology Broker

Ontology Broker is the central distribution of ontology, either component parts or in whole, to third-party ontology consumers. This pattern is most useful in enterprise middleware applications that provide interoperability services for a wide range of disparate sources with some level of dynamism. By utilizing broker architecture for ontology distribution, applications may dynamically assemble subsets of larger models into purpose-built smaller models that provide useful concepts and vocabulary terms used in other interoperability processes.

Although this pattern does not directly relate to semantic interoperability of disparate data sources, it is an architectural solution that enables the dynamic supply of ontology to tools that do.

The value of this pattern emerges when multiple ontologies are used in an enterprise-wide environment to resolve semantics from a number of sources. Multiple smaller ontology component models can be assembled on demand rather than requiring the use of several discrete, large, and possibly difficult to maintain, conceptual models.

Class

Behavioral pattern—The value of this pattern is embodied by the resulting behavior of the component or ontology within the overall system or process architecture.

Intent

- To provide a centralized facade for the distribution of model components
- To enable the simpler componentization of ontology parts
- To create a layered ontology architecture for composite vocabularies

Motivation

- Data architects have proven over time that the creation and maintenance of large monolithic ontologies, data dictionaries, and vocabulary standards are prohibitive—this pattern provides a repeatable framework for the componentization of model parts.
- The deployment of multiple middleware environments within a large enterprise causes a fragmented information infrastructure that is difficult to unify with monolithic ontologies.

[6] http://jodi.ecs.soton.ac.uk/Articles/v01/i08/Doerr/

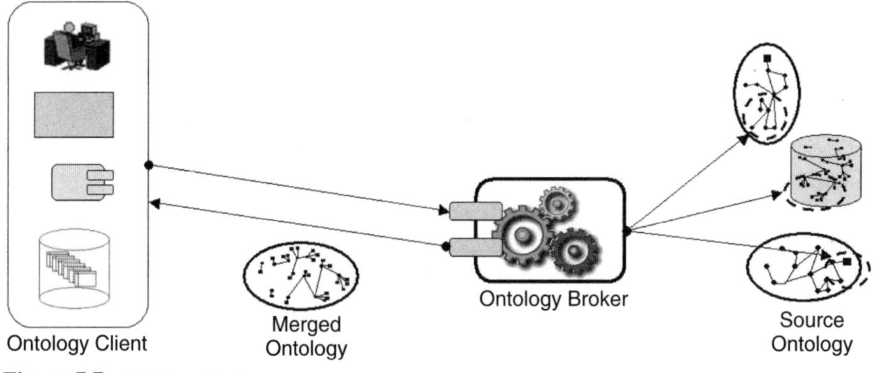

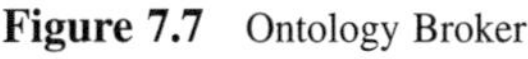
Figure 7.7 Ontology Broker

Applicability

Use the Ontology Broker pattern when

- Multiple ontologies must coexist in order to provide vocabulary resolution across discrete applications.
- A large monolithic ontology is undesirable or impossible.
- A single ontology or class of ontologies is expected to undergo change during the evolution of the systems collaboration.

Structure

- An application broker handles requests and responses for model components.

Participants

- Client applications
- System architects
- Registry services

Consequences

- Smaller, more manageable model components
- Greater use of indirection in ontology distribution
- Less time spent managing monolithic conceptual models

Implementation

- Sandpiper Software has implemented a component-based ontology platform that closely mirrors the Ontology Broker pattern.[7]

[7] For more information, see: www.sandsoft.com.

- Either a custom implementation or off-the-shelf tools provide the brokering capabilities. Typically used in conjunction with a registry and a taxonomy to provide directory services for the ontology components, the Ontology Broker is a dispatch point for models used elsewhere to align disparate application semantics.

Decoupled Blueprint

This pattern involves the manual use of ontology by architects and designers as a way to derive new insight into an applications domain and scope. Typically a static, decoupled model will provide details regarding various constructs and/or behavior in applications and data artifacts that can be used later to implement the system more effectively.

Examples of this kind of use can be found in the rise in popularity of the ontology concept during the late 1980s and early 1990s—especially within large aerospace organizations and the US federal government. A great deal of time was devoted to producing models of domain, systems, parts, and assemblies with the expectation that the activity of modeling would produce knowledge for the engineer who was later tasked with designing a functional system to automate some process.

A key characteristic of this type of activity was the completely decoupled nature of the ontology from the applications, which unfortunately resulted in file cabinets full of ontologies that were outdated almost immediately.

Were it not for the intangible value this pattern creates for those who create the models, Decoupled Blueprint would likely be considered an antipattern. However, many organizations may well benefit from the knowledge gained from a modeling effort without necessarily implementing a set of tools that make the ontology useful in a design or run time context.

Class

Behavioral pattern—The value of this pattern is embodied by the resulting behavior of the component or ontology within the overall system or process architecture.

Intent

- To document knowledge about a domain, organization, or process
- To reveal new insight into a domain or problem space
- To create model of reality for a given scope
- To educate project members about the nature of the domain in which they are working

Motivation

- In large organizations dealing with complex problems, it is often the case that no single resource knows very much about a broad scope of the problem domain—creating a significant problem for architecture teams charged with solving interoperability problems that span many domains.

- The risks associated with the lack of documentation and knowledge reuse for large organizations are well documented.

Applicability

Use the Decoupled Blueprint pattern when

- There is a lack of documented knowledge about a significant problem space that one or more project teams are expected to create software applications for.
- There is no sponsorship for more sophisticated tools, which actually link the ontology to design time or run time tools and processes.
- The project team is unfamiliar with their domain.

Structure

- A team creates models for other team members to review and learn.

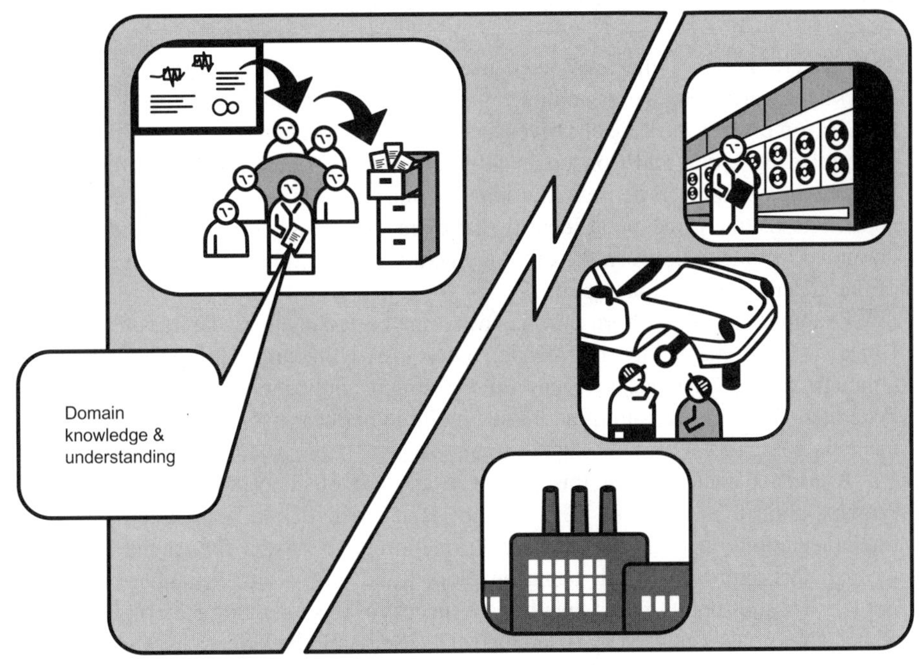

Figure 7.8 Decoupled Blueprint

Participants

- System architects
- Data modelers
- Domain experts

Consequences

- More educated team
- Documented (but static) knowledge about the problem domain
- New information about organizational assumptions and operating practices

Implementation

- This pattern requires no physical implementation in a computer. Because its purpose is decoupled from any automated objectives, it merely requires paper, pencil, and a trained modeler.
- There was widespread use of this pattern in the US federal government during the later 1980s and early 1990s—primarily making use of object role modeling (ORM) and entity relationship ER diagrams.

Model-Driven Federated Governance

This pattern involves the automated use of ontology to implement and control diverse software components and information resources within a holistic IT ecosystem. In some ways the Model-Driven Federated Governance pattern combines the Decoupled Blueprint and Conceptual Identity patterns. Like Decoupled Blueprint, this pattern's scope can be multiple domains and multiple systems, subsystems, or applications. Like Conceptual Identity, there are well-defined linkages, both procedural and technical, between the ontology being developed and the implementation of the underlying IT infrastructure.

Examples of this pattern can be found in the architectural definitions of the Object Management Group's Model Driven Architecture (MDA). However, although the MDA tends to focus on the single application, the Model-Driven Federated Governance pattern addresses multiple applications existing in different business contexts.

A core characteristic of this pattern is the use of conceptual models as the primary control mechanism. This in turn is used to derive application-specific implementations that can be dynamically reconfigured on the fly. At the time of writing, this approach is still being developed and only a small number of early implementations point toward the ultimate success of this emerging pattern. In many respects the goals espoused by federated MDA, Semantic Web Services, and service grid communities can be viewed as a realization of this pattern.

Class

Behavioral pattern—The value of this pattern is embodied by the resulting behavior of the component or ontology within the overall system or process architecture.

Intent

- To automate application-specific implementations
- To shift the emphasis of software engineering toward models and away from algorithms
- To provide dynamic reconfigurability for middleware tasked with the interoperability of highly complex IT environments
- To reduce the amount of time required in the programming of new features and capabilities for application interoperability scenarios

Motivation

- Keeping IT systems on pace with business change is very expensive.
- Reprogramming interfaces and routines when interoperability requirements change is time consuming, inefficient, and error prone.

Applicability

Use the Model-Driven Federated Governance pattern when

- Frequent changes to the business are causing rising maintenance costs for the infrastructure that provides interoperability among enterprise systems.
- The chaos of a disorganized middleware infrastructure is becoming impossible to manage.
- The business requires more adaptive infrastructure to respond to opportunities and to maximize existing market advantages.

Structure

- A centralized platform infrastructure provides hooks into interfaces from a wide variety of systems, subsystems, and applications for the purposes of reconfiguring both process and data flows on the fly.

Participants

- System architects
- Domain experts
- Middleware tools

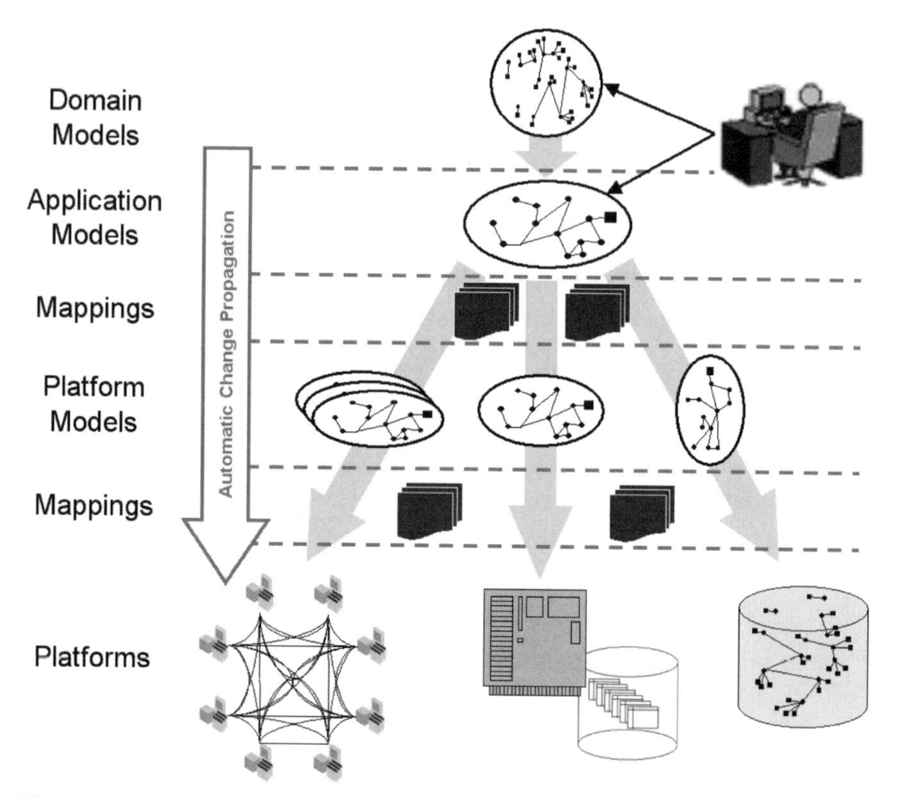

Figure 7.9 Model-Driven Federated Governance pattern

- Enterprise applications
- Subsystems and databases

Consequences

- Less time spent manually reconfiguring system interoperability
- Fewer costs associated with keeping pace with business change
- More capitalized opportunities due to faster response time when reconfiguring internal and external interoperability interfaces

Implementation

- The implementation of this pattern can take many forms. Custom-defined tools to orchestrate service availability can be purpose built, or off-the-shelf software can be purchased and installed to manage specific technologies such as Web Services. Future generations of tools will make use of ontology as the central modeling interface for managing services, data, and process throughout the enterprise. However, implementations with more than basic capabilities are likely to come from custom-developed solutions.

- The commercial implementations of IT management software from Relicore, Troux, and Collation make use of ontology and other kinds of models to support configuration management of enterprise IT system services.[8]
- The Object Management Group's Model Driven Architecture (MDA) has yet to see a full operational software implementation, but a key aspect of their planning and design—federated MDA—does implement the essential features of this pattern.
- Semantic Web Services, as envisioned by the sponsors of the OWL-S working group within the W3C, is also closely associated with this pattern—as it involves the dynamic (re)configuration and composition of services in support of a truly dynamic integration infrastructure.
- This pattern may also be realized in the development of composite functional applications—as envisioned by Delphi Group, Oracle, Sybase and others.

Design Practice: Distributed Ontology Blackboard

One possible future ontology design pattern is the distributed ontology blackboard. Unlike the other patterns discussed in this chapter, the authors are unaware of a software implementation of the distributed blackboard pattern—therefore, we've chosen to discuss it as a sidebar.

To better understand the scope and consequences of this pattern, it will be useful to revisit some the observations we've made about actual organizations earlier in the book. Different user communities (groups, organizations, sub-organizations) exist throughout an enterprise or agency—they each operate with different missions, objectives and tasks. Likewise their perspectives, concerns, and interests are variable and specific to each community. These perspectives themselves frequently array into hierarchies of different abstraction levels and aggregations that define a group and their associated tasks. Unlike traditional classification ontologies, operational units will oftentimes have the need for highly empirical ontologies that reflect instance data that is tightly-coupled with actual real-world events or objects.

The distributed ontology blackboard pattern intends to operationalize typical classification ontology, as well as robust empirical ontologies that contain current or near-future values about a mission or business process. This pattern will lead to a more robust situation analysis—sort of a state-machine for a business, unit, battlespace or any other arbitrarily defined domain.

The distributed blackboard pattern relies on the use of a populated ontology as a real-time container for context-independent information about a broad domain space. In turn, the broad world-view information is disseminated and interpreted into specialized local views through notifications and inference tools. This pattern differs from the Logical Hub and Spoke pattern in two important ways:

Continued

[8] Release 1.0, February 2003.

- How the Ontology is Used—In the Logical Hub and Spoke pattern, the ontology is primarily a schema-level device for reconciling schema and semantic conflicts among distributed sources. Conversely, the Distributed Blackboard is itself a populated schema with active rules and both schema and instance data modeled in the ontology.
- How the Information is Distributed—Whereas the Logical Hub and Spoke is primarily a request/reply type distribution, the Distributed Blackboard is a publish/subscribe model. Distributed community members (human or machine) are able to subscribe to aspects of the ontology information that is relevant to them and get notified when there are changes to those ontology components.

Crucial to a successful Distributed Blackboard pattern is the use of various semantic tools to enable adaptive behavior from the ontology and its component parts. Machine-learning algorithms, inference tools and model-based mappings all enable the transparent discovery and movement of information through the logically incompatible views that exist on the edges of community. In particular, since the Distributed Blackboard pattern will rely on both classification and instance level inference capabilities, it will be one of the few patterns that explicitly requires the use of both Abox and Tbox inference capabilities—which respectively enable instance level and class level deductions on data sets.

Simply put, the Distributed Blackboard attempts to use ontology to mimic the way distributed information exists in the real world—it is imperfect, incomplete and sometimes contradictory—while maximizing the software system's ability to adapt to high volumes of information input and improve the coherence and effectiveness of that information in resource constrained environments. Ultimately it will allow both human users and machines make use of real time global data flow on their own terms and aligned with their own ontologies.

Logical Hub and Spoke

The Logical Hub and Spoke is the behavioral use of ontology to synthesize, aggregate, and augment data from multiple disparate applications or software components for the purpose of re-presenting that data in a new context that is meaningful to the other application(s). Often when people first become familiar with the role of ontology in digital systems this is the pattern that is intuitively obvious.

Because a given ontology is a domain abstraction, it is initially easy to grasp how specialized data representations, when associated with their abstract concepts via ontology, can be reassembled into a relevant view for users. Despite the intuitive understanding of this capability, the actual implementation of this pattern requires a high degree of formalism in mapping operational schemas to a context-free ontology that accounts for a significant shift in abstraction levels. This formalism is not trivial.

Soapbox: Semantics Need Not Be Standardized!

Most of the literature about ontology implies, or explicitly states, that an ontology is a *shared* vocabulary meaning. Indeed, it is true that in many cases ontology is best shared at a deep vocabulary level. However, this doesn't scale well. As the "users" of the ontology increase, so do their unique contexts. Too many users and the vocabulary fails. Instead, the better way to use ontology is as a pivot for mediating between vocabularies. This way many unique interfaces can map to the same ontological structures, regardless of programming language, and reach agreement on mutually shared lingo: all the advantages of a hub without a central point of failure; data fidelity without rigid, grandiose, and granular ontological commitments.

When successful, the resultant pattern may effectively eliminate the necessity for programmatic algorithms to transform, aggregate, and augment the semantics of data as it passes between contexts, thereby saving both time and money in the development and maintenance phases of an enterprise interoperability program.

Class

Behavioral pattern—The value of this pattern is embodied by the resulting behavior of the component or ontology within the overall system or process architecture.

Intent

- To enable model-driven resolution of semantic data conflicts
- To lower costs associated with the production and maintenance of application programs that exclusively deal with the structural and semantic representation of data between multiple contexts
- To eliminate the classic $n^2 - 1$ problem with associating data schemas within a community of applications existing inside different contexts

Motivation

- Developing and maintaining IT interoperability is highly expensive and time consuming.
- As the number of interfaces and data schemas increase in a community, the number of programmed components to convert data between them rises exponentially, causing dramatic negative consequences for budget and efficiency.
- The problem of data semantics among the contexts of different applications is fundamentally a human problem, which can only be addressed in one of two ways, through application programming logic or through models—

models are a better alternative because they can remain free of programming details and retain a high degree of conceptual integrity over time.

Applicability

Use the Logical Hub-and-Spoke pattern when

- The scope of a community requiring interoperability rises beyond five participant systems (~24 unique transformation points).
- The data sets in an interoperability community are large and/or very complex.
- Changes in the community will likely dictate frequent changes to the interfaces, data, and transformation points.

Structure

- A central ontology provides an abstraction layer for disparate sources of underlying data. A rich formalism in the model-based mappings is required to maintain the semantic integrity of the data as it moves through different contexts.

Participants

- Data modelers
- Domain experts
- Model-based mapping tools (custom or COTS)

Consequences

- Fewer interfaces to maintain
- Elimination of coded transformation routines between interoperable systems

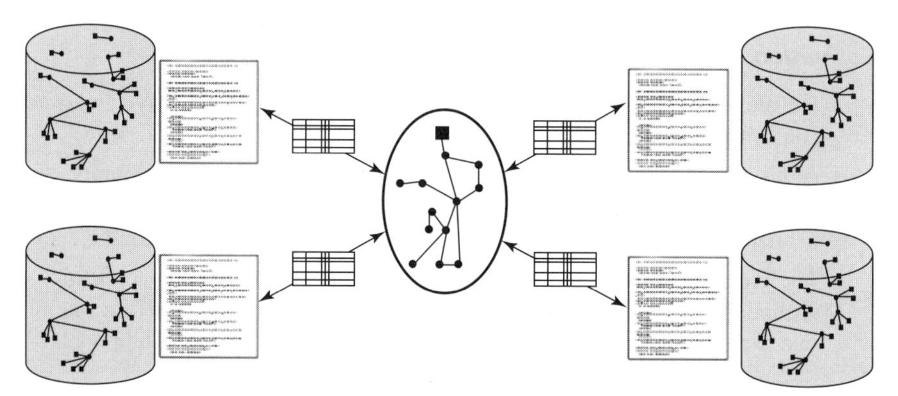

Figure 7.10 Logical hub and spoke

- Associated reduction in time and money for the development and maintenance of interoperable systems.

Implementation

- A number of robust implementations of this approach have been accomplished in the academic realm and only a few in the commercial realm. There is a high degree of confusion and overlap in the industry about less sophisticated approaches to this pattern that simply involve the programming of scripts or components to associate an implementation schema to the conceptual ontology, but these approaches do not offer the substantial reduction in either interfaces or programming time and look the same only in component diagrams.
- Modulant Solutions has implemented a version of this pattern utilizing older modeling languages like Express and a well-developed proprietary mapping specification dubbed "context mapping."[9]
- The University of Karlsruhe's FZI Research Center for Information Technologies has created a framework named "Mapping Framework for Distributed Technologies in the Semantic Web" that provides a sophisticated design, based on modern modeling languages, for capturing both semantics and context among distributed applications.[10]
- Ontopia provides a tool set for implementing industry standard Topic Maps in an enterprise environment.
- Unicorn provides both design and run time tools that implement the Logical Hub-and-Spoke pattern.

Perspectives—Logical Hub of Hubs

We were tempted to add an additional pattern, *Logical Hub of Hubs*, to describe the concept of linking several Logical Hub and Spoke communities through ontology bridges. However, because we have only heard this concept discussed in theory, and not seen it in practice a single time (much less three times) we have chosen to exclude it here. However, it is worthwhile to mention that by loosely coupling the hub ontology involved in several Logical Hub and Spoke communities, via semantic mapping models or some other sort of ontology bridge or transformation scheme, a tremendous potential for interoperability exists. If there is in fact overlap between ontologies that mediate various communities, it is possible that a single coupling between those two ontologies could create many more connections among participating enterprise systems. Concerns with this theoretical approach would certainly include the loss of data semantics and/or context as they move through more than one ontology as well as performance issues related to additional hops, transformations, and interfaces. Nonetheless, the concept could prove quite valuable when linking already established niche communities to a greater whole, for instance, supplier alliances in manufacturing and antiterrorism alliances among state law enforcement organizations.

[9] For more information, see: www.ontologyx.com.

[10] Knowledge Transformation for the Semantic Web Proceedings 2002.

Smart Query

This pattern involves the use of ontology as a conceptual query interface for diverse sources of information. More targeted search retrieval can be achieved by using the ontology as a query target because it searches based on context, concept, and inference-rather than simple word matches or popularity algorithms.

Capabilities like this will manifest themselves in a number of different kinds of solutions, including natural language search and response, context-aware help, enhanced internet search, resource locators, and agent-based searches.

Limitations with this pattern exist because content will have to be associated with concepts and context in order for full capabilities to become apparent. Although this may still be attractive for specific applications, it is prohibitive for the Internet as a whole. Nonetheless, for enterprise interoperability scenarios the use of this pattern may well be a highly desirable alternative—particularly because the process of tagging old content ensures the long-term utility of that same information. For an enterprise, work to utilize this pattern will be an effective investment in its information infrastructure.

Class

Behavioral pattern—The value of this pattern is embodied by the resulting behavior of the component or ontology within the overall system or process architecture.

Intent

- To substantially improve the quality of results returned from software-based search of structured and unstructured content
- To reduce the amount of time wasted on sifting through irrelevant search results
- To create a viable search interface for autonomous software agents that can enable more intelligent autonomous behavior from the agents

Motivation

- Information retrieval can consume a significant amount of time and resources in information-intensive businesses.
- Widespread and popular search methods are semantics free, creating great difficulty in finding narrowly targeted searches.
- The rate of new digital information being created is rising exponentially—more data will be created in the next three years than in all of recorded history[11]—indicating that current search-related problems will only worsen over time.

[11] Information Integration: A New Generation of Information Technology, M. A. Roth, *IBM Systems Journal* Volume 41, No. 4, 2002.

Applicability

Use the Smart Query pattern when

- A significant portion of organizational time is spent on data search and retrieval.
- Accuracy and quality of search results have a direct impact on business objectives.
- Increases in the amount of organizational data are expected to rise significantly in the coming years.

Structure

- An ontology of domain concepts provides the basic concept search capabilities, which can be combined with more traditional text-matching searches. The search framework must reflect classifications explicitly made between the data and the ontological concepts in the model. The underlying instance data may then be retrieved on demand and in context with the search request.

Participants

- Data modelers
- Domain experts
- Query engines
- Inference engines

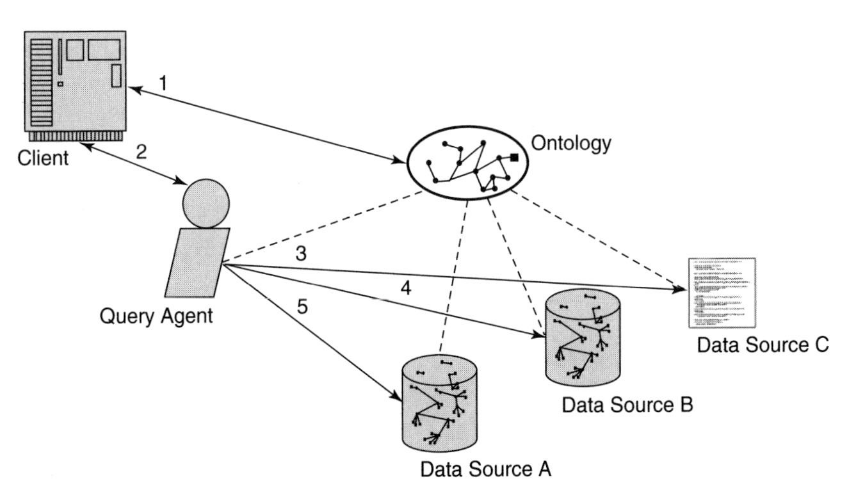

Figure 7.11 Smart Query

Consequences

- Substantial work to associate existing data with newly explicit semantics and context
- Reduced time required to find targeted, hard-to-find data through standard search mechanisms
- Increased quality of search results because they are presented in the context of the searcher

Implementation

- A wide variety of approaches can be taken to implement this pattern. Most approachable is the basic Semantic Web model put forth by the World Wide Web Consortium. Using RDF-based OWL ontology associated with tagged content will provide the core linkages between instance data and the conceptual model. Additional approaches, focused on semistructured and structured data, provide mapping capabilities that can weave traditional metadata values into richer classification schemes provided by ontology. In this way, search across structured, semistructured, and unstructured data can be accomplished.

Herding Cats

Herding Cats occurs when a large community process, characterized by many classes of heterogeneous data experts from related industries, attempts to reach agreement on a homogenized set of industry terminology. The process of creating ontology from a large set of experts from loosely associated domains has been a common way of defining vocabulary standards and ontology for some time.

As far back as the 1960s large community processes were established to define data exchange formats and reach agreements on terminology. Today, the results of these efforts are mixed. In all cases, and as with language itself, the vocabulary has changed—resulting in more committee work. But in some cases the committees have settled on the development of high-level conceptual ontology that governs the more frequently changing implementation standards.

In all but a few cases, this pattern should be seen as an antipattern and avoided. Instead, consider the Structure Reuse and Creation by Automation patterns as alternatives.

Class

Creational pattern—The value of this pattern is manifested during the process of defining a set of concepts, relations, and constraints to comprise a given ontology.

Creational antipattern*—The value of this antipattern is to serve as a warning for others who believe that it is a solution to the problem of defining concepts, relations, and constraints for a given ontology.

Intent

- To specify a set of concepts that can be utilized in a machine-processable environment for the purposes of vocabulary alignment
- To seek input from a wide range of experts who can contribute important facts about the relationships, constraints, and values of domain concepts

Motivation

- Creating ontology with appropriate levels of abstraction is difficult.
- When only a few people are used to generate a domain ontology, key aspects of the model may be limiting for future applications.

Applicability

Use the Herding Cats pattern when

- A team is working in a wholly new domain with few preexisting ontologies.
- Preexisting ontologies are determined to be at the wrong level of abstraction to be useful.
- Tools, or relevant data sets, do not exist that can automatically generate a set of ontological concepts.

Structure

- Not much structure to discuss here, only long, problematic meetings!

Participants

- Ontology modelers
- Domain experts

Solution Consequences

- A generalized conceptual model with a machine-processable set of abstractions with sufficient breadth to cover a given domain.

* Note—We have listed this solution as both a pattern and an antipattern because in some rare cases an entire ontology may need to be generated from scratch; thus it is useful to discuss the consequences of that approach here.

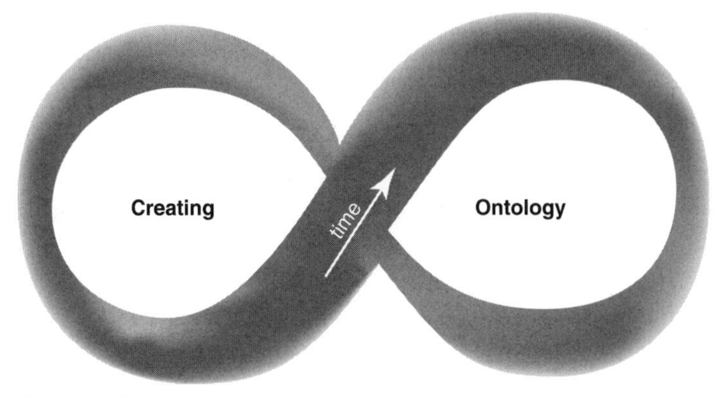

Figure 7.12 Herding Cats

- Note: Many groups (internal and external bodies) also produce low-level implementation-oriented data vocabularies and codify them as a standard for a given domain—this characteristic is purposefully omitted as a solution because it results in antipattern consequences.

Antipattern Consequences

- Expect a reasonably large ontology developed in this way to take several years of quarterly meetings—this timetable can be slightly accelerated in corporate environments where significant mandate and budget have been issued.
- Expect the first several iterations of the ontology to have major structural flaws that need to be adjusted as the implementation details (of the applications that use it) get worked out.

Implementation

- ISO 10303 Standards for the Exchange of Product Data (STEP) undertook a multiyear program involving hundreds of international participants to develop a series of standards at both the application and domain levels.
- ebXML is a standards body dedicated to the development of electronic business data. This group took several years to develop both application- and domain-level models.
- Health Level 7 (HL7) has produced several reference information models (RIMS) for the health care industry that are characteristic of domain ontology.

Structure Reuse

The Structure Reuse pattern of ontology creation consists of the careful selection and assembly of concepts in previously available ontology. As with all good architecture approaches, creating an ontology should heavily rely on theft[12] (the use of reference implementations) to speed results, encourage patterns, and ensure high-quality results.

When creating ontology by Structure Reuse, the process would involve seeking out domain models that have already been created for your domain and then using them as a basis for delivering your own ontology—which could differ quite significantly from the original source of your model. Typically the process would rely more on the reuse of classification structures than the reuse of actual class name labels. Often the most important part of the new model is the depth of its model dimensions, not the vocabulary of its data entities.

For the purposes of creating a new ontology there are frequently many existing taxonomy or data models available. For example, if tasked with developing an ontology to mediate semantic conflicts in an automotive parts middleware environment, why not begin with STEP Part 214, STEP IR 109, and the OAG Canonical Model as a starting point? All three of these industry models have the benefit of thousands of person-hours of development and hundreds of implementations. Most industries have several highly developed and publicly available data models or ontologies.

Class

Creational pattern—The value of this pattern is manifested during the process of defining a set of concepts, relations, and constraints to comprise a given ontology.

Intent

- To specify a set of concepts that can be utilized in a machine-processable environment for the purposes of vocabulary alignment
- To reuse work that may have already been done in modeling a particular domain

Motivation

- Creating ontology with appropriate levels of abstraction, and with appropriate domain coverage, is difficult.
- If previously existing domain models are not consulted, an ontology development team runs the risk of repeating mistakes or missing opportunity to benefit from more elegant solutions.

Applicability

Use the Structure Reuse pattern

[12] Grady Booch.

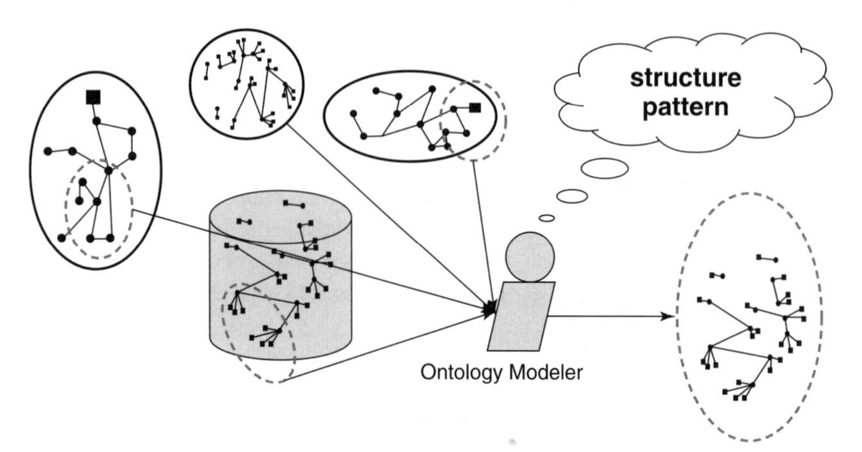

Figure 7.13 Structure Reuse

- Whenever possible!
- When a team is working in a known domain space with previous ontology and model implementations that are in public domain or can be licensed.

Structure

- Seek out all possible sources of previous conceptual model development.

Participants

- Ontology modelers
- Domain experts
- Preexisting ontology

Consequences

- Speedier ontology development
- Better-quality initial draft ontologies
- More legwork tracking down ontology
- Differing abstraction levels among models that must be corrected manually

Perspectives—Automated Ontology Creation

One additional pattern not included here, *Creation by Automation*, is used to describe the practice of creating ontology from the use of algorithms that read volumes of data and then make guesses at their ontological categories. However, because the authors have only ever witnessed this kind of technology produce fairly simplistic taxonomy—not the kind of ontology we are generally referring to in this chapter—it seemed inappropriate to identify it as a pattern. It should be noted that advances in pattern recognition software may well reach a point where they can produce fairly sophisticated ontology, but also that no pattern recognition software will ever be able to infer patterns that are not directly manifested in the data—such implicit patterns are quite common.

REPRESENTATION AND LANGUAGES

No universally accepted standard for ontology representation exists. Today there is a plethora of language specifications, graphical notations, and natural language notations that all provide valid alternatives for modeling ontology. Additionally, the usefulness of ontology may also manifest in simple whiteboard diagrams with little or no formalism. But for the purposes of semantic interoperability in enterprise systems, certain basic criteria should be considered:

- **Processability**—the ability of the representation language to be efficiently processed by programs tasked with inferring truth values or providing interoperability
- **Accessibility**—the market penetration and familiarity of the representation language throughout the industry and among software professionals
- **Usability**—the ease with which new users can adopt and become proficient in the use of the representation language
- **Expressiveness**—the ability of the representation language to capture unambiguous semantics relevant to a given domain
- **Life cycle coverage**—the scope of the representation language's utility throughout the development life cycle (e.g., design, architecture, implementation, requirements, testing)

A simplistic review of how some of the key ontology representation candidate languages stack up against these criteria is provided in Table 7.2.

A chart such as this can be informative, but also misleading. For example, to developers and architects looking to maximize the semantic content of an information infrastructure, the processability and expressiveness of representation languages like OWL and KIF trump all other factors. Likewise, although a representation language such as XML has wide market penetration and a high degree of usability, the

Table 7.2 Data Table—Comparison of knowledge representation languages

	OWL	UML	RDF	ERD	KIF	XML	Express
Processability	5	2	5	4	4	3	3
Accessibility	1	5	2	5	2	5	2
Usability	3	4	1	5	2	5	3
Expressiveness	5	3	5	2	5	1	2
Life cycle	2	5	2	2	2	1	2

sheer limitations of the XML schema's semantic content severely compromise its utility as a true ontology language.

For the remainder of this section, individual descriptions will be provided for languages with at least a medium level of support for expressiveness. Shorter summaries will also be provided for interesting, but noncore, ontology representation candidates.

Web Ontology Language (OWL)

The Web Ontology Language is the rapidly rising superstar of the ontology world. It is a product of the union between two historic efforts: DAML and OIL. DARPA Agent Markup Language (DAML) is primarily a frame-based knowledge representation system, whereas the Ontology Inference Layer (OIL) is a language to support logical reasoning about models. Because the two parallel efforts were highly complementary, they joined the specifications into a DAML+OIL specification in 2000. Since then the natural evolution of DAML+OIL has continued, but as the time neared for inclusion in the World Wide Web Consortium (W3C) it took on a new name: Web Ontology Language.

OWL combines the strengths of a very OO-like representation system with a highly formalized logic uniquely amenable to efficient operations performed by inference engines. These description logics (DLs) are the foundation of what makes the OWL approach highly decisive and unambiguous.

Unambiguous and decisive truth values are at the core of why OWL is so important. Unlike representations based on first-order logic and/or frames, the OWL system will always guarantee an answer to any question you pose to the ontology—with obvious benefits to enterprise users.

Counterpoint: On the Value of OWL

Although the authors of this book firmly believe in the ultimate success of OWL, they recognize that it comes at a cost and that it is not without drawbacks. The designers of OWL largely forsake concerns about the complexity of the language in order to exact its significant descriptive capabilities. OWL's computational, technical, and conceptual complexity make it overkill for many software users—even most software engineers.

> The longtime imbedded traditions of frame-based modeling (OO or ER) are at odds with many of the core principles that underlie the OWL requirements (especially the description logics based component). Style issues like (a) how abstractions are modeled, (b) subclass relationships, (c) language property usage and scope, and (d) decidability and completeness requirements all contribute to a steep adoption curve and some potential long-term barriers. So, despite OWL's advanced capabilities, the cost of adoption may still be too high for many with only basic needs.

In frame-based environments, like OO systems, the modeler can make arbitrary assertions about classes and relationships without being forced to clarify those assertions. In a description logics-based system, the modeler starts with primitive concepts and then goes about combining them in interesting ways to create defined concepts—which are always reducible to the primitives. In this way, ambiguity is avoided and the ontology can be built up with increasing complexity while still enabling inference engines to traverse the model and infer facts about data.

Strengths

- Effective union between frame-based systems and logic-based systems.
- Decideability—A query always produces an answer, or, put another way, an algorithm should always terminate.
- Expressiveness—The OWL model is highly capable of describing real-world relationships and complex associations.
- Processability—OWL models, and the inference engines that operate on them, can be processed very efficiently and with great speed.

Weaknesses

- OWL is a new standard with little market penetration; significant examples of OWL implementations are primarily in the academic world.
- Because of its requirements for formalism and its different paradigm of model development, OWL is unintuitive for the vast majority of data modelers who already have experience with object or relational approaches.
- OWL was developed as a solution for ontology representation and processing at the machine level; therefore, it has no graphical modeling constructs and few features designed to make it approachable by humans.

Implementation

In August of 2003 the W3C approved OWL as a candidate recommendation, which means that more reference implementations are required before proposing the recommendation. As it currently stands, the OWL specification embodies three distinct sublanguages:

- **OWL Lite**—a lightweight implementation that focuses on classification hierarchy and basic constraints for ontology construction. The expectation is that OWL Lite will provide an easier reference model for tools and a quicker migration path for content owners who wish to convert taxonomy and simple thesauri into the OWL Lite format.
- **OWL DL**—the core OWL implementation that provides maximum expressiveness while still providing computational completeness and decideability. The DL part of OWL DL refers to its usage of the description logic paradigm for ensuring that computational completeness.
- **OWL Full**—a layer that enables modelers to inject ontological assertions into RDF documents without being restricted by syntactic constraints, which guarantee computational completeness. This is the most expressive part of OWL, but it makes no guarantees about the decidability of the resulting ontology.

The following example provides the basic OWL Lite syntax that would be used to describe a part of an ontology that says that "employees are people that work for at least one company."

Code—OWL Lite Syntax: Employees are People that Work for a Company

```
<owl:Class rdf:ID="Employee">
  <rdfs:subClassOf rdf:resource="&mammal;Person"/>
  <rdfs:subClassOf>
    <owl:Restriction>
      <owl:onProperty       rdf:resource="#worksForCom-
      pany"/>
      <owl:minCardinality
              rdf:datatype=
              "&xsd;nonNegativeInteger">1</owl:minCar-
              dinality>
    </owl:Restriction>
  </rdfs:subClassOf>
  . . .
</owl:Class>
```

Although this is just a simplistic example, it is easy to see how the rigorous attention to generalization and associations would provide a sound basis for programmatic evaluation of assertions in this language. Unique to OWL is the creation of first-order classes for associations that are independent of the objects that they are associated with. For example, the association `worksForCompany` is an anonymous class that exists as a property of employee, but that same association could be used elsewhere were it given a class name. This is just one way that OWL differs from UML as a modeling language.

Unified Modeling Language (UML)

Although the roots of UML can be traced as far back as 1990, the unification and standardization of the language didn't begin to occur until 1996. Since then, the industrialization of the language has led us through several revisions to the standard and widespread adoption by development teams throughout many industries.

A direct comparison between UML and OWL is not quite fair because the languages were really developed to solve different problems. UML was developed as a graphical notation to assist the full life cycle development of complex object-based software applications, whereas OWL was developed as a model syntax designed to be formal and expressive enough to offer unambiguous notation for the run time processing of ontology. In short, UML is for humans and OWL is for machines.

However, UML class diagrams have a great deal of overlap with OWL precisely because they are also intended to model concepts and associations found in the real world. This begs the question: Can UML be an effective ontology modeling language?

As it turns out, OWL and UML are much more complementary than competitive. Because UML is popular and familiar to modelers all over the world, it makes sense to consider UML as a useful tool for people to model ontology. Because OWL is formal and expressive, it makes sense to consider OWL as the most useful tool for machines to process ontology.

In fact, work is being done on several fronts to examine ways in which the UML specification can be modified to fully support the more formal syntax of OWL in UML's visual environment. Some tools today already provide plug-ins to popular UML modeling tools so that they can output OWL syntax. But a standardized way of supporting this capability has not yet been reached.

Confusing the matter, the XML Metadata Interchange (XMI) standard is an XML-based format that provides syntax to the full-range UML capabilities. Can a UML- and XMI-based solution provide ontology support for run time applications? The answer is Yes, but not as well as OWL. Because XMI doesn't change any of the underlying assumptions and approaches to UML (it only provides a syntax for it) the issues with UML's ambiguity and lack of formalism still apply—making it less capable of providing finite answers to all questions posed to a given ontology.

Strengths

- Unmatched marketplace acceptance as a visual modeling standard
- Mature standard with many stable revisions and innumerable implementations
- Developed as a human interface to modeling ontology

- Well-understood, even intuitive approach toward modeling domain classifications and associations

Weaknesses

- Frame-based system not grounded in formal mathematical soundness—results in ambiguity that is difficult to resolve when applying algorithms to the resulting models.
- Its origins as a human approach toward modeling have resulted in the development of a common machine-processable syntax as an afterthought—XMI.

Implementation

Options for use of UML as a run time ontology tool can take several forms:

- Proprietary tool syntax (Rose, Together, etc.) to support run time processing
- Export as XMI to support run time processing
- Export as OWL to support run time processing

Of these alternatives, the most desirable is probably the OWL export, but it is the least-developed approach.[13] Implementations that utilize the XMI or proprietary syntax are more common, but they result in issues with the processability of the resulting models—to overcome these issues additional metadata constructs would be required in the model itself to provide cues for an external engine, such as an inference engine, to provide deterministic results from traversing the models.

Let's consider a small part of an ontology that asserts "managers are employees employed by a company." The XMI representation would look like:

Code—XMI Syntax: Managers are Employees Employed by a Company

```
<XMI version="1.1" xmlns:UML="org.omg/UML1.3">
    <XMI.header>
      <XMI.model xmi.name="Company"
           href="Department.xml"/>
      <XMI.metamodel xmi.name="UML" href="UML.xml"/>
    </XMI.header>
  <XMI.content>
    <UML:Class name="Company" xmi.id="Company"/>
    <UML:Class name="Employee" xmi.id="Employee"/>
    <UML:Class name="Manager" xmi.id="Manager"
```

[13] See Chapter 6, Metadata Archetypes, for a discussion of OMG activity for OWL-based UML support.

```
      generalization="Employee"/>
    <UML:Association>
      <UML:Association.connection>
        <UML:AssociationEnd name="employees"
          type="Employee"/>
        <UML:AssociationEnd name="works for"
          type="Company"/>
      </UML:Association.connection>
    </UML:Association>
  </XMI.content>
</XMI>
```

Once again, this is a simplistic example, but it serves to highlight several things about this type of model construct. Unlike OWL, the associations are local associations only—they are inherently tied to two first-order classes via two end point definitions. Likewise, the UML construct doesn't provide for rigid formalism in the use of associations and leaves rigor to the human modeler. On the positive side, simply glancing at the model tells humans much about the ontology in a very intuitive way—which OWL cannot.

Resource Description Framework (RDF)

RDF is the historical precursor to OWL and one of the first purpose-built modeling facilities to provide an ontology layer as an overlay to web content. With its beginnings as the Platform for Internet Content Selection (PICS) within the W3C, RDF was geared toward web content from the onset.

Much of what has already been presented about OWL also applies to RDF. But OWL both extends and restricts RDF in important ways. RDF has little or no support for advanced ontology modeling features such as cardinality constraints, defined classes, equivalence statements, and inference capabilities, among others. But RDF can also be used more flexibly than OWL Lite, which has formal and constrained rules about how to build the ontology. This means that RDF can be more flexible to use but less decisive in its results.

Knowledge Interchange Format (KIF)

KIF was developed in the early 1990s to provide a robust mechanism for knowledge exchange between computer applications. It was not intended to be an internal representation for applications (as with entity-relationship models) or to be interacted with human modelers (as with UML). Instead its focus was to provide an extremely expressive notation, based on first-order predicate logic, to encapsulate data and declarative semantics as they moved over the wire from one application to another.

Currently, KIF is the de facto standard for discussing ontology and logic-based standards among the researchers who develop them. It is a very robust language that

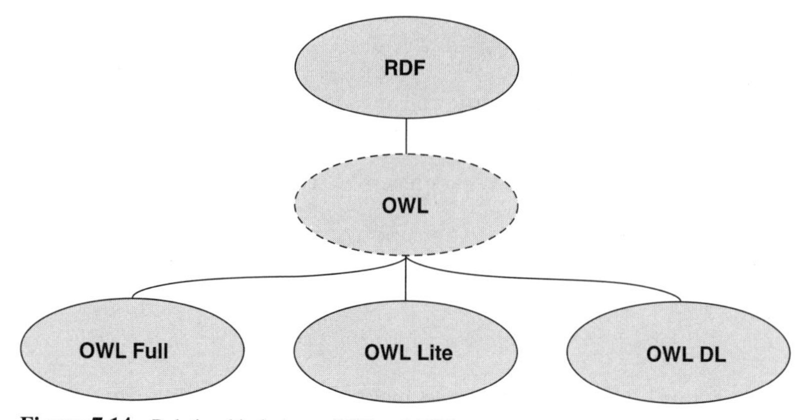

Figure 7.14 Relationship between RDF and OWL

can effectively describe all things that can be said or mathematically expressed. Unfortunately, this expressiveness comes at a cost. It is not intuitively grasped by nonmathematicians and is processing intensive for machines.

A simple example of KIF to assert "employees work for companies" would be the following:

```
(exists ((?x Employee) (?y Company)) (works_for ?x ?y))
```

KIF is also the foundation for many other knowledge representation languages. Most notably, the conceptual graphing (CG) techniques being developed by John Sowa rely a great deal on KIF as a foundation for the graphical notation—all CGs can be converted to KIF and vice versa.[14]

Other Modeling Formats

There are many examples of other data structures supporting ontology development. By and large, they are much less expressive and provide only minimal semantics for the application to work with. In many cases, with Express, Entity-Relationship, and XML specifications, for example, the modeler would be required to model his or her own semantic models into the ontology itself. This is quite possible, but it lacks the portability of languages that directly support some form of semantic rigor in the language itself. In some cases the semantic rigor is not included at all.

Examples in development of this approach are given by organizations like Standards for the Exchange of Product Model Data (STEP), Open Applications Group (OAG), and Applied Data Resource Management, which have each developed domain ontology in formats like Express, XML, and ERD, respectively.

[14] John Sowa, *Knowledge Representation: Logical, Philosophical, and Computational Foundations.* Brooks/Cole 2000.

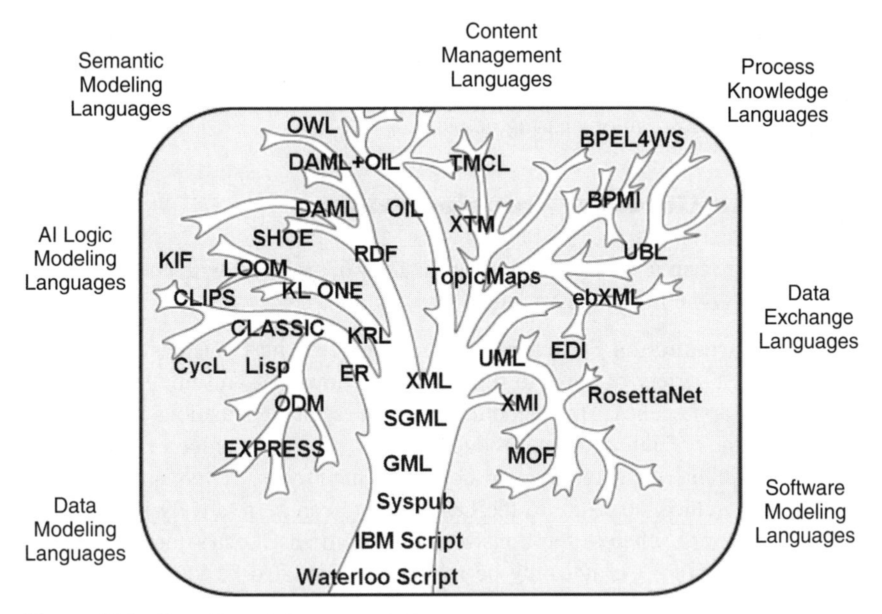

Figure 7.15 Tree of knowledge representation

However, the clear path forward is for knowledge-intensive systems to begin using one of the first four structures mentioned (OWL, UML, RDF, KIF) as the basis for ontology development and processing.

ONTOLOGY TRANSFORMATION

Last, but certainly not least important, in the discussion of ontology is the topic of ontology transformation. The idea of transformation can mean one of many things, including transformation of syntax, transformation of semantics, and transformation of structure. In each of these cases there are a good deal of theoretical problems and automated solutions are still in their infancy.

An indication of the significance of this topic is the formation and success of an entire conference series on the subject. The Knowledge Transformation for the Semantic Web[15] conference was held in Lyons, France during the 2002 proceedings on Artificial Intelligence. This popular conference dealt exclusively with topics pertaining to transformation, change, and mappings among discrete ontology.

Why Bother Transforming Ontology?

The reason why transformation is so important is that as ontology, as well as schema in general, continues to proliferate throughout the Internet and inside the enterprise,

[15] Full proceedings may be found at: www.isbn.nu/1586033255.

we will increasingly be asked to move data from one model to another. As with everything, performing these functions manually, or with code, is an alternative. But finding ways to automate the process is crucial to the dynamic functionality required by a large-scale information-sharing network.

Issues with Ontology Transformation

Much like the semantic conflicts discussed Chapter 6, problems with ontology transformation can come in several varieties.

- **Transformation of Structure**—Within a given syntax, such as XML, there may be a variety of ways to represent the knowledge, resulting in different schema approaches. This modeling freedom results in situations in which the same kinds of information, product data for example, can be expressed in a multilevel hierarchy or a pseudo-object-oriented format. Transformation from the hierarchical structure to the OO structure can be relatively painless, but still arbitrary, whereas the transformation from an OO structure to a hierarchical structure can not only be arbitrary, it can also be lossy. The loss of information about the structure is equivalent to the loss of important semantics about the intended usage of the data. Not only is automating this process difficult in some cases it can be impossible.
- **Transformation of Syntax**—For certain ontology representation languages this issue can be considered trivial; for example, transforming ontological syntax from OWL to KIF or from KIF to conceptual graphs is fairly straightforward. But transforming ontology from syntax with radically different semantic constructs can be quite problematic, for instance, OWL to UML or KIF to entity-relationship syntax. Some languages simply do not support similar features—and thus transformation can also be lossy.
- **Transformation of Semantics**—Typically one would think of preserving semantics when transforming ontology, not transforming them. But in some cases the actual meaning of concepts may need to change in order to support business rules about the equivalency or difference of certain concepts. In these cases the machine would certainly require additional cues beyond what was already modeled in the ontology in order to correctly perform the transformation. Approaches to semantic and contextual ontology mapping can solve this, whereby humans would direct the mapping process.

Solutions for Ontology Transformation

Today, most work for transforming ontology is still done manually or by custom-written transformation code—which results in brittle, tightly coupled interfaces between ontology representations. Alternatives to these tightly coupled techniques are scarce, many still coming from academia.

Standards-Based Solutions

- Topic Maps—Although the Topic Map specification is more suited to hierarchical data structures with simple context/scope, this mapping standard may be used to transform basic ontology from one form to another.

Academic Solutions

- MAFRA[16]—Ontology Mapping Framework is a conceptual framework and methodology for creating "bridge ontologies" that relate source and target schemas in a loose manner while allowing separate software components to transform the ontologies at run time.
- X-Map—An MIT Sloan effort led by Ashish Mishra and Michael Ripley takes an XML centric approach toward solving the semantic interoperability problem among XML documents.[17]

Proprietary Solutions

- OntologyX—uses a "context model" and robust "master ontology" as the baseline for generating mappings between other ontologies and basic XML schemas while maintaining the original precision of the vocabulary meanings.[18]
- Modulant—uses a context-based information interoperability methodology (CIIM) to facilitate context-sensitive mapping among ontology and schema.[19]
- MIOSoft—has developed a "context server" that unifies customer data and relationship management business rules across disparate applications and information silos; their mapping technology is a context-sensitive object-oriented approach.[20]

This subject of ontology transformation is very rich. Mature solutions will soon appear that take into account the complete range of semantic conflicts that could occur between ontologies, as well as the issues that arise from different representation languages and structures. Until then, we can keep writing XSL\T- and Java-based transformation programs.

[16] For more information, see: mafra-toolkit.sourceforge.net.

[17] *Automated Negotiation from Declarative Contract Descriptions*, Ashish Mishra, Michael Ripley, and Amar Gupta, MIT Sloan School of Management.

[18] For more information, see: www.ontologyx.com.

[19] For more information, see: www.modulant.com.

[20] For more information, see: www.miosoft.com.

FINAL THOUGHTS ON ONTOLOGY

This chapter has presented an overview of the current state of ontology work throughout the industry, why ontology works, and patterns of ontology structure, behavior, and creation. It should be clear by now that ontology is indeed a multifaceted solution to a wide range of problems. The relative maturity of ontology usage in the industry is still quite low, but organizations everywhere are becoming ever more committed to semantic approaches using ontology.

There is no tractable alternative to ontology for solving the vast array of issues and problems that businesses are faced with today. The dramatic acceleration of the rate at which new digital information is being created, coupled with the increasing fragmentation of data standards, data formats, schema structures, application models, and development approaches, signals a rapidly approaching information crisis. This crisis could cripple businesses that rely on information to drive their business, or, for companies who invest in semantic interoperability, it could spur innovation and lead to competitive market advantage.

Chapter 8

Multimodal Interoperability Architecture

KEY TAKEAWAYS

- New adaptive infrastructure approaches are at hand today.
- Multiple "modes" of semantic solutions are required for enterprise problems.
- Multiple enabling technologies are required for adaptive/dynamic architectures.
- The complexity of enterprise infrastructures is not solved with simple solutions.
- The benefits of a semantic interoperability architecture are significant and tangible.

Why confound an already obtuse phrase like "semantic interoperability architecture" with a word like "multimodal"? Because it accurately describes the best overall solution approach. Semantic technologies, as we have seen in previous chapters, are wide and varied. They operate at different architectural layers and provide differing reasoning services. This complexity is the natural state of things. Semantic solutions for interoperability, and the architectures we must use to describe them, apply to different modes of operation, different service interfaces, different technologies, and different function points—thus requiring a multimodal solution.

A useful analogy is in software security technology. Although many architects are tempted to draw a box on a diagram and label it "security," anybody who's ever truly built a large E-commerce system knows that security is not just one thing—it has to be accounted for at every layer. Security touch points like network, data, application, and functions like authorization, authentication, realm protection, and rights management span systems, architecture layers, networks, and data repositories. Likewise, a semantic interoperability architecture must cut across other enterprise systems to be most successful at creating an adaptive and dynamic service execution framework.

This chapter will first address the purposeful omission of a simple and intuitive semantic architecture picture. Second, a general multimodal semantic architecture, describing technology goals and various solution views, will be presented. Special attention will be given to model views, component diagrams, and sequence

Adaptive Information, by Jeffrey T. Pollock and Ralph Hodgson
ISBN 0-471-48854-2

diagrams as a means to communicate the needed interoperability functions. The final part of the chapter will dissect a hypothetical product life cycle management (PLM) solution case and provide some details as to its implementation and advantages.

A SIMPLE SEMANTIC ARCHITECTURE—WHY NOT?

A natural reaction of many technically savvy people who have just learned about the promise of semantic technologies is to try and fit it into their current understanding of enterprise architecture. This is natural. However, it is also problematic. As mentioned in the introduction to this chapter, semantic technologies are not simply a box in an architecture diagram; instead they are part of the framework and interfaces between existing layers.

Typically, software architects are familiar with the kinds of *n*-tier application models and middleware architectures depicted in Figure 8.1:

Unfortunately, it seems quite easy to simply put a box in a layer and call it a "semantic engine"—a black box that magically solves semantic conflicts among applications.

If these simplistic notions of semantic capabilities were indeed accurate, they would imply that a "semantic engine" is a new kind of tool that can plug into an existing vendor offering. This couldn't be farther from the truth. In fact, the revo-

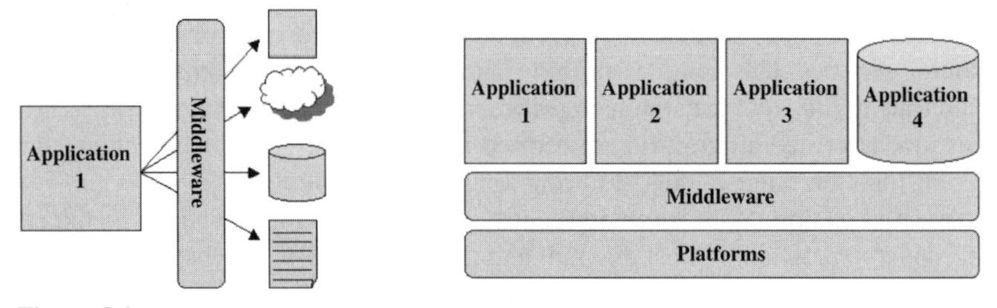

Figure 8.1 Typical n-tier and middleware application architectures

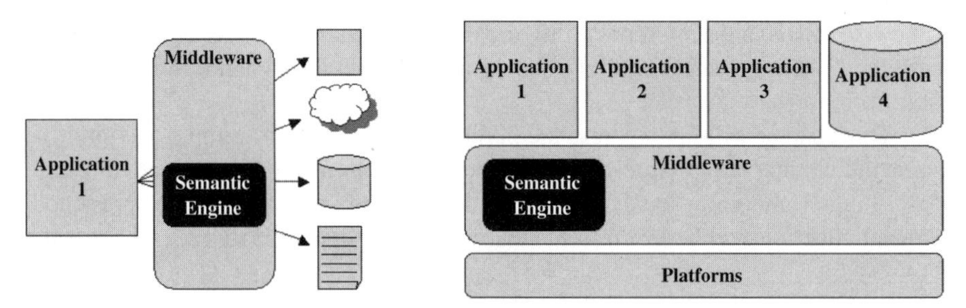

Figure 8.2 A simple, but inaccurate, depiction of a semantic architecture

lution around computable data semantics is so revolutionary that most traditional middleware vendors are unable to incorporate the technology into their platforms at all—because it would require a wholesale rethinking about how to represent information inside their tools.

Other times this "semantic engine" is depicted with enterprise applications, functional systems, or even at the data layer. Regardless of where it is placed, it always falls short of depicting anything meaningful about a complete semantic architecture. This is because of the inherent complexity of how the schema architecture interacts with physical components and the many locations (e.g., middleware, adapters, applications, databases, desktop) where the semantic services might be rendered. This complexity is a necessary evil because of how complex our software infrastructures actually are. We, as software engineers, cannot avoid that fact. But the overall complexity of the semantic architecture is largely irrelevant to all but a few alpha geeks—precisely because once it is in place, the rest of the enterprise users are interacting with simple models relevant in their own narrow domains. This is yet another reason why a semantic architecture is a long-lived investment and not a passing fad.

GENERAL MULTIMODAL SEMANTIC ARCHITECTURE

One of the more difficult exercises an architect can undertake is to describe an architecture outside of a very specific problem context. Architectures only make sense to the degree that they can be manifested in something concrete. However, given the newness of semantic technologies, and the capabilities they may provide, it will be useful to examine the "semantic architecture" at a general level of analysis. Treatment of the semantic architecture at an abstract level should enable us to talk about the overall goals and principles of a semantic approach and indirectly highlight advantages that can be accomplished by applying them.

The more difficult aspect of this treatment will be to provide a visual representation of the architecture that is simultaneously informative, accurate, and relevant to a broad problem space. This will be accomplished by providing a functional decomposition of the core capabilities in a semantic interoperability architecture and then describing relationships among its parts in a conceptual view. Finally, a generic component view will be assembled to visually depict the various elements of a multimodal approach to semantic architecture planning.

Architectural Modes

As discussed at the beginning of this chapter, a semantic architecture must function in different modes to achieve end-to-end adaptive and autonomic behavior. These modes correspond to aspects of software architecture such as layers and interfaces—wherever brittleness may impede dynamic behavior. Table 8.1 describes these modes, the spot in an architecture where they apply, and a short description of how a semantics-aware solution may be put in place.

Table 8.1 Data Table—Semantic architecture modes of operation

Architecture Mode	Application Area(s)	Use of Semantics
Information Mode	Data Persistence Metadata Persistence Domain Models	Annotated semantic descriptions and maps allow runtime components to repurpose information on-the-fly without human inputs
Interface Mode	Service Interfaces Bean Interfaces Object Interfaces Procedure Interfaces	Annotated interfaces allow runtime components to dynamically discover, interpret and assemble functions without human inputs
Transaction Mode	Transaction Monitors Process Monitors Process Orchestration Workflow Engines	Annotated transaction models allow runtime components to convert and repurpose procedures adaptively without human inputs
Security Mode	Authorization Authentication Certificates/Signatures Audit Reconciliation	Annotated security policy models allow security tools to apply general policies to specific software components and data sets without human inputs.
Network Mode	Resource Management Mobile Networks Protocol Management Utility Configuration	Annotated network models enables management software to monitor and control diverse system-wide network resources without human inputs

Key Technology Goals

Design goals are a way of setting a stake in the ground that creates parameters for a final system vision. Having multiple, and sometimes conflicting, goals can create circumstances in which a sacrifice has to be made for one goal in order to achieve a second. To fully understand the technical vision it is appropriate to identify and prioritize all the goals for the implementation and design of the overall system architecture. These technology goals must also be measurable, so it is important to use metrics to measure progress toward the goals.

The matrix in Table 8.2 identifies a list of technology goals, describes the rationale behind the goals, and identifies a metric that can be used to measure successes. Because semantic technologies are a particular technology component of a larger business system, the priority of certain technology goals can be stated as: "This goal is [crucial, important, optimal] for a semantic architecture." In this way, the focus of semantic architectures can be described by the operational goals that they attempt to accomplish.

The overarching technical goal, which can be seen in three of the four crucial goals mentioned in Table 8.2, is to manage change better. Semantic technologies rely on a number of technical innovations to better prepare an architecture and

Table 8.2 Data Table—Semantic interoperability architectural goals

Goal	Rationale	Metric	Priority
Dynamic and Adaptive	A dynamic architecture responds in real time to changing conditions by generating its own instruction sets on the fly when encountering unforeseen circumstances. Key design considerations include: dynamic queries, mappings, transformations, and service discovery.	In a given community of known applications, measure the costs associated with bringing a new application on-line (or changing an existing application's interface) and measure the costs associated with making the new system interoperable with the old ones.	Crucial
Information-Centric	An information-centric architecture separates the information from the application in highly portable components. This model-driven information is used to drive the implementation of run time code and also provide contextual cues for interpreting semantics.	Two metrics: (1) evaluate the costs of rewriting the total number of interfaces systems and (2) determine the cost of redeploying the entire system in a new program or platform language.	Crucial
Modular	A modular architecture creates a flexible structure for adding and removing functional and technical capabilities with minimal impact to the rest of the system. Key design considerations include: loose coupling of components, service-oriented interfaces, and the use of open technology protocols and standards.	Total costs associated with reworking code or reconfiguring functions from the existing system(s) when adding new application or middleware functionality.	Crucial
Scaleable	A scaleable technical architecture (physical and logical) allows a system to grow with minimal or no changes to application code as the user base expands.	Evaluate costs associated with doubling the throughput capacity of the entire system.	Crucial
Autonomic	An autonomic architecture has built-in mechanisms for configuring, optimizing, generating (code & schema), securing, and repairing itself.	Introduce a massive technical infrastructure failure and evaluate how much the total cost of downtime and repairs impacts the bottom line.	Important
Reusable	A reusable architecture reduces the amount of development that must be done to support future enhancements and modular plug-ins. It may also reduce the total effort of initial implementations as well by reducing redundant work.	Measure the amount of duplicate work that must be done when adding new functionality to the site.	Important
Development Speed	Development Speed is an architectural consideration because it is possible to meet all other goals and still architect a system that takes far too long to develop.	Measure the impacts of architecture decisions within the context of "level of effort"—computing that into FTE equivalency, and thereby costs.	Optimal

system infrastructure for unforeseen future changes—which in turn result in lower costs and more timely information.

Semantic Interoperability Architecture Principles

Unlike other aspects of technology planning and design, the principles used to guide software architecture should be enduring and seldom changed. Semantic interoperability capabilities are consistent with, and indeed reinforce, a number of widely accepted rules and guidelines that comprise architecture principles. The following principles[1] provide both motivation and consistency for the practical application of emerging technologies that enable semantic interoperability.

- **Separate Architecture Concerns**—Software applications should separate the "what" from the "how" both internal and external to program execution.
- **Technology Independence**—Software applications should remain independent of specific technologies and technology platforms.
- **Responsive Change Management**—Changes to the information and technology infrastructures should be implemented in a timely manner.
- **Tolerance of Technical Diversity**—Controlled management of technical diversity in the enterprise environment should be undertaken to minimize the nontrivial costs associated with maintenance of multiple environments.
- **Planned Interoperability**—Software and hardware should conform to models and standards that promote greater transparency among data, applications, and technology.

And finally, Michael Daconta succinctly articulates the principle most closely associated with the classic architectural edict, "Separate the concerns"—and its usefulness in understanding the role of enterprise data—in his declaration of data independence.

Declaration of Data Independence[2]

- Data is more important than applications.
- Data value increases with the number of connections it shares.
- Data about data can expand to as many layers as there are meanings.
- Data modeling harmony is the alignment of syntax, semantics, and pragmatics.
- Data and logic are the yin and yang of information processing.
- Data modeling makes the implicit explicit and the transparent apparent.
- Data standardization is not amenable to competition.
- Data modeling must be decentralized.

[1] Adapted from The Open Group Architecture Principles at: www.opengroup.org.

[2] Michael Daconta, Semantic Technologies for eGov Presentation 2003 at: www.daconta.net.

- Data relations must not be based on probability or luck.
- Data is truly independent when the next generation need not reinvent it.

These principles, and especially the principle of data independence, are the motivation behind the rise of semantic architectures. They imply shortcomings in current off-the-shelf technical solutions and offer guidance for those who wish to develop alternative architectural solutions.

Functional Decomposition

Now that we have identified our goals and motivating principles for a semantic interoperability architecture, it is time to answer the question, "What shall it do?" Despite the myriad of complex answers to this question, the authors have opted to start at the highest level and answer this question as simply as possible.

The significant elements of this high-level functional decomposition are agents, services, and semantics. Roughly speaking, these also equate to the user (agent), the medium (service), and the information (semantics). This functional breakdown, albeit simple, is a necessary starting point for the architectural discussion because it makes the first stab at separating the various concerns within the framework. Unlike other functional models, this architectural framework will clearly delineate physical connectivity and protocol concerns from information models and metadata artifacts. This underscores several of the core tenets we described previously. Let's examine this functional breakdown a little further.

- **Agents**—This functional layer is intended to capture any entity that can initiate a query or post information to services on the framework. An agent may be a human, a software application, or an intelligent software agent—among others. Functionally, the agent is usually a proxy for some other interested target, like a software system, a database, an analytical tool, or even another human.

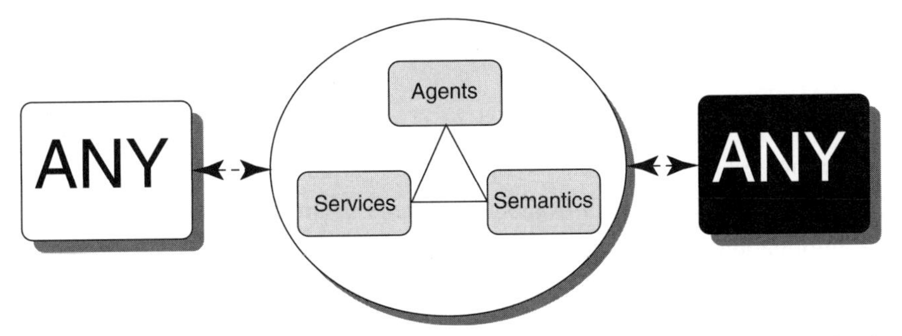

Figure 8.3 A semantic architecture will enable the meaningful communication of information from any technology component to another by enabling agents to use previously unknown dynamic services and data semantics without the intervention of human designers or programmers.

- **Services**—This functional layer represents anything that may provide information, functionality, or some other type of utility to agents within the semantic architecture. Services can be thought of as functionally similar to WSDL specifications in Web Services but need not be a web service or XML-based protocol. The Global Grid Forum has prepared specifications for Grid services, dynamically activated services, which most closely matches the functions provided for in this layer. Likewise, the Semantic Web Initiative is working on specifications suited for this layer.
- **Semantics**—This functional layer is comprised of anything that may be treated as data or metadata. Back in Chapter 6 we defined metadata as consisting of data patterns, syntax, structure, referents, context, and rules. Appropriately, this functional layer provides for the logical storage of data, schema, mappings, ontology, and business rules. In practice, as we will see later, this layer is a *logical* layer, which means that the physical location of the artifacts and components are of only secondary importance in this scheme. The value lies in the interoperability that is enabled by having them all related to one another inside logical communities or domains.

Although this is a simple functional decomposition, it serves the purpose of separating the primary concerns of the architecture. As we move into some more specific examples the delineations set forth here will continue to serve as the "lenses" through which we see the rest of the architectural components.

Conceptual View: Dynamic and Autonomic Capabilities

A semantic interoperability architecture should be capable of discovering and effectively utilizing new information and services without intervention from humans. The following conceptual architecture describes how components in a processing framework might interact to accomplish those goals.

Key Components

- Agent—either human or software initiator for queries, posts, or service requests
- Protocol—the medium over which basic communication occurs
- Service—a software component that offers various capabilities to agents
- Directory—taxonomy of available services and information resources
- Information—anything that contains context, metadata, and content
- Models—explicit or implicit organization of metadata and content

The conceptual model depicted in Figure 8.4 can be summarized in the following series of statements about the relationships among components in the model:

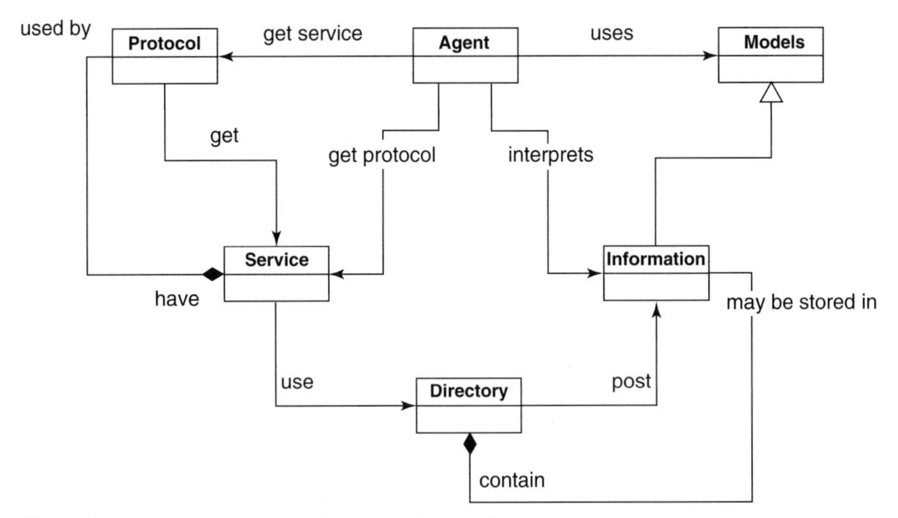

Figure 8.4 Conceptual semantic interoperability architecture

- An agent must use protocols to get services.
- A service can provide new protocol information to an agent.
- Services use directories to post information.
- Information contained in directories may describe services.
- Agents use models of information.
- Information described by models assists agent tasks.

Another way to describe the conceptual view is to utilize a sequence diagram to describe how an agent might discover new services and information and begin to use them without direct prior knowledge of either their semantics or context.

Of course, the sequence diagram in Figure 8.5 represents a conceptual view of sequence and shouldn't be interpreted as a low-level sequence diagram with actual classes and operations. The intent of this procedural view is to demonstrate how an agent, acting on behalf of a software system or a human, can first discover a set of services, then meaningfully determine what those services do (to be understood within the context of the agent), initiate a communication with a new service, and finally, interpret the content that the new service is providing by comparing the new content to already established models known to the agent.

In this way, agents acting on behalf of something else can dynamically adapt to their operating environment and community systems in addition to performing operations that configure, heal, or protect themselves without the involvement of an application programmer.

However, for this conceptual architecture to function, key semantic services must be in place.

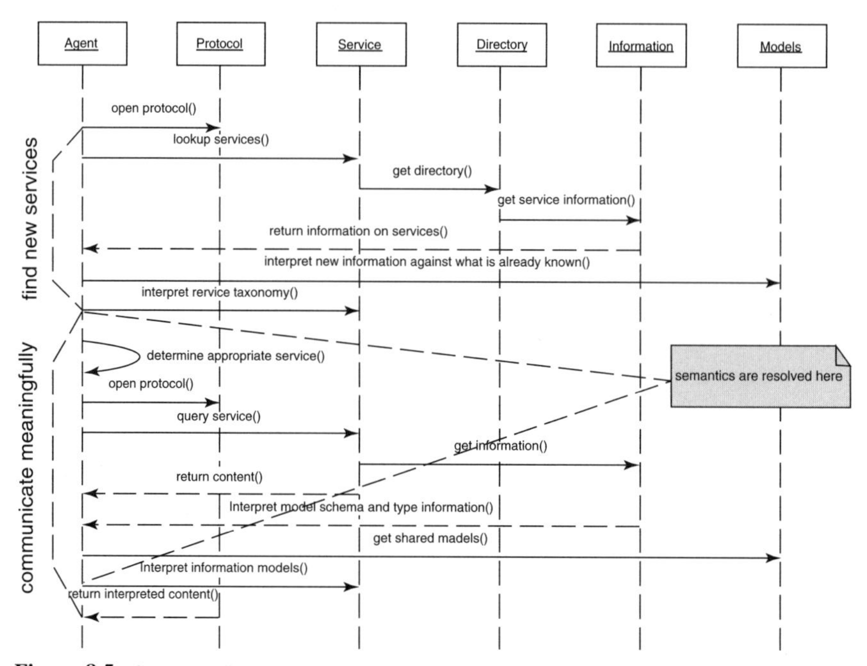

Figure 8.5 Sequence diagram of conceptual interaction among components

Component View: Multimodal Semantic Interoperability Architecture

More practical to the discussion of semantic interoperability architectures is the component view, which describes the software services that enable the kind of interaction previously described in the conceptual view. The architecture described in this book is termed a multimodal semantic architecture because it does not prescribe a single best approach for handling enterprise semantics. Instead, a key assumption is made that there are many ways to approach the semantics problem. As discussed in Chapter 4, there are different conceptions of what enterprise semantics are—none of which is exclusively correct—and thus different software-based approaches for handling them. One consequence of this is that no single vendor offers an end-to-end tool set with a complete range of multimodal semantic capabilities.

Figure 8.6 offers a simple component view of the key services that comprise a multimodal semantic architecture. These services are true services—different agents may utilize them on demand as needed. Additionally, there are no built-in single points of failure whereby a single service instance is required by all agent systems. Likewise, different implementations of each service may provide alternate functionality and varying degrees of capabilities. This is precisely what makes this type of service-based semantic architecture so powerful.

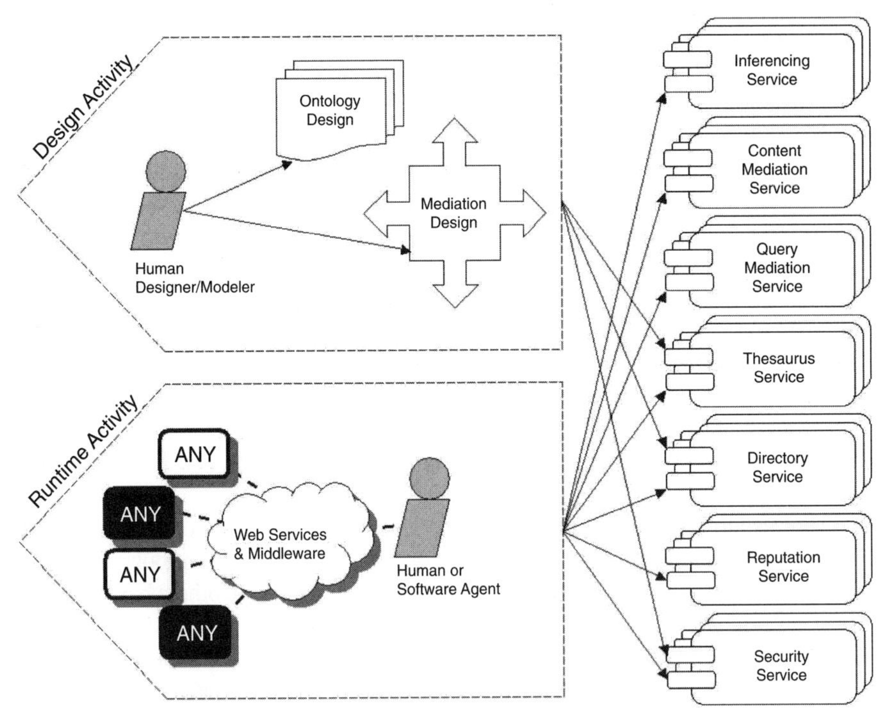

Figure 8.6 Static component architecture—key services

Figure 8.6 does not display any sequence information about how the components are used, only dependencies among the design time and run time work flows. An explanation of each service and capability is provided below.

Ontology Design

This capability is achieved through various software tools. Ontology modeling tools differ tremendously in implementation and provide a widely different set of capabilities throughout the life cycle. Typical life cycle stages for ontology design include creating, populating, validating, deploying, maintaining, and evolving. A full-featured tool set should include some capabilities in each of these areas.

Tool vendors still choose to support ontology in different schematic languages. Although some have already adopted newer formats like OWL and RDF, others continue to rely on relational models or Express notation. Some of the more popular tools in this capability area include Protégé 2000, Cerebra Construct, OntoEdit, Coherence, and Stanford's Chimera.

Key Enabling Technologies

- Resource Description Framework (RDF)
- Web Ontology Language (OWL)

- Description Logics (DLs)
- XML Rule Initiative (RuleML)
- Ontology Definition Metamodel (ODM)

Mediation Design

Mediation design is the activity that creates semantic mappings among schema, federated query structures, and enterprise communities of similar context. It is the mediation design capability that enables designers to model a loosely coupled logical architecture among diverse information resources. Typically the ontology plays a central role in the mediation design by providing a pivot point for documenting and describing, in machine-processable ways, how data semantics and context overlap. Sometimes, as with inferencing, the mediation design is not critical because alternate mechanisms for determining semantics can be relied on—such as language axiom bridges.

Tool vendors are taking fairly different approaches to mediation design. Some are using topic maps as a focal point for mediation and have created tools that align schema semantics through the use of topics and scope, whereas others have created robust proprietary approaches for aligning semantics through thesaurus mappings. Still others have created robust model-based mapping approaches that model the relationships among schema while accommodating divergent contexts. Some of the popular tools offering mediation design capabilities include Protégé 2000 TMTab, MioSoft Context Server, Intelligent Topic Manager, Modulant Contextia, and Contivo Analyst.

Key Enabling Technologies

- Topic Maps
- XML Namespaces 1.1
- Many other proprietary and academic technologies

Inferencing Service

A run time inferencing service is a pretty straightforward concept. This capability is an on-the-wire service that can accept models and queries as input and provide informational responses back to an agent as a response. As previously described, inference engines support various model logics that are often not compatible with each other. One popular standard model type is the web ontology language (OWL), which in its description logic form is 100% reliable and can offer high-performance inferencing.

Different agents can use an inferencing service differently. For example, whereas one agent may be interested in querying a model for a certain concept and its associated metadata, another agent may wish to use the inference service to merge two ontologies and evaluate their overlap. Inferencing is one of the most important tools for deriving implicit semantics from ontology models. Some popular tools offering inference capabilities include Cerebra Server, RDF Gateway, TRIPLE, and OpenCyc.

Key Enabling Technologies

- SOAP
- Inference Algorithms (public or proprietary)
- Description Logics (DLs)
- Knowledge Representation (KR) Languages

Query Mediation Service

Query mediation presents one of the most difficult challenges in the semantic interoperability and Semantic Web visions. The need is unambiguous. Querying for content across widely divergent repositories (XML, relational, HTML, object, word processing files, etc.) requires drastically different query mechanisms (text matching, SQL, X-Query, binary decomposition, etc.) that are not easily centralized. A query mediator would allow information that has been semantically mapped into an ontology to be queried from a single point and in a single query construct—regardless of the native structure of the underlying information.

The value for agents is undeniable. On finding interesting services, with different query interfaces, the agent could appeal to shared ontologies and construct a query based on the semantics of the ontology. A query mediator would evaluate the semantic and context mapping metadata and construct a query in the native structure allowable by the service. Functional capabilities that must exist in the query mediation layer include query routing, query decomposition, parallel plan generation, subquery translation, and results assembly. Although no vendor currently offers these capabilities in a federated query environment for heterogeneous sources, BEA's Liquid Data Product makes strides in this area and several university sponsored software tools have also tackled the challenge.

Key Enabling Technologies

- SOAP
- XQuery 1.0
- XPath 1.0
- XML Rules Initiative (RuleML)
- Proprietary query mapping approaches

Content Mediation Service

A core capability for the semantic interoperability architecture lies in its ability to mediate various content structures and business contexts into a local view for agents.

The content mediation service is a high-powered data aggregator that can repurpose information into context-sensitive views without human-generated application code. Ontology is the logical hub of this service. Content mediation would typically implement the ontology pattern "Logical Hub and Spoke" and use the ontology as the core mediation schema for a variety of different schemas that were mapped into it.

Critical to this implementation is the ability of the content mediation service to interpret the schema and mapping models produced by the mediation design component. This would allow the run time service to dynamically mediate on the fly without requiring humans to remap or implement a programmatic solution for the decomposition of instance data—and recomposition into other formats. As with all optimal mediation design artifacts, the ontology would not necessarily represent a "shared meaning" among targets and sources; rather, it would provide a common pivot for the semantic maps and context specifications.

Key Enabling Technologies

- SOAP
- Mediation Design (see above)
- Mediation Algorithms (public or proprietary)
- XML

Thesaurus Service

In Chapter 4, the topic of semantics as synonyms was covered. The thesaurus service is an implementation of this capability. Essentially this service can be thought of as a synonym and antonym repository for data vocabulary terminology. Unlike some implementations of thesaurus utilities (like Contivo's or the Institute of Computer Science in Greece) that focus on the design time use of a thesaurus as a mechanism to implement the "Design Moderator" pattern, the thesaurus service conceived of in this scenario is a run time service that provides on-the-fly insight into data meaning by cross-referencing synonyms and antonyms. In this capacity a thesaurus service would provide crucial links in the resolution of unknown data semantics for agents that are attempting to resolve new schema relationships in newly discovered models.

Key Enabling Technologies

- SOAP
- XML
- Resource Description Framework (RDF)
- Web Ontology Language (OWL)

Directory Service

The directory service is a pretty straightforward concept, essentially the same as envisioned by the UDDI Web Services framework. Like UDDI, the directory service would be partitioned into a number of different "books"—yellow pages, green pages, blue pages, white pages, etc.—to accommodate different types of services. Then lower-level taxonomy would provide categorization for the services. Agents browsing the directory could identify one or more interesting services by their classification scheme and relevance to their need. Additionally, a directory service would need to identify the type of schema that a service used for its content and what ontologies those schema were mapped to. This information would enable the agent to base decisions off of the likelihood of further successful communication without the assistance of human programmers.

Key Enabling Technologies

- SOAP
- Universal Description, Discovery and Integration (UDDI)
- Web Services Description Language (WSDL)
- Web Ontology Language—Services (OWL-S)
- OASIS ebXML Registry/Repository

Reputation Service

Once agents begin to utilize services without human intervention it will become necessary to build in automated trust mechanisms to ensure quality communications. A reputation service, much like the EBay reputation service, would rank service providers on a range of quality metrics like uptime, schema compatibility, information quality, and other relevant business measures. This is a crucial capability once we begin to wean ourselves off point-to-point brittle connection strategies. The shift toward dynamic capabilities requires a more comprehensive set of services that include reputation management.

Key Enabling Technologies

- SOAP
- Web Ontology Language—Services (OWL-S)
- Proprietary reputation algorithms

Security Service

As with any enterprise architecture, security controls should be at every layer. Security issues must be addressed from the hardware all the way to human resources—no single point of failure should be allowable. Within the context of semantic interoperability, a security service would mirror the capabilities offered by Web Services application firewall vendors (like Westbridge Technologies) but offer additional data-level security measures.

Typical Web Services Security Concerns

- Authentication
- Authorization and access (service level)
- Single sign-on
- Encryption
- Nonrepudiation and signatures
- Denial of service and replay attacks
- Buffer overflow attacks
- Dictionary attacks

For security to be effective in a dynamic and collaborative environment additional measures would have to be taken to identify trust and security levels for fairly granular components.

Semantic Interoperability Security Concerns

- Authorization and access (data level)
  - Role-based hierarchical (e.g., rank or employment level)
  - User-based
  - Third-party validation
- Realm and scope protection schemes
  - Data level
  - Schema level
  - Metadata access
- Secure access requests
  - Mechanisms for agents to efficiently request access from service owners

These types of security mechanisms are intended to provide a loosely coupled resource for identifying third parties and determining which data assets they may reliably see based on their credentials.

Logical View: Models, Maps, Schema, and Logics

A logical view is intended to communicate the interrelatedness of the logical, nonphysical elements of the architecture. In Figure 8.7, the relationships among information are demonstrated. Conceptually, information is simply related from schema to schema. However, there are a number of ways to relate schema to schema. In Figure 8.7 several ways of associating schema are displayed at once. In the uppermost part of the figure schema are related in a point-to-point manner with a mapping. The central relationship, shown by the "content mediation" line, is from schema to ontology to ontology to schema—with model-based mappings acting as the intermediary between the ontology models and the application schema. Finally, in the lower part of the figure, a schema is related to an ontology that

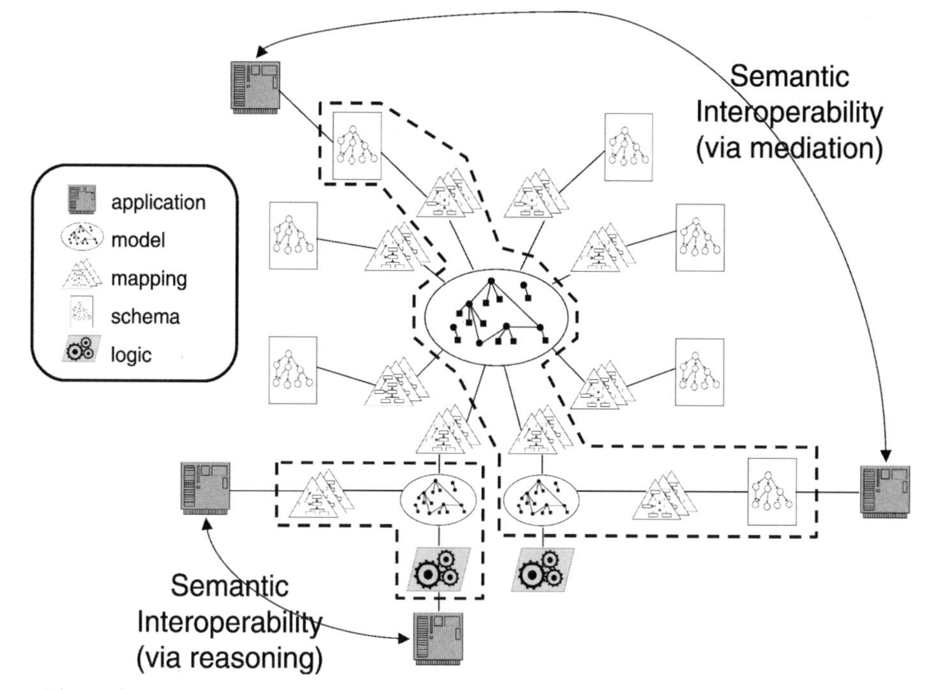

Figure 8.7 Logical architecture of models, mappings, schema, and logics

contains description logics, thereby enabling semantic interoperability through inference capabilities.

When reading this model, it is helpful to imagine what kinds of artifacts are actually involved in the logical structure.

- **Application**—This could be an application of any type or business function. Technically, the only prerequisite is that the application contains information that may be represented by some sort of schema. Even in applications without structured content, a schema that accurately describes the unstructured content may be created and used in an interoperability scenario like this.
- **Model**—This is intended to represent a conceptual model, or ontology. There is no prerequisite as to the type of ontology model; therefore, OWL-based models that contain description logics or UML-based models structured in XMI format are relevant to this logical view.
- **Mapping**—Here we mean "model-based mapping" to differentiate basic matrix mappings from mappings that account for application context and more complicated structures in the data. Examples of these type of mappings include topic maps and several proprietary and academic mappings like

Modulant's context maps, Portugal Polytechnic's MAFRA[3] approach, and MIT's X-Map[4] framework.

- **Schema**—Any application schema. Used here to specifically represent schema about implementation data, process, or business rules.
- **Logic**—Any set of logics that accompanies an ontology to provide inference capabilities for machines to traverse the model and infer new facts. Description logics is the most common logics to fulfill this purpose.

Process View: Dynamic Service Discovery

The first step required for dynamic interaction of services is to provide for the discovery of services and service descriptions. This is a fairly well known problem with solutions provided for in the Open Grid Service Architecture and the Semantic Web Services Initiative. The purpose of covering this topic here is to highlight the issues that arise when different taxonomies of service descriptions are used and to describe how that may be solved with semantic interoperability capabilities.

When looking up services there may be few problems so long as the directory you are using and the services you happen to be looking for are all using the same taxonomy of service classification. But much like how the phone's Yellow Pages vary from city to city, the taxonomy used to describe services may vary from community to community. These differences in taxonomy structure are easily accommodated with Topic Maps, which can relate hierarchical classification structures to one another.

A content mediation service may be used to execute Topic Map instructions at run time and formulate new lookup queries based on a mediated view of the classification taxonomy. This would enable services to use directories with previously incompatible taxonomy for service listings.

The sequence diagram in Figure 8.8 offers a conceptual view of how this would work. An actual implementation of this scenario would provide for a more optimized, and therefore different, sequence diagram than the one presented here.

Process View: Dynamic Service Collaboration

Once a set of services is identified that are suitable for follow-on queries, the actual task of collaboration begins. Unlike Web Services or Service Grid schemes, the goal of semantic interoperability is to enable communication across services where there

[3] Alexander Maedche, Boris Motik, Nuno Silva, Raphael Volz, "*MAFRA—A MApping FRamework for Distributed Ontologies in the Semantic Web.*" Proceedings of the Workshop on Knowledge Transformation for the Semantic Web (KTSW) 2002. Workshop W7 at the 15-th European Conference on Artificial Intelligence, July 2002.

[4] Wang, David, "*Automated Semantic Correlation between Multiple Schema for Information Exchange.*" Massachusetts Institute of Technology Department of Electrical Engineering and Computer Science master thesis, June 2000.

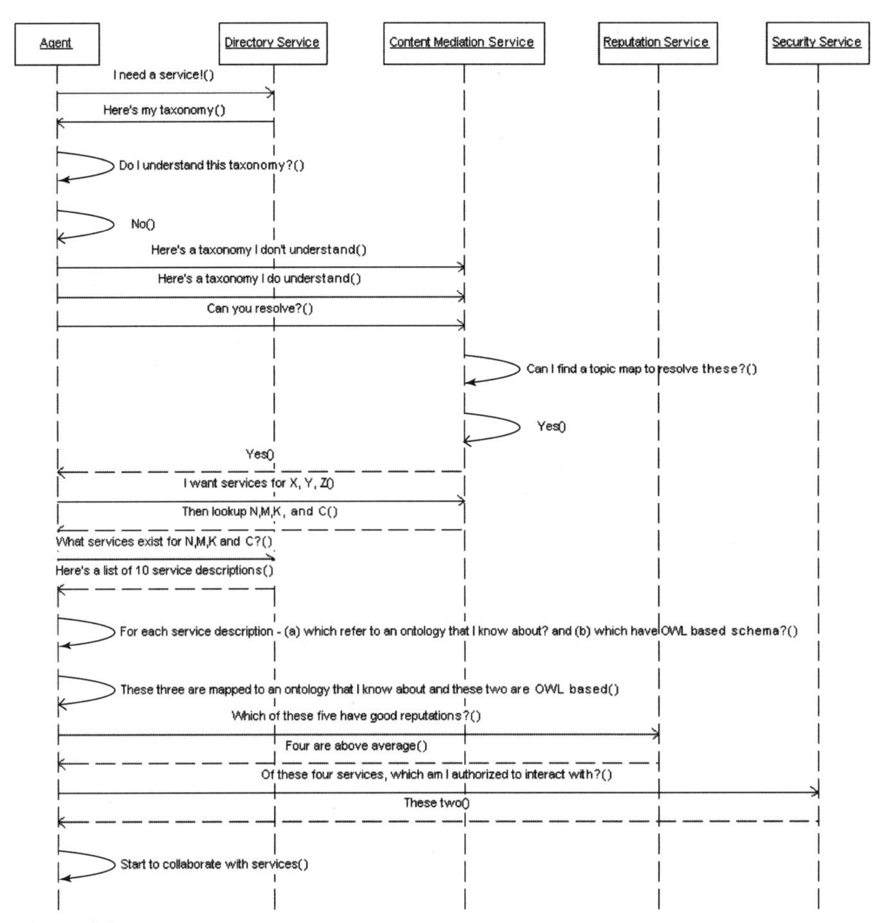

Figure 8.8 Dynamic discovery of new services via a previously unknown directory

is no standard schema or shared XML models. This is accomplished via a network of models, schema, thesauri, and mappings that provide routes for semantic engines to make reliable assertions about data meanings that were previously unknown.

The example in Figure 8.9 describes how two services may be utilized without previously knowing their internal workings, query interfaces, or data formats. The first example service, "New Service FOO," contains a data schema that is queried like an SQL X-Query system and returns a non-OWL-based data or schema. Therefore, the type of content mediation used for this service does not require an inference capability. The second service, "New Service BAR," does contain OWL-structured data and schema and does make use of the inference service capability.

This sequence diagram has been highly simplified, and many technical omissions may be observed; however, its intent is solely to provide a conceptual, process-

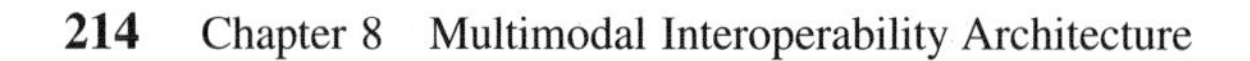

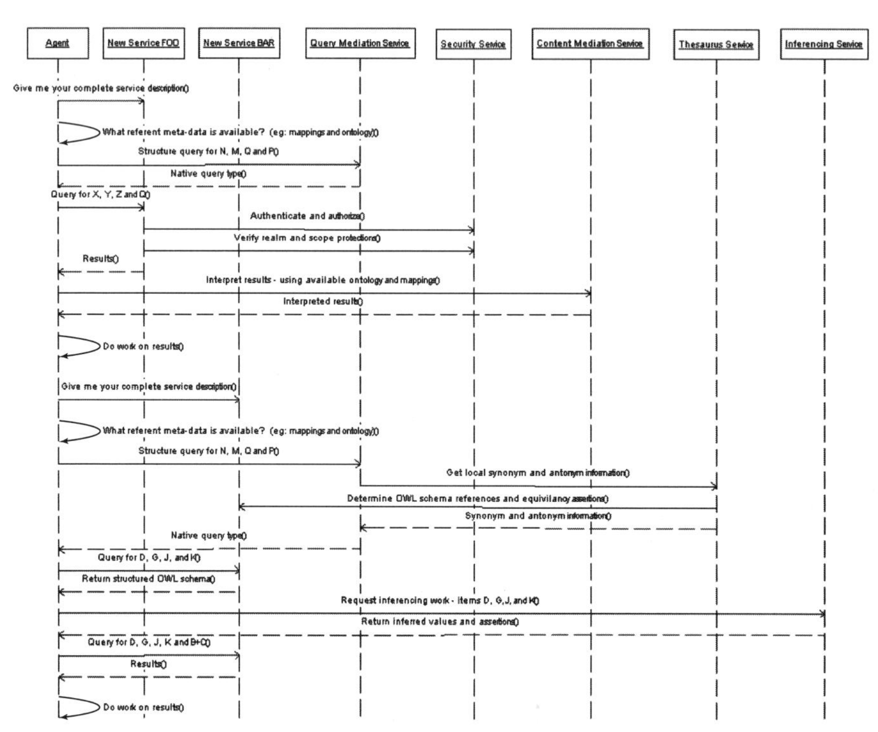

Figure 8.9 Dynamic collaboration of previously unknown services and content

oriented view of how these multiple steps would be accomplished to provide service-based semantic interoperability among previously unrelated services.

EXAMPLE DYNAMIC INFORMATION HUB: PLM INTEROPERABILITY

It has been stated that the lack of product data interoperability is the Goliath facing today's global supply chain,[5] which imposes at least $1 billion per year on the automotive industry alone.[6] The following example architecture will offer some insight into how to build an infrastructure that will dramatically improve the interoperability of product data and product life cycle applications.

Use Case View: Machine-to-Machine Product Data Exchanges

In this scenario Goliath CarMaker, a top automobile manufacturer, has decided to upgrade its PLM interoperability infrastructure. After years of constant investment

[5] EPM Express Way, January 2001 ["PDM Interoperability", Emery Szmrecsanyi].

[6] RTI/NIST Report, March 1999 ["Interoperability Cost Analysis of the U.S. Automotive Supply Chain"].

and maintenance, its legacy infrastructure was still inefficient and expensive to maintain. Goliath's old environment was comprised of different technologies that were brittle and inflexible to change. Their technical components were customized EAI platforms that required hand-written adapters and integration scripts. Goliath would spend millions to tweak its interoperability infrastructure each time a PLM system was upgraded or reconfigured to optimize business process or add new suppliers.

To launch its new semantic interoperability infrastructure Goliath has opted to focus on a particular auto platform and several of its key suppliers. Goliath's internal PLM system is a legacy environment that is highly customized to its internal processes and supply chain systems. BigParts supplies wheel and brake assemblies for the automaker and uses EDS's TeamCenter PLM application. Although TeamCenter is capable of outputting standard data formats, like OAG, Goliath's highly customized systems require specialized data inputs throughout the supply chain process. Acme supplies electrical systems to Goliath and uses MatrixOne as its internal PLM environment.

One of the key goals for Goliath is to reduce the time and expenses related to bringing on new suppliers, so a recently signed supplier of cooling systems has been scheduled to be brought on-line after the initial deployment.

The first two use cases to be implemented are Update Release Status and Add New Supplier. One dependency on the Add New Supplier use case is for the Supplier Parts Update to occur. In total, three use cases will be considered for the scope of this architecture discussion.

Unlike most use case scenarios, the actors in these use cases are enterprise PLM systems. Therefore, this overall interoperability scenario is fundamentally a

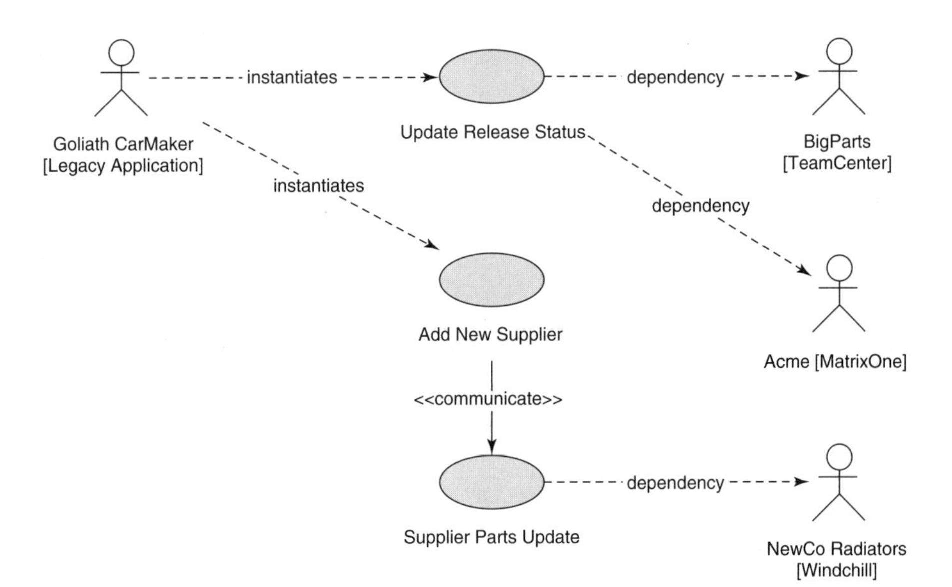

Figure 8.10 Product lifecycle management actors and use cases

machine-to-machine exchange environment. The initiators of these use cases can be system events or time itself. Both of the key use cases revolve around the updating of key PLM data in the correct context and view of each of the systems. Important data exchanged in this process includes parts master data, document master data, Goliath-specific attributes from other ERP systems, product structure, and transformation matrices for the parts.

Key Assumptions

As with any initial architecture, certain key assumptions must be made to constrain the possible technology choices. The following list represents the assumptions we are working with here:

- Data sources and targets already exist.
- Exchange schema and data formats are already in place.
- Connectivity protocols must be capable of passing through fire walls.
- Standards will be used as much as possible.
- Data and process information should be as up to date as possible.
- Updates to data and process information can be done on demand.
- Consistent processing of product data will be guaranteed.
- The infrastructure operates on a protected subdomain and is not public.

Component View

The component view, Figure 8.11, is a straightforward representation of a service grid architecture. Multiple components share a common protocol of communication and utilize a common set of directory services to find and communicate with one another. Key services include directory, security, content mediation, query mediation, reputation, and basic grid service utilities (audit, assessment, reporting, transport management, provisioning, and monitoring).

This type of physical infrastructure provides a loosely coupled overlay to existing brittle middleware. This coexistence of architectures can be visualized in the NewCo and Goliath diagrams, whereby each of those systems participate in other enterprise networks. A simple Web Services adapter allows the pass through of native communication protocols (like CORBA over IIOP or Tibco Rendezvous). Another element of this architecture is the protected nature of its deployment. Rather than deploying on a public HTTP domain, the entire deployment exists on a dedicated subdomain implemented by Goliath and routes through several fire walls that provide packet-level encryption and network security.

Logical View

A logical view typically shows the relationships between the logical layers of the architecture. Here we've used the logical view to show the relationships among the

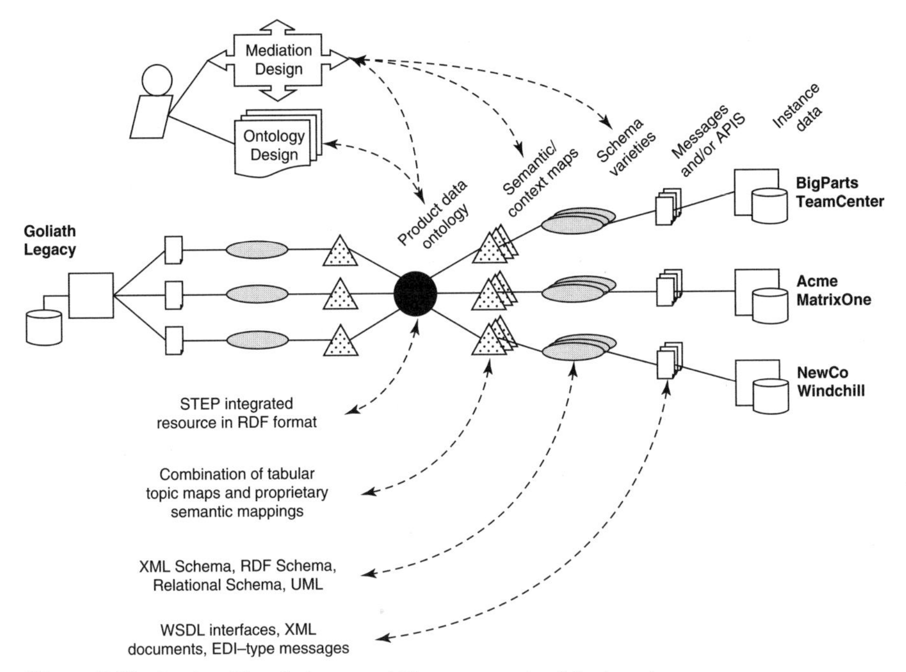

Figure 8.11 Product lifecycle interoperability component architecture view

information artifacts and the logical connections between them. From the system architecture perspective, the physical location of these artifacts is largely irrelevant; internally, the system uses URIs to locate and retrieve the artifacts on demand. The service-based registry tracks the file names, locations, and associations among the files and provides these capabilities to any service that has permission to use the directory.

This logical view is closely aligned with the decomposition of metadata presented in Chapter 6. The source applications, MatrixOne, Windchill, etc., provide access to the raw data. Their syntactic metadata is encapsulated by the interfaces the source applications present in the form of messages, XML documents, and service-oriented APIs. These syntactic metadata constructs are in turn modeled and constrained by the structural schema that govern them—in the form of XML schema, relational schema, and other model-based representations of the data. Structural schemas are related to one another and to conceptual models via referent metadata. It is the referent metadata, in the form of semantic and model-based mappings, that provide a means to capture local application context and relate it to other views of the information. These divergent views of information are resolved via a conceptual domain representation, ontology, that provides a pivot point for the reinterpretation of the data meaning in other forms.

Functionally speaking, these artifacts are used as guides for services in the semantic interoperability architecture to perform their work. At run time these services use the metadata at each level to provide enough expressiveness so that they

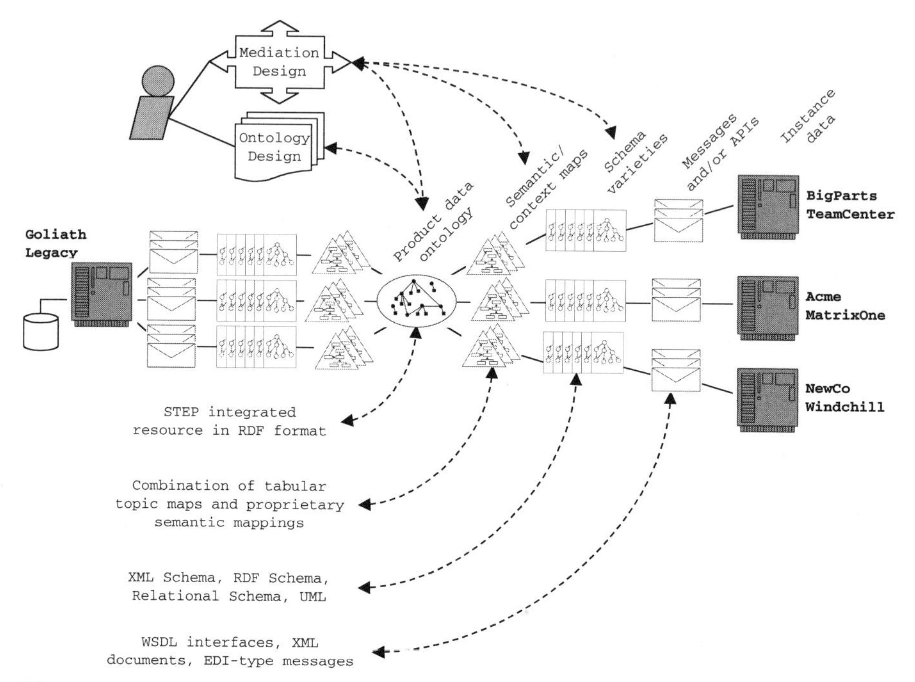

Figure 8.12 Product lifecycle interoperability logical architecture view

can operate independently of human programmers. A key logical underpinning of this approach places primacy on models, and therefore the software designers, over application code, and therefore software programmers. This fundamental shift in focus will drive industry toward more dynamic infrastructures and heighten the importance of software modelers and designers in enterprise environments.

Architecture Patterns

The structure of this product life cycle semantic interoperability architecture leverages a number of architectural patterns. Some of these patterns are traditional software architecture patterns, whereas others belong to the emerging class of semantic interoperability and ontology patterns.

- **Service-Oriented**—the use of well-defined contracts and interfaces among distributed software components to provide services within a broader system context
- **Federated Data Access**—the use of distributed data in a common system of systems environment without the use of a central repository for homogenized data representations

- **Dynamic Collaboration**—a service-oriented architecture that functions in a dynamic fashion without previously negotiated, brittle, one-off interfaces among the services
- **Ontology Broker**—the use of a service to broker required ontology components, from other ontology parts, dynamically in response to outside service requests
- **Logical Hub and Spoke**—the use of ontology as a central pivot point for diverse information models, schema and mappings
- **Smart Query**—the use of ontology as a query mediation reference that enables queries to be structured in a local context while still having global reach

Infrastructure Benefits

Nobody in business advocates technology for technology's sake. In this semantic interoperability architecture scenario Goliath CarMaker had very specific goals—to combat the expenses related to maintaining its supplier networks. By deploying a semantic interoperability architecture for product life cycle management (PLM) Goliath would easily accomplish those cost-savings goals along with several other important benefits.

- **Improved Return On Investment**—Infrastructure is the only technology component that can actually improve ROI for other, noninfrastructure, software applications—it does so by accelerating the usefulness of enterprise software and ensuring the survivability of those applications when things change.
- **Reduced Operational Overhead**—By making the interoperability infrastructure more dynamic, fewer dedicated human resources are required to operate the environment.
- **Improved Fact-Finding Capability**—When an interoperability infrastructure is programmed in code, it can do little to move beyond its own algorithms—conversely, a dynamic infrastructure that is model driven can discover facts that software programmers may never have thought of.
- **Reduced Maintenance Costs**—By making the interoperability infrastructure more dynamic, fewer problems will arise when interfaces change and when business drivers inevitably force reconfigurations to the system.
- **Improved Real-time Capability**—Real-time capabilities are not just the speed of transactions; a dynamic interoperability infrastructure enables business to add new capabilities in real time—without weeks and months of development.
- **Reduced Deployment Time**—The time required for bringing new systems on-line is dramatically reduced when interfaces and data structures can be discovered and utilized on the fly.

Insider Insight—Sandeep Maripuri With an Engineer's Perspective on Semantic Technologies

So what does this mean to us software engineers? On paper all of these semantic techniques seem impressive, but we engineers are a generally skeptical lot: six years ago we doubted the potential of XML as a cure-all for platform neutral data representation, five years ago we wondered if the EJB specification was just going to add bloat to our code-bases, and most recently whether or not the automagic promise of UDDI was really going to allow us to discover random services to appropriately solve runtime needs. Are semantic technologies really going to improve our ability to deliver technology, or are they just another good idea?

Like many other software engineers, I've cut my teeth building integration points between systems—writing complex custom integration components to transport data between flat files or XML documents and database-centric applications. And like many of you, I graduated to building enterprise-grade multitier client-server applications. And, unfortunately, like many of you, I have struggled with the overall notion of achieving true separation between logical architecture layers. Throughout these various software engineering practices, the basic challenges remain the same: making data transparent and transient amongst various containment mechanisms.

As engineers and architects, we invest a lot of time and effort building robust persistence mechanisms between our domain models and datastores and detachable transfer APIs between our domain models and presentation or integration layers, attempting to isolate architecture layers for maximum decoupling. But we effectively struggle to avoid shotgun changes—in a typical user-driven application, if we change the domain model, do we really escape from changing our data model, persistence mechanism, transfer objects, and presentation models, or do we just make five changes instead of two? Is separation actually achieved, or do we effectively just write a lot of API-level transformations within our code in addition to the necessary persistence transformations between our object models and data models?

This kind of code and program logic is often trite and uninteresting—the patterns are well known, and it is primarily a question of rote execution. And true, there are frameworks to available to aid in data transience: object/relational mapping and persistence frameworks, XSL/T publishing frameworks, and so on. But again, shotgun changes aren't avoided: In fact, they are exacerbated. Instead of just changing code, we now additionally update mappings, stylesheets, and deployment descriptors. Linkages across architectural layers are all point-to-point—decoupling isn't actually achieved, undermining their purpose.

Semantic interoperability removes a lot of these uninteresting and redundant programming cycles from the development process. Mapping the meanings contained within a domain model, database model, presentation model, and integration/data transfer schema to a common ontology allows us to describe each model individually and separately from the models with which it interacts. We are able to describe the essence of each model in a simple manner; changes to a single model are decoupled and are not forcefully propagated to other models. Choose the same ontology to back each model description, and true architectural separation is achieved.

This allows us to concentrate on application logic and business rules, delegating data and API-level transformations to a neutral third party that will translate information between model instances seamlessly. And because the ontological associations embed contextual and semantic information, these translations are more accurate than brittle code. Project cost is reduced, bugs are eliminated, and engineers' time is concentrated in areas where they are providing the most value, allowing for greater functional implementation in similar time frames.

But perhaps more importantly, semantic modeling techniques can be extrapolated to allow engineers to overcome many of the issues our projects regularly endure: understanding a system solution or functional decomposition contained in a single subject matter expert's head, or third parties conforming to transfer schemas in structure but not in meaning. These can easily be overcome by contextual analysis of current and planned systems—the metainformation regarding data structures, business rules, and data element meaning can be explicitly captured and explained using a base ontology, thereby eliminating the engineer's need to wade through pages of outdated data dictionaries, functional specifications, or making wild assumptions. This enables us to create domain models that more accurately reflect true business needs and application requirements.

These technologies as a whole are not going to solve each and every computing issue, but they are going to solve those persistent engineering issues that plague us with their dreariness, redundancy, and complexity for minimal benefits achieved. These semantic capabilities will free us to concentrate on software engineering problems that provide the most value for our stakeholders, and indeed, ourselves, which is decidedly a good thing.

Sandeep Maripuri
Director of Engineering, Product Development
Modulant Solutions

FINAL THOUGHTS ON SEMANTIC INTEROPERABILITY ARCHITECTURES

As we move forward further into the information age the importance of ubiquitous information sharing is ever more pressing. Without a dramatically new paradigm for managing the complex networks of data that have emerged in past decades—not to mention the vast archives of data not yet created—we will drown in meaningless text, facts, and words.

Software architectures that will successfully respond to this problem will not be simple or one-dimensional—there is no magic universal adapter for data. However, the progress made in a range of technical solutions provided by standards bodies, universities, and commercial software providers is rapidly converging into a useful, elegant, and repeatable solution. Semantic interoperability is no longer a dream; it is the inevitable result of better engineering. Unlike many technical "point solutions," a successful interoperability architecture will be multimodal in nature—covering an array of problem spaces on different axes and with different core approaches.

Chapter 9

Infrastructure and E-Business Patterns

KEY TAKEAWAYS

- Infrastructure patterns describe how semantic architectures are deployed.
- Infrastructure patterns should be applied in context of business model schematics.
- Adaptive infrastructures derive from newly emerging technology configurations.
- Four basic rationales exist for why to deploy semantic architectures: monitor policies, generate code, mediate conflict, and govern complex systems.

Semantic interoperability infrastructure should be a crucial topic of discussion for business people today. Unlike the Semantic Web architecture, semantic interoperability is focused on providing a technology infrastructure for enterprises—not the Internet as a whole. The next generation of technology infrastructures for business will have built-in adaptability. They will provide a dynamic foundation on which enterprise applications can be deployed and reused in many environments. This foundation will enable business information to flow seamlessly among applications regardless of its origin and destination. This foundation will also enable technologists to configure new services and transactions in real time—without having to code and recode interfaces again and again.

This kind of vision is not as far-fetched as some would believe. This chapter will start with a description and comparison of some semantic interoperability patterns. These patterns will focus on how emerging semantic technologies can be positioned in the enterprise environment and leveraged for improved interoperability. Next, various atomic business schematics will be presented and evaluated in light of the semantic interoperability architecture patterns. This should provide a business context for the patterns for architects to evaluate how they might be used in their business models.

As Figure 9.1 shows, the purpose of this chapter will be to discuss the infrastructure patterns and show their linkages to business model schematics.

Adaptive Information, by Jeffrey T. Pollock and Ralph Hodgson
ISBN 0-471-48854-2

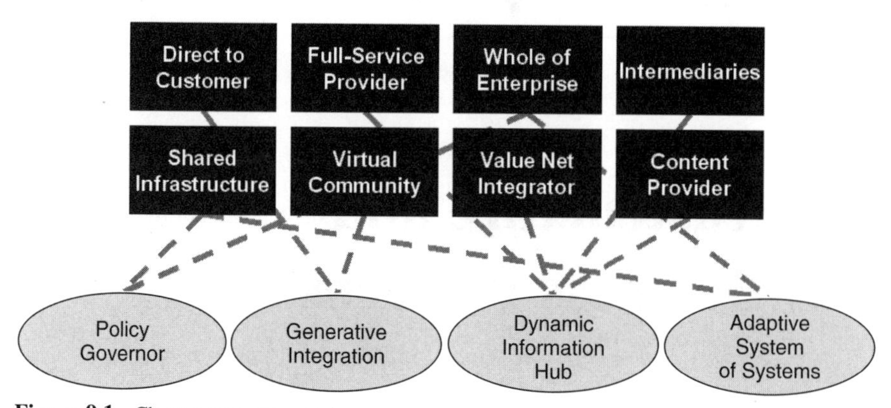

Figure 9.1 Chapter concepts overview

SEMANTIC INTEROPERABILITY INFRASTRUCTURE PATTERNS

Infrastructure patterns are blueprints for end-to-end application infrastructure designs. They encompass a range of software tiers and usually involve more than one functional application. Their value lies in being able to repeat successes by defining successful approaches and examining new ways to deploy infrastructure.

Semantic interoperability is inherently an infrastructure issue. In the past it was relatively easy to follow a three-tier or *n*-tier approach for application development and be done with it. However, as the importance of information access and transparent business process continues to rise, there is utmost value in a solid infrastructure plan that maximizes information visibility and reduces access barriers to silos of data, process, and behavior. These patterns will provide insight into these areas.

As opposed to architecture patterns, infrastructure is really about the foundation of enterprise applications. Infrastructure is the thread that keeps software chaos from becoming unmanageable; it is the cornerstone that provides stability for a potentially fragile array of one-off software applications with different rules and behavior.

By talking about infrastructure, we specifically mean to address concerns about information and connections between functional software applications. It is the authors' belief that the evolution of enterprise infrastructure will enable distributed applications on many platforms to be managed from a virtual console of models and model relationships. These models will drive ubiquitous data visibility and adaptive application connectivity.

Infrastructure Pattern Summary

Like Web Services, EAI (enterprise application integration), and the Semantic Web, the result patterns discussed here are focused on providing frictionless interoperability among software applications. The four patterns presented here represent innovative approaches for combining the strengths of Web Services, EAI, and Semantic Web technologies into a cohesive infrastructure that demonstrates value to enterprise architects and business management.

In their essential form, the four patterns discussed here are fundamentally about visibility into chaotic environments. Policy Governor provides visibility (and some control mechanisms) into enterprise IT networks and services. Generative Integration provides visibility into enterprise data and organizes it in ontology and taxonomy. Dynamic Information Hub and Adaptive System of Systems go a bit further and complement the visibility of data, process, and services by providing run time execution environments to build out dynamic and adaptive response mechanisms in the infrastructure itself.

No other technologies on the market today, or on the horizon for tomorrow, offer the promise of semantic technologies. The promise to provide frictionless information flows from application to application has been with the IT industry for years. Only now, with the advent of emerging technologies and the rise of computing horsepower, is the environment ripe to tame the problems that have plagued us for years.

Table 9.2 provides a snapshot of the strengths and capabilities of these semantic interoperability infrastructure patterns. They are contrasted with Web Services, EAI, and Semantic Web technologies to provide context—not to imply that there is an either/or decision point. In fact, the four patterns introduced here rely quite heavily on technologies from more traditional infrastructure approaches. It should become apparent throughout this chapter that these new interoperability patterns

Table 9.1 Data Table—Infrastructure pattern summaries

Pattern	Description	Maturity
Policy Governor	The Policy Governor uses agents and ontology for the purpose of monitoring and controlling the configuration and distribution of protected resources.	High
Generative Integration	Generative Integration uses thesaurus or ontology in a design time capacity for the purpose of generating transformation code that will unite disparate data schemas. Typically, the ontology is considered to be a shared vocabulary.	High
Dynamic Hub and Spoke	Dynamic Hub and Spoke uses ontology or thesaurus in a run time capacity to mediate transformations and queries among disparate data sources. Typically, the mediating ontology is not considered to be a shared vocabulary.	Moderate
Adaptive System of Systems	Adaptive System of Systems is a synthesis of the previous patterns. It combines policy governance and monitoring with code generation and dynamic conflict resolution. It is a completely model-driven framework.	Low

Table 9.2 Data Table—Infrastructure pattern capabilities overview

Key:
○ Unfulfilled
◖ Partially Fulfilled
● Fulfilled

				Semantic Interoperability			
	Web Services	Traditional EAI	Semantic Web	Policy Governor	Generative Integration	Dynamic Hub and Spoke	Adaptive System of Systems
Message Delivery	●	●	○	○	○	◖	●
Business Process	◖	●	◖	○	○	◖	●
Metadata Sophistication	◖	○	●	◖	◖	●	●
Semantic Conflict Resolution	○	○	◖	○	◖	●	●
Distributed Component Governance	○	◖	○	●	○	●	●
Code Generation	○	○	◖	○	●	◖	◖
Adaptive Service Negotiation	◖	◖	○	◖	○	◖	●
Dynamic Information Manipulation	○	◖	●	◖	○	●	●
Loosely Coupled	●	◖	●	○	○	●	●
Service Oriented	●	◖	○	◖	○	●	●
Information-Centric	○	◖	●	◖	●	●	●
Standards Led	●	○	●	○	○	◖	◖
Overall Maturity	◖	●	○	◖	●	◖	○

represent interesting ways to combine established technologies—not reinvent new protocols, specifications, or topologies.

Landscape of Approaches

A common theme for each of these new patterns is to establish designs and architectures that enable businesses to move toward adaptive capabilities and away from resource-intensive, largely manual coding efforts. As such, a common approach is the use of models to drive code generation, run time services, and information visibility and alignment across distributed systems. One way to envision the differences among these approaches is to map the model-driven capabilities on a graph. Figure 9.2 shows the relative sophistication of each approach along two axes: (1) model-driven connections, which signify service- and application-level connections and process control mechanisms, and (2) model-driven information, which signifies the interrelatedness of data and the availability of metadata to assist with data (re) interpretation.

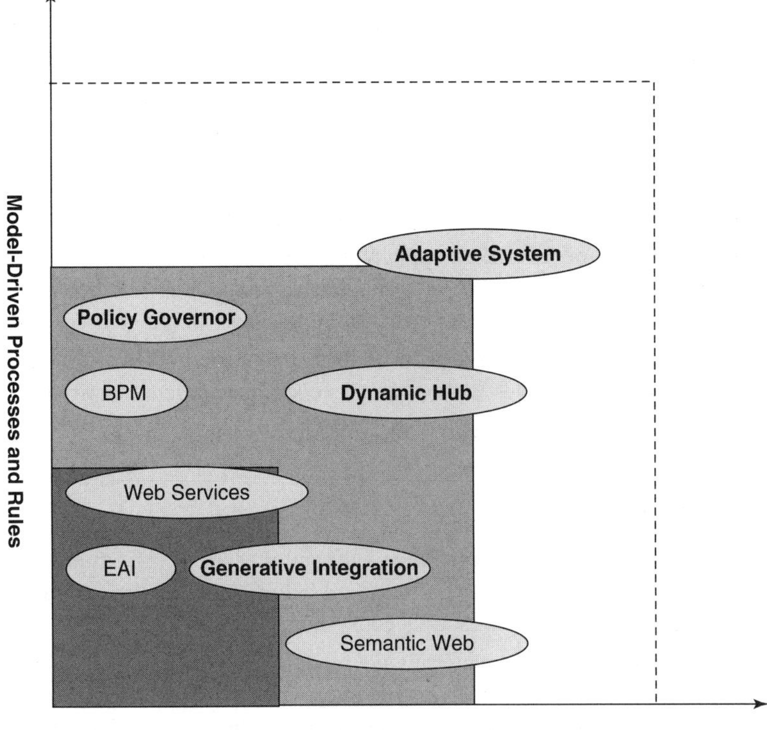

Figure 9.2 Model-driven pattern capabilities

Policy Governor

The Policy Governor pattern consists of the use of agents and ontology for the purpose of monitoring and controlling the configuration and distribution of protected resources. A Policy Governor can provide a central view of the network services and applications to assist the management and planning of IT initiatives, but the most important feature of the Policy Governor pattern is the ability to govern and guide individual services and application interfaces in their connections to other IT resources. It is this more active governance capability that is a characteristic of other policy-driven applications such as property rights management and security system monitoring.

When to Use

Use this pattern when complex rules and interdependencies dictate an otherwise difficult-to-manage and chaotic network of protected resources. These resources could include IT applications and systems, digital media, or any protected resource, for example, distributed services or software applications belonging to different logical infrastructures with separate configuration management concerns that still reside on the same network infrastructure.

Forces

- Increased complexity of policy interdependence and the lack of tools to successfully manage or predict the impact of changes in those complex rules networks
- The rising diversity of networked applications and services on the enterprise infrastructure causing configuration management and process control roadblocks that can result in exorbitant costs to the business
- Continued proliferation of data formats and schema types for XML and other data representation languages with variant levels of semantic expressivity
- Rising interconnectedness of diverse information systems

Ontology Patterns Employed

- **Model-Driven Federated Governance**—the automated use of ontology to implement and control various constructs and behaviors in multiple applications and data artifacts in a holistic IT ecosystem.
- **Decoupled Blueprint**—This pattern is the manual use of ontology by architects and designers as a way to derive new insight into an applications domain and scope. (This ontology pattern is included for completeness: Many application vendors generate models of the IT environment without making explicit connections for behavior.)

Technology Services Utilized

- **Control Center**—The control component provides the management console for service and application configurations, sometimes including process control mechanisms.
- **Security Service**—The security component provides on-the-wire application fire wall services and offers additional data-level security measures.
- **Directory Service**—The directory service provides an interface for resource identification and service descriptions.
- **Inference Service** (optional)—The inference service is an on-the-wire service that can accept models and queries as input and provide informational responses back to an agent as a response.

Logical Pattern Model

Figure 9.3 is intended to represent the use of ontology to model policies and rules about interfaces, protocols, system attributes, status monitors, and other technology

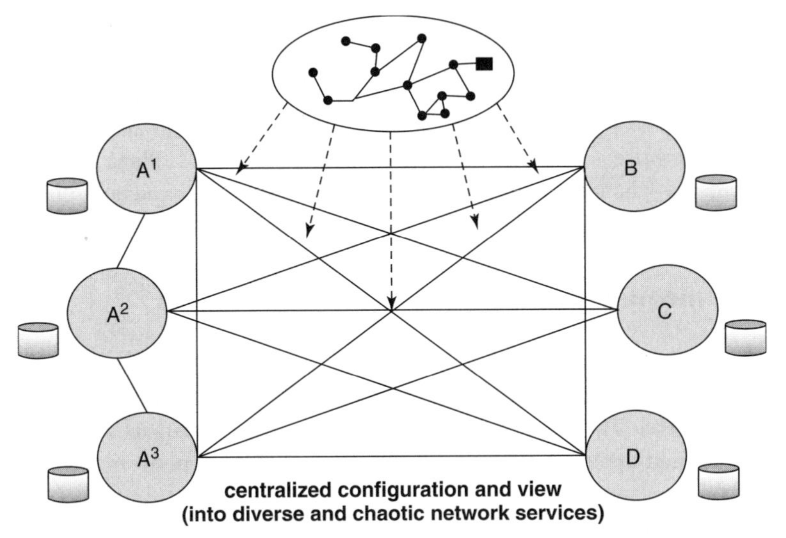

Figure 9.3 Active model infrastructure pattern logical model

configurations. In turn, the ontology would be controlled via a dashboard utility that allows for visibility and on-the-fly adjustments to the controlled environment.

Enabling Technologies

The following list of technologies is representative and not exhaustive:

- Ontology—provides a conceptual schema for describing network applications
- OWL-S—enables the dynamic reconfiguration of web service nodes
- SWRL—provides loosely coupled but highly formal rule expressions
- UDDI—a Web Services directory framework
- BPEL—provides a schema and a topology for process execution

Known Implementations

Although every effort has been made to verify the accuracy of the descriptions provided herein, errors, omissions and changes to the vendor tool or implementation are bound to occur. The authors heartily recommend that you contact the vendor directly with any questions regarding its technology.

- **Collation**[1]—Collation's platform provides a broad range of network and service capabilities that range from the discovery of services to the construction of topology maps in different views and change tracking of enterprise

[1] For more information, see: www.collation.com.

services. The use of various topology maps provides an active monitoring capability that Collation uses to target the network management niche.

- **Collaxa**[2]—Focusing on a different layer of the architecture, Collaxa provides a BPEL (business process event language) for the management and configuration of enterprise services in business process work flow. Collaxa uses the BPEL process schema as the basis for monitoring, managing, and controlling service connections among enterprise Web service deployments.

Generative Integration

Generative Integration is a pattern of semantic interoperability infrastructure that relies on the design time use of thesaurus or ontology for the purpose of relating disparate data schemas. The infrastructure consists of the code artifacts generated by the design tool and deployed to the run time components that perform the data transformations.

Typically, this pattern is an incremental step forward from traditional EAI- and BPM-type tools, which have far less sophisticated information-handling capabilities but quite sophisticated messaging, transaction, application management, and process management software built in.

When to Use

Use this pattern when an already deployed message infrastructure is in place and a low-cost solution is required to augment the messaging infrastructure with a less expensive approach for managing enterprise information.

Forces

- Inflexibility in the rapid configuration of run time data manipulation programs within integration architectures
- Continued proliferation of data formats and schema types for XML and other data representation languages
- Rising interconnectedness of diverse information systems

Ontology Patterns Employed

- **Design Moderator**—Design Moderator is the design time use of a cross-application ontology to align vocabularies among multiple disparate applications or software components.
- **Conceptual Identity**—This pattern is the use of an explicit ontology as the design centerpiece for the development of a single software application.

[2] For more information, see: www.collaxa.com.

Technology Services Utilized

- **Mediation Design** (thesaurus-based)—thesaurus-based mediation design is the activity that creates semantic mappings among schema and taxonomy by providing synonym and antonym mappings. The results can then be used to autogenerate transformation routines for run time software components.
- **Thesaurus Service**—This service would provide crucial links in the resolution of unknown data semantics (vis-à-vis synonyms and antonyms) for software components integrating with new services.
- **Security Service**—The security component provides on-the-wire application fire wall services and offers additional data-level security measures.
- **Directory Service**—The directory service provides an interface for resource identification and service descriptions.
- **Inference Service** (optional)—The inference service is an on-the-wire service that can accept models and queries as input and provide rich semantic responses back to an agent as a response.
- **Data Miner Service** (optional)—The data miner service is a typical machine learning analytics package that can operate on quantities of data to find previously unknown patterns.

Logical Pattern Model

Figure 9.4 is intended to depict a logical integration scenario. In this case, six software applications are joined to a common physical hub with common transport protocols. Internal to the hub, individual point-to-point transformation routines are stored in the form of scripts, code, or mappings. These transformation routines have been generated from a design time tool and downloaded to a production environment.

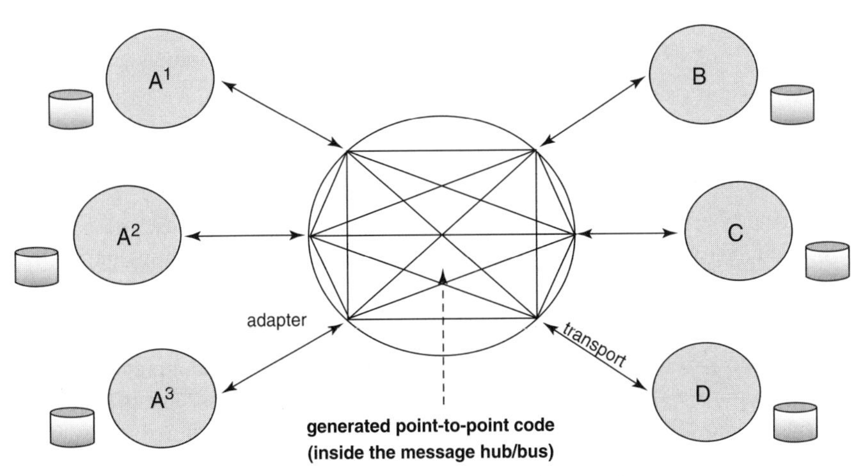

Figure 9.4 Generative integration infrastructure pattern logical model

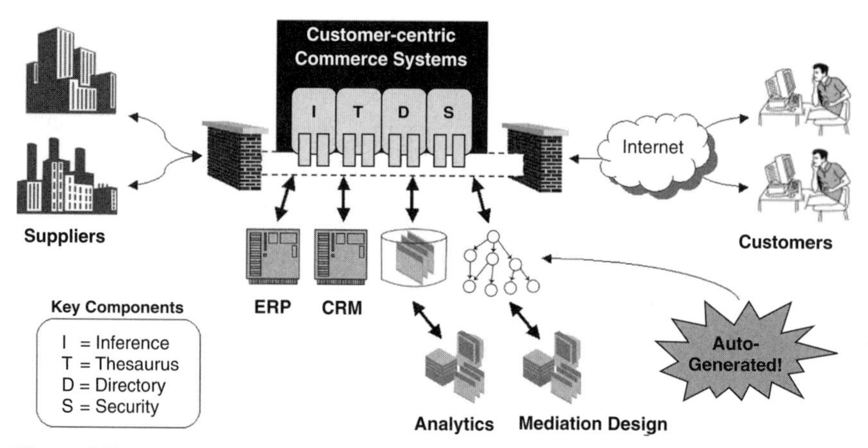

Figure 9.5 Generative integration infrastructure pattern component model

Example Architecture Component Model

The component model in Figure 9.5 represents a service-level view of the same scenario. In this case, the system functions are deployed as on-the-wire Web Services that can be messaged at any time. Note that the deployed transformation artifacts are generated and deployed into the run time environment via the mediation design tools.

Enabling Technologies

The following list of technologies is representative and not exhaustive:

- XML—supplies intuitive data representation for thesaurus development
- Ontology—provides a conceptual schema for describing network applications
- XSL/T—provides a platform-neutral XML transformation language
- RDF—core ontology representation language

Known Implementations

Although every effort has been made to verify the accuracy of the descriptions provided herein, errors, omissions, and changes to the vendor tool or implementation are bound to occur. The authors heartily recommend that you contact the vendor directly with any questions regarding its technology.

- **Contivo**[3]—the Contivo technology approach leverage a decoupled thesaurus that actually evolves with every new mapping of synonyms. The thesaurus is referenced during design time, where mappings are constructed. Overlap is

[3] For more information, see: www.contivo.com.

detected among like schemas (that use similar targets in the thesaurus) and used to drive the generation of transformation code in a number of useful formats—most notably XSL/T.

- **Unicorn**[4]—the Unicorn model leverages RDF-based ontology as the target for design time mapping of synonyms and more complex data structures. Once its workbench has been used to build the mappings, exports can be accomplished to a number formats—including SQL and XSL/T.

Dynamic Information Hub

The Dynamic Information Hub pattern mitigates the deployment of code-based point-to-point routines for manipulating data as it moves between contexts and semantic spaces. The logical configuration is much the same as in the Generative Integration pattern. However, rather than using the mappings to generate code—which is then deployed on other platforms—the mappings themselves are deployed to the run time environment. By deploying the mappings themselves, unexpected transformation configurations can be handled. The downside of this approach is that an "engine" for interpreting the mappings, data, and ontology at run time must also be deployed and configured to work with the underlying messaging infrastructure. This pattern assumes a service-oriented approach, enabling the engine to simply be configured as an on-the-wire service, but other approaches (such as hub and spoke or message bus) can also be leveraged.

When to Use

Use this pattern when a large number of data sources, with fairly complex disparate data formats, need to be made interoperable—and there are minimal constraints regarding the use of additional run time services to interpret and transform data in the broader middleware ecosystem.

Forces

- The need to integrate data in previously unforeseen ways, on demand and during run time operations
- Continued proliferation of data formats and schema types for XML and other data representation languages
- Rising interconnectedness of diverse information systems

Ontology Patterns Employed

- **Logical Hub and Spoke**—This pattern is the behavioral use of ontology to synthesize, aggregate, and augment data from multiple disparate applications

[4] For more information, see: www.unicorn.com.

or software components for the purpose of re-presenting that data in a new context that is meaningful to the other application(s) and their context.

- **Conceptual Identity**—This pattern is the use of an explicit ontology as the design centerpiece for the development of a single software application.
- **Smart Query** (optional)—This pattern is the use of ontology as a conceptual query interface for diverse sources of information.

Technology Services Utilized

- **Mediation Design** (model-based mappings)—Mediation design is the activity that creates model-based semantic mappings between schema and ontology. The resulting artifacts can then be used in on-the-fly dynamic transformation routines for run time software components that were previously incompatible. Likewise, the mediation design tool set would allow modelers to define generic query structures that can later be decomposed in platform-specific routines.
- **Content Mediation**—Content mediation is a service with the ability to interpret design artifacts (from the mediation design tool set) to mediate various content structures and business contexts into a local view for human or software agents. A key characteristic of this service is its ability to perform dynamically and without brittle point-to-point scripts or program code.
- **Query Mediation Service**—This service would allow information that has been semantically mapped into an ontology to be queried from a single point and in a single query construct—regardless of the native structure of the underlying information.
- **Security Service**—The security component provides on-the-wire application fire wall services and offers additional data-level security measures.
- **Directory Service**—The directory service provides an interface for resource identification and service descriptions.
- **Reputation Service**—This service would rank suppliers on a range of quality metrics like uptime, schema compatibility, information quality, and other relevant business measures.
- **Data Miner Service** (optional)—This service is a typical machine learning analytics package that can operate on quantities of data to find previously unknown patterns.

Logical Pattern Model

The logical model in Figure 9.6 represents the internal use of ontology (or taxonomy) as a run time operational pivot point for various data formats. Note the absence of any point-to-point routines—instead, the applications context is modeled into the ontology directly and used to facilitate data transformations on the fly.

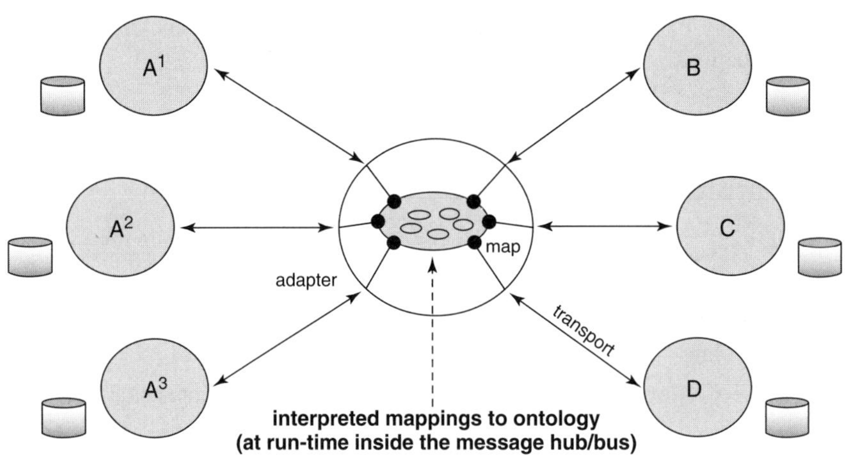

Figure 9.6 Dynamic information hub logical model

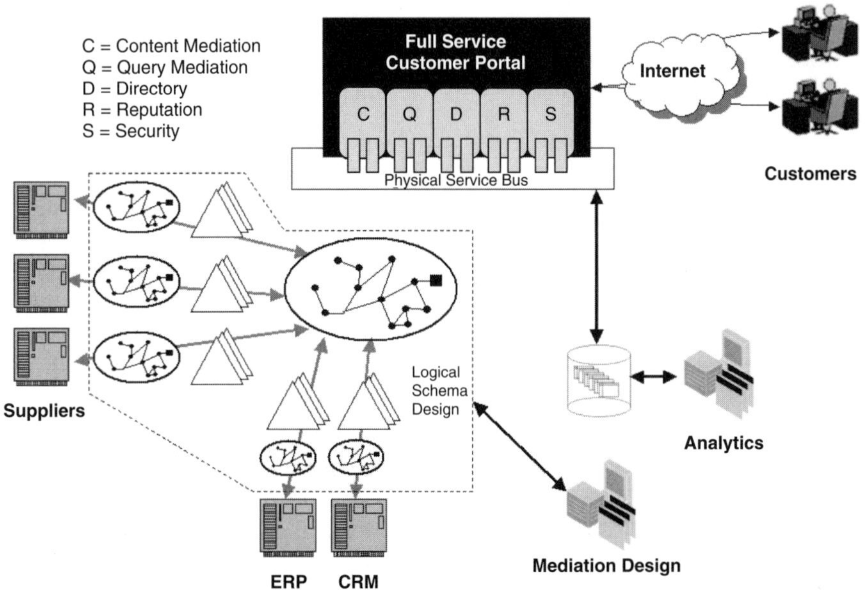

Figure 9.7 Dynamic information hub component model

Architectural Component Model

Figure 9.7 depicts how the logical schema architecture is an overlay to the physical service architecture. Although the physical services are deployed in a traditional buslike manner, the schema associations and logical links provide a network of semantic metadata that can be used to drive transformation and data re-presentation in a wide array of target software applications—without prior knowledge about which applications will ask for what data.

Enabling Technologies

The following list of technologies is representative and not exhaustive:

- Ontology—provides a conceptual schema for describing network applications
- RDF—core data representation language
- OWL—provides logic capabilities alongside a restricted RDF set
- Topic Maps—context-sensitive schema mapping capabilities
- X-Query—flexible query language capable of alternate representations
- RuleML—provides constraint logics around RDF and XML models

Known Implementations

Although every effort has been taken to verify the accuracy of the descriptions provided herein, errors, omissions, and changes to the vendor tool or implementation are bound to occur. The authors heartily recommend that you contact the vendor directly with any questions regarding its technology.

- **MioSoft**—The MioSoft Customer Context Server is an object-oriented system that manages context as a means of aligning business semantics about customer data. It uses a graphical user interface that allows analysts to identify and mange the contexts in which customer data is leveraged throughout the enterprise and aligns the semantics of that metadata so that interoperability between the various customer data repositories can occur. While this technology does not leverage any of the newer OWL or RDF capabilities for semantics, the object-oriented approach provides a rich mechanism for mediation between disparate information meaning, rules, data cleansing, and analytic components.
- **Modulant Solutions**[5]—the Modulant approach relies on a neutral model for the reconciliation of semantics at design time. During run time, requests can be made of the Contextia server for on-demand transformations without the necessity of code generation or prewritten scripts and programs. Modulant provides capabilities for STEP compliant data models in the Express format and can import and export application data in a number of formats.

Adaptive System of Systems

The adaptive system of systems pattern is in some sense a synthesis of the previous patterns. By combining the governance of distributed services (Policy Governor),

[5] For more information, see: www.modulant.com.

enabling more sophisticated code generation for distributed services (Generative Integration), and providing on-demand data manipulation for unforeseen situations (Dynamic Information Hub), the Adaptive System of Systems pattern may accomplish a truly organic technology infrastructure.

An important note is that this pattern can subsume a number of semantic technologies and make use of their strengths in both run time and design time environments. Therefore, certain optional component technologies—such as the Data Mining Service—can be deployed to augment core pattern strengths without directly constraining the patterns' usage scenarios.

Forces

- The need to create an infrastructure capable of specifying its own integrations on demand without the intervention of programmers for connections, process, or information
- Rapid development and integration demands continuing to pressure enterprise middleware into faster and faster deployment scenarios
- Continued proliferation of data formats and schema types for XML and other data representation languages
- Rising interconnectedness of diverse information systems

Ontology Patterns Employed

- **Model-Driven Federated Governance**—the automated use of ontology to implement and control various constructs and behaviors in multiple applications and data artifacts in a holistic IT ecosystem
- **Ontology Broker**—the central distribution of ontology, either component parts or in whole, to third-party ontology consumers
- **Logical Hub and Spoke**—the behavioral use of ontology to synthesize, aggregate, and augment data from multiple disparate applications or software components for the purpose of re-presenting that data in a new context that is meaningful to the other application(s) and their context
- **Smart Query**—the use of ontology as a conceptual query interface for diverse sources of information
- **Conceptual Identity**—the use of an explicit ontology as the design centerpiece for the development of a single software application.

Technology Services Utilized

- **Inference Service**—This service is an on-the-wire service that can accept models and queries as input and provide informational responses back to an agent as a response.

- **Mediation Design** (multiple)—In a Whole of Enterprise environment the mediation design component would include both the thesaurus and model-based semantic mapping capabilities.
- **Content Mediation**—This is a service with the ability to interpret design artifacts (from the mediation design tool set) to mediate various content structures and business contexts into a local view for human or software agents. A key characteristic of this service is its ability to perform dynamically and without brittle point-to-point scripts or program code.
- **Query Mediation Service**—This service would allow information that has been semantically mapped into an ontology to be queried from a single point and in a single query construct—regardless of the native structure of the underlying information.
- **Thesaurus Service**—This service would provide crucial links in the resolution of unknown data semantics (vis-à-vis synonyms and antonyms) for software components integrating with new services.
- **Security Service**—The security component provides on-the-wire application fire wall services and offers additional data-level security measures.
- **Directory Service**—The directory service provides an interface for resource identification and service descriptions.
- **Reputation Service**—This service would rank suppliers on a range of quality metrics like uptime, schema compatibility, information quality, and other relevant business measures.
- **Data Miner Service** (optional)—This service is a typical machine learning analytics package that can operate on quantities of data to find previously unknown patterns.

Logical Pattern Model

Figure 9.8 represents the synthesis of previous pattern advantages. Models can be used to drive every aspect of the infrastructure's interoperability by combining the ad hoc adaptability of the Dynamic Hub and the active governance of the Policy Governor. Interfaces, policies, and rules are managed in models as well as information and data content.

Architectural Component Model

The component model in Figure 9.9 depicts the fact that although each system physically belongs to a services bus, it is the logical association of models and schema that ultimately drive the application's participation, policies, and data portability.

Enabling Technologies

- See Policy Governor, Generative Integration, and Dynamic Information Hub patterns.

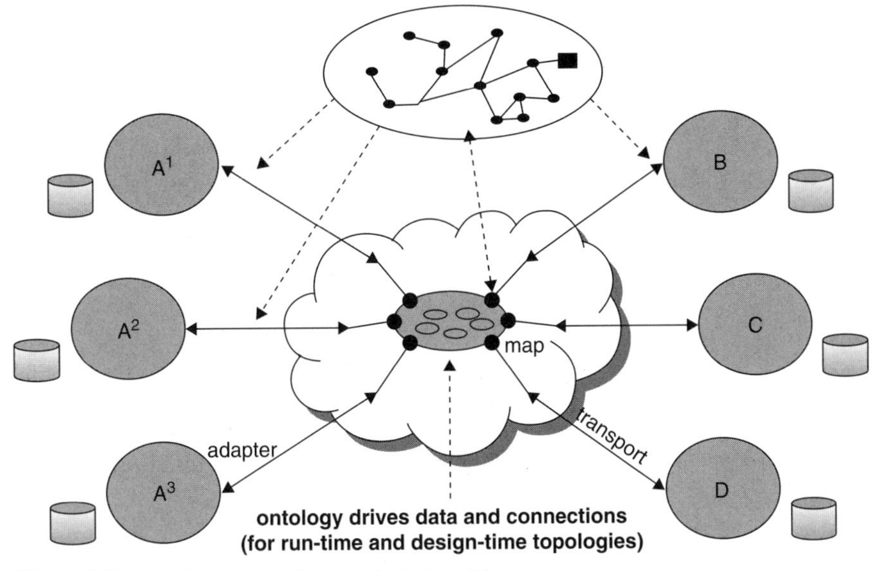

Figure 9.8 Adaptive system of systems logical model

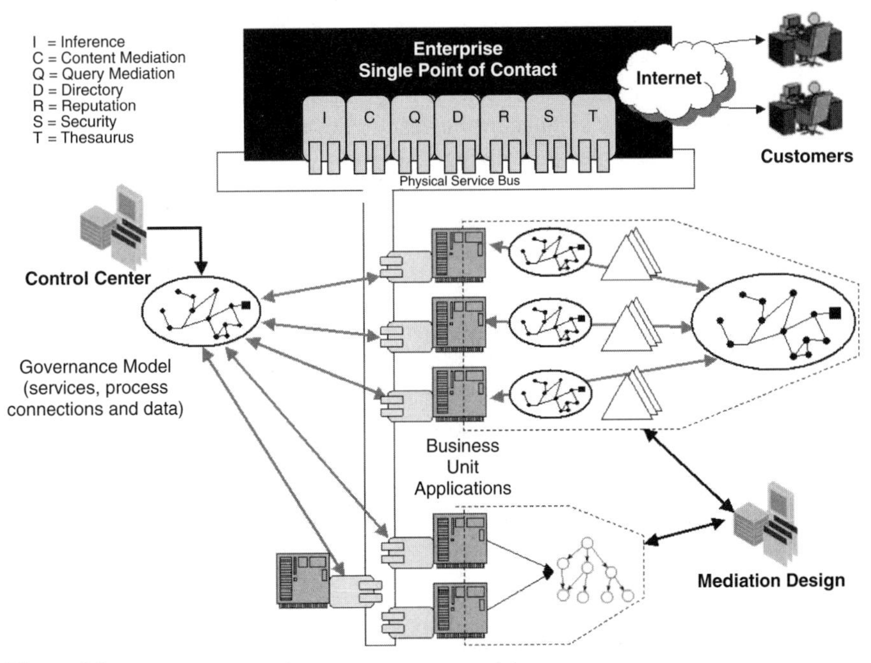

Figure 9.9 Adaptive system of systems component model

Known Implementations

Implementations demonstrating the range of capabilities for this pattern are non-existent at the time of writing. However, it would not have been included as a pattern if it were not a vision that is commonly espoused among software futurists. In fact, a number of analysts and standards organizations are rapidly aligning around the adaptive system of systems paradigm.

- **Model-Driven Architecture**[6]—The Object Management Group's vision for the MDA, more specifically the federated MDA, is a reference design model for this type of approach. The MDA relies on models for every layer of enterprise architecture to drive dynamic, platform-specific implementations that can be far more flexible than non-model-driven approaches.
- **Service Grid**[7]—The Global Grid Forum's architecture for grid services closely mirrors the dynamic service component articulated as part of this pattern. Their designs will be a crucial aspect of future development toward these goals.
- **Semantic Web Services**[8]—This form broadly means the combination of adaptive service interface negotiation and dynamic information manipulation. As a product of DAML-S and OWL-S Semantic Web service conceptions, the term may sometimes mean adaptive service interface negotiation only.
- **Enterprise Nervous System**[9]—Gartner Group's vision of the enterprise nervous system has provided a description of capabilities and characteristics that are necessary to drive these adaptive capabilities in the enterprise. Although Gartner classifies a broad range of applications that contribute to the nervous system architecture, it readily acknowledges the importance of emerging technologies in realizing this vision.

INFRASTRUCTURE PATTERNS IN BUSINESS CONTEXT

This section will provide an overview of the different ways a generalized multimodal architecture can be configured and applied to enterprise computing problems. Whereas the previous sections covered a general conceptual architecture, this section will decompose various electronic business models and provide a schematic for how to leverage semantic technologies within them.

E-Business schematics provided by Peter Wiell and Michael Vitale in their book *Place to Space*[10] will be used as a basis for this decomposition. Their breakdown

[6] For more information, see: www.omg.org.

[7] For more information, see: www.ggf.org.

[8] For more information, see: www.swsi.org.

[9] Gartner IT Expo, Orlando 2002, Roy Shulte presentation "Enterprise Nervous System Changes Everything." Presentation may be found at:
http://symposium.gartner.com/docs/symposium/itxpo_orlando_2002/documentation/sym12_16k.ppt.

[10] Place to Space, Peter Weill, HBR 2001.

provides sufficient coverage to describe the breadth of applications for semantic technologies in actual business models. These schematics also enable a more detailed discussion of architecture trade-off analysis for each business model. Furthermore, this approach will enable us to revisit the semantic architectures in the context of capability case studies presented in Chapter 10.

Business Schematic Summary

Table 9.3 Data Table—Business schematic summary

Business Model	Description[a]	Infrastructure Pattern
Direct to Customer	Provides goods or services directly to the customer, often bypassing channel members.	Generative Integration
Full-Service Provider	Provides a full range of services directly to the customer, often bypassing traditional channel members.	Dynamic Information Hub
Whole of Enterprise	Provides a firmwide single point of contact, consolidating all services provided by a large multiunit organization.	Adaptive System of Systems
Intermediaries	Brings together buyers and sellers by concentrating information.	Dynamic Information Hub
Shared Infrastructure	Brings together multiple competitors to cooperate by sharing common IT infrastructure.	Adaptive System of Systems
Virtual Community	Creates and facilitates an on-line community of people with a common interest, enabling interaction and service provision.	Generative Integration
Value Net Integrator	Coordinates activities across the value net by gathering, synthesizing, and distributing information.	Dynamic Information Hub
Content Provider	Provides content (information, digital products, and services) via intermediaries.	Dynamic Information Hub

[a]Place to Space, Peter Weill, *HBR* 2001.

Direct to Customer

Companies employing this model provide goods or services directly to the customer, often bypassing channel members. Buyers and sellers typically interact directly over electronic channels—such as the Internet—to exchange information, money, and sometimes even goods and services.

Example Companies

- Dell
- RealNetworks
- Gap
- CDNow

Who Owns What?

	Customer Relationship	Customer Data	Customer Transaction
Direct to Customer	✓	✓	✓

Business Model Schematic

Unlike traditional business models, the Direct to Customer model does not rely on distributors, dealers, or any other intermediaries.

Semantic Technology Opportunities

Challenges and opportunities identified in this section are specifically chosen from a broader set of business concerns because they are most likely to benefit from semantic technology architectures in some capacity.

- Increased demand for sophisticated logistics management presents opportunities to leverage information technology in streamlining processes.
- Channel management demands present opportunities to use technology for timely data retrieval.
- Customer service demands present opportunities to develop integrated extranet and ERP systems for improving customer access to information (such as inventory and product information).
- Service provider goals of owning customer relationships present opportunities to extend new services and information to customers.
- Demands to improve product information present opportunities to connect with supplier information systems and provide customers with more accurate data.
- Service provider challenges to better market and prospect customer data present opportunities to make better use of customer data and buying trends.

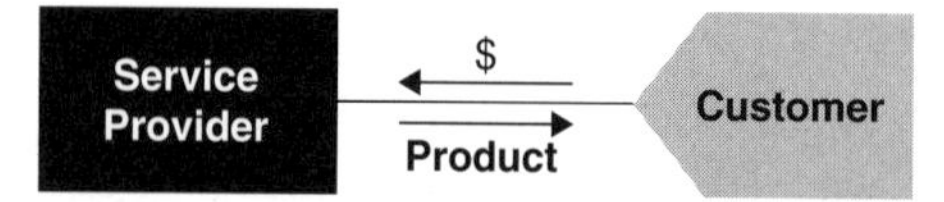

Figure 9.10 Direct to customer business model schematic

Generative Integration Infrastructure Pattern

The overall approach toward applying semantics technologies in this business model is to leverage the Generative Integration Architecture Pattern, which in turn relies on the core dictionary and thesaurus semantic interoperability approach. A technically complex semantic interoperability deployment is not required because of the relative simplicity of this business model schematic. The Generative Integration pattern will take advantage of the benefits of automated script generation for backend interfaces—enabling greater connectivity to both supplier and internal information systems. In turn, customers can take advantage of better customer service, information access, and improved process.

A complementary, but secondary, approach could be to take advantage of the Pattern Analysis or Inference semantic patterns to mine customer data and discover new relationships and trends.

Capability Case Solution Example

Additional details about capability cases and the applicable solution stories may be found in Chapter 10. The capability case that applies to this business model and architecture pattern is the following:

- European Environment Agency Content Registry

Full-Service Provider

Companies employing this business model schematic typically provide a full range of services directly to the customer, often bypassing traditional channel members. Although this schematic is similar to the Direct to Customer model, it differs in the scope of information, services, and product that are provided to the customer—typically engaging the customer with more robust capabilities. These enhanced capabilities usually demand greater service provider interaction with a range of third-party suppliers, with value-added consolidation, on behalf of their customers.

Example Companies

- Prudential Advisor
- Barnes and Nobel
- ChemNet
- GE Supply

Who Owns What?

	Customer Relationship	Customer Data	Customer Transaction
Full-Service Provider	✓	✓	✓

Business Model Schematic

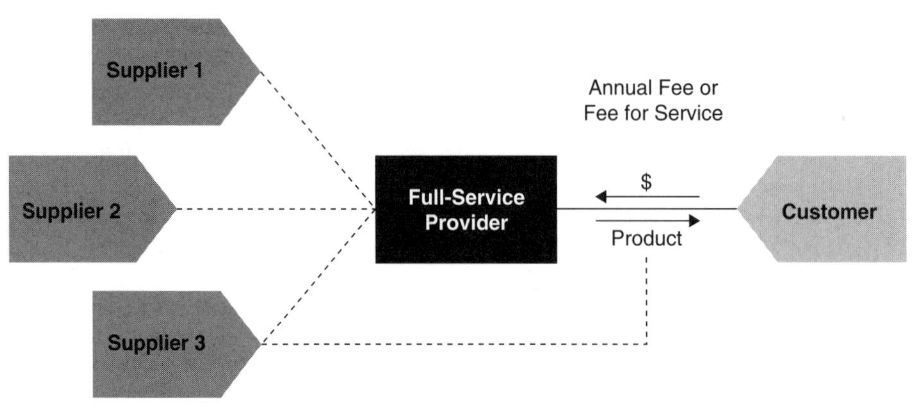

Figure 9.11 Full-service provider business schematic

Semantic Technology Opportunities

- The role of full-service provider presents an opportunity to excel at customer relationship management practices, demanding a holistic view of the customer to successfully impact sales channels.
- Inventory management and product logistics present both a challenge and an opportunity—the number of suppliers in this model will exceed the Direct to Customer schematic, and the demand for value-added services will drive innovation in the value chain.
- The technology infrastructure demands on this type of schematic can become fairly complex—a number of opportunities exist to reduce costs and improve efficiency in key technical areas such as:
  - Transaction processing
  - Business unit integration
  - Global corporate integration
  - Supplier integration

Dynamic Information Hub Infrastructure Pattern

Because of the additional complexity in supporting a wide range of suppliers and the demand to provide value-added services in addition to core supplier services, the infrastructure demands are more significant than Direct to Customer. Therefore, a more robust Dynamic Information Hub pattern will be employed here. This architectural pattern will enable the full-service provider to create a wide array of business partnerships with value-added suppliers without fear of a prohibitively expensive infrastructure to maintain. Because the information hub has embedded autonomic capabilities it can evolve and adapt to changing messages, XML documents, and database interfaces with minimal input from technical staff.

Depending on the extent of investment in a semantics-based approach, additional steps can be taken to further automate the technical message services—particularly if Web Services are leveraged.

Whole of Enterprise

The Whole of Enterprise schematic is a firmwide single point off contact, consolidating all services provided by a large multiunit organization. Typically this kind of schematic would be employed within an organization where multiple operational models are being leveraged—confusing customers and suppliers as to who their points of contact are and what processes they are a part of. Because of the decentralized nature of government services, this schematic applies to federal agencies as well.

Example Companies

- Federal government
- Colonial Limited
- Ford
- Chase Manhattan Bank

Who Owns What?

	Customer Relationship	Customer Data	Customer Transaction
Whole of Enterprise	✓	✓	✓

Business Model Schematic

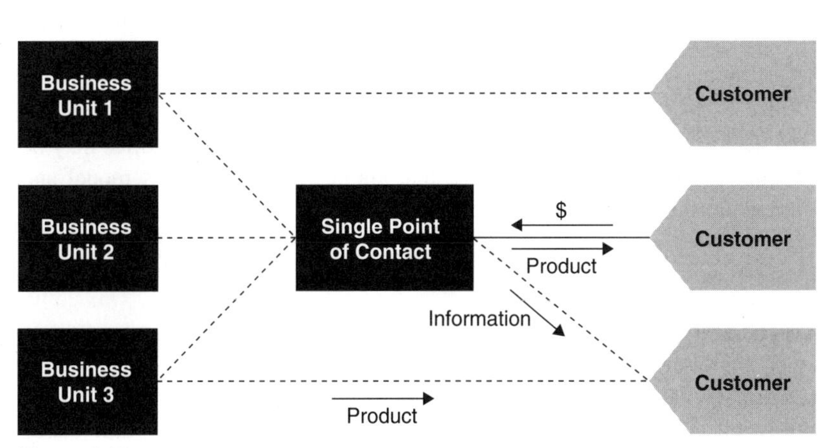

Figure 9.12 Whole of enterprise business schematic

Semantic Technology Opportunities

- The challenge of providing an integrated single point of contact provides an opportunity to streamline and reengineer business processes—especially those that impact a customer's experience of the enterprise.
- Undertaking the commitment to provide a Whole of Enterprise experience to customers presents a significant infrastructure management challenge and the opportunity to deploy interoperable solutions.
- Providing a unified face to customers across the enterprise creates a data management challenge that is likely at a much larger scale than previously undertaken—this presents an opportunity to separate the data- and information-level concerns from the underlying technical infrastructure to provide greater flexibility and deployment options.
- The vast number of ERP and other enterprise systems that are likely to already exist provide the opportunity to roll out a firmwide interoperability program for business applications in multiple units.

Adaptive System of Systems Infrastructure Pattern

Because the Whole of Enterprise business schematic could easily result in a dizzyingly complex network of interconnected IT systems, it is imperative to have a control mechanism for that complexity. The Adaptive System of Systems architectural pattern provides for a federated governance schema that allows analysts to control the run time behavior, configuration, and maintenance of enterprise systems in a largely model-driven capacity—thereby simplifying the overall complexity by abstracting data formats, service interfaces, and component configurations to models.

The actual deployment of this pattern would be expected to differ widely among organizations—depending on their systems and complexity of interaction among services. Additionally, the semantic technology services that are deployed would vary significantly because of the broad scope of this type of semantic deployment.

Capability Case Solution Example

Additional details about capability cases and the applicable solution stories may be found in Chapter 10. The capability case that applies to this business model and architecture pattern is the following:

- US Air Force Enhanced Information Management

Intermediary

The Intermediary category consists of portals, agents, auctions, aggregators, and other service providers that exist between customer and supplier. The value-added services that they provide and the number of buyers and sellers who are involved in the exchanges help define the types of intermediaries that exist. Intermediaries make their bread and butter by getting paid for the transactions that they facilitate between buyers and sellers.

Example Companies

- EBay
- ESteel
- Priceline
- Yahoo

Who Owns What?

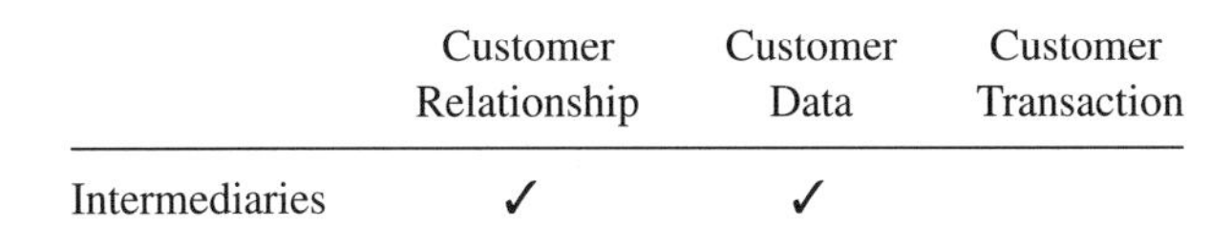

	Customer Relationship	Customer Data	Customer Transaction
Intermediaries	✓	✓	

Business Model Schematic

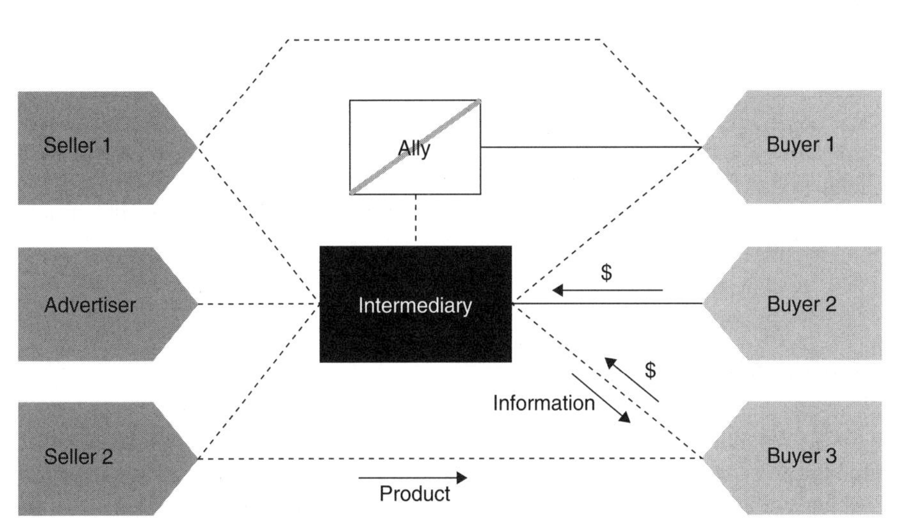

Figure 9.13 Intermediary business schematic

Semantic Technology Opportunities

- The challenge to keep pace with growth demands creates an opportunity to leverage flexible and dynamic infrastructure services that can scale rapidly.
- Of crucial importance, the need to bring customers back and own the relationship creates an opportunity to improve the utility of customer data that is tracked in a variety of software applications.
- Difficulty in keeping tabs on buyer, seller, and advertiser information creates an opportunity to leverage knowledge management systems that can classify,

categorize, mine, and manipulate data from the intermediaries constituents—in turn, this can be used to profile customers and drive innovative marketing programs.

Dynamic Information Hub Infrastructure Pattern

The intermediary business schematic can take many forms, for example, consumer-to-consumer applications such as EBay and Yahoo or business-to-business applications like ESteel. For the purposes of discussing an architectural approach, we will cover the more complicated of the scenarios—business-to-business intermediaries. In these cases the Dynamic Information Hub is the most pragmatic architecture to apply.

The Dynamic Information Hub is the best selection because it enables the IT environment to evolve and adapt to changing information from either the sellers or the buyer. From the intermediary's perspective, where accurately representing critical product data from a variety of different sellers is crucial, the ability to have the system respond on its own to internal product information representations, new XML formats, and new message types will dramatically lower the maintenance costs for keeping pace with change.

Capability Case Solution Example

Additional details about capability cases and the applicable solution stories may be found in Chapter 10. The capability case that applies to this business model and architecture pattern is the following:

- Ford e-Hub

Shared Infrastructure

The Shared Infrastructure schematic is marked by a high degree of cooperation among competitors. Typically, a single firm will host a shared infrastructure for joint owners and third-party suppliers—who are not part of the ownership structure. This infrastructure is sometimes developed as a response to a dominant market force or to meet demand for new services that is not currently being met. Like the Intermediary schematic, payment may either go directly from customers to suppliers or be routed through the shared infrastructure. Unlike the Intermediary schematic, the Shared Infrastructure Schematic is not driven by a desire to own the customer relationship; instead, value is derived by efficiencies of scale and cooperation.

Example Companies

- Covisint (automotive)
- Amsterdam Internet Exchange (ISPs)
- SABRE (travel and airline)

Who Owns What?

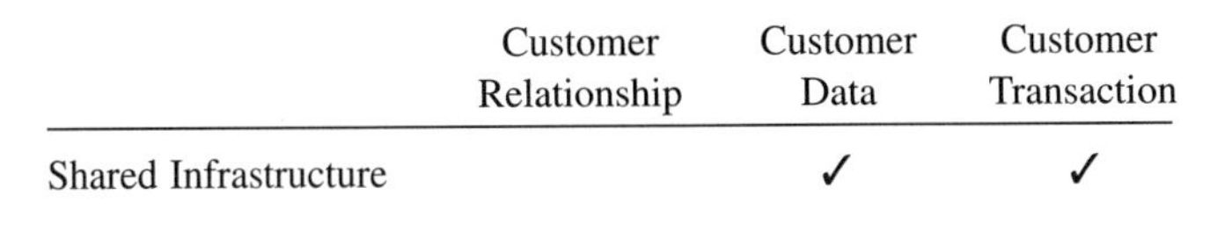

	Customer Relationship	Customer Data	Customer Transaction
Shared Infrastructure		✓	✓

Business Model Schematic

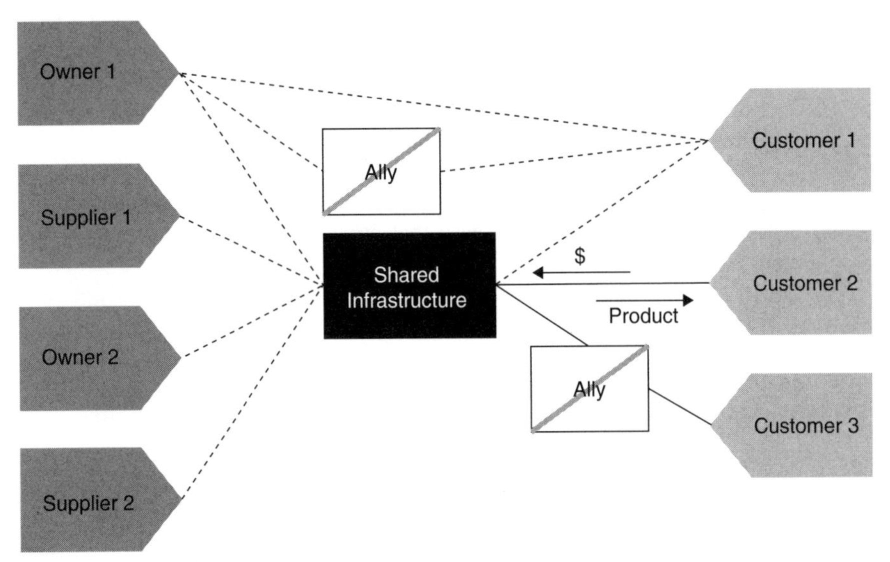

Figure 9.14 Shared infrastructure business schematic

Semantic Technology Opportunities

- The challenge of high cooperation on technical infrastructure creates an opportunity to provide a highly flexible system that can easily accommodate new owners and suppliers into the infrastructure.
- Cross-organizational commitments provide the opportunity to establish standards and policy boards to ensure that the shared architecture and standards are sufficiently meeting the needs of participants.
- The challenge to ensure that no single organization becomes dominant can create an opportunity to build an open infrastructure framework that does not place unfair burden (for adoption) on any organization that is part of the alliance.
- The sheer diversity of IT platforms and systems that may need to interoperate with the shared infrastructure is an opportunity to build a highly adaptable infrastructure to reduce expenditures related to maintaining external interfaces.

Adaptive System of Systems Infrastructure Pattern

The Adaptive System of Systems pattern is an optimal fit in this scenario because of the diversity of systems, business units, and cultures that may choose to align themselves together around a shared infrastructure. As the shared infrastructure grows and evolves, the costs associated with changing interfaces, data formats, and infrastructure adapters can become prohibitive and erode the efficiencies generated by the cooperative business model.

In practice, a centralized, model-based paradigm for managing services, data, process, and diverse technical platforms will greatly improve efficiency and adaptability for participants in the shared infrastructure.

Capability Case Solution Example

Additional details about capability cases and the applicable solution stories may be found in Chapter 10. The capability case that applies to this business model and architecture pattern is the following:

- Cargo Airlines System Integration

Virtual Community

The Virtual Community schematic provides a channel for the fundamental need for humans to communicate with like-minded people. One of the last pure-play internet business models, the Virtual Community schematic often combines with other schematics, such as Full Service Provider or Content Provider, to augment their revenue base. As a stand-alone model, Virtual Communities almost always base their structure on the customer segments they are servicing. As with the Full Service Provider model, stickiness and retention are the key to sustainability and typically drive the value of a community's advertising real estate.

Example Companies

- The Motley Fool
- Planet Out Partners
- Friendster
- Always On Network

Who Owns What?

	Customer Relationship	Customer Data	Customer Transaction
Virtual Community	✓		

Business Model Schematic

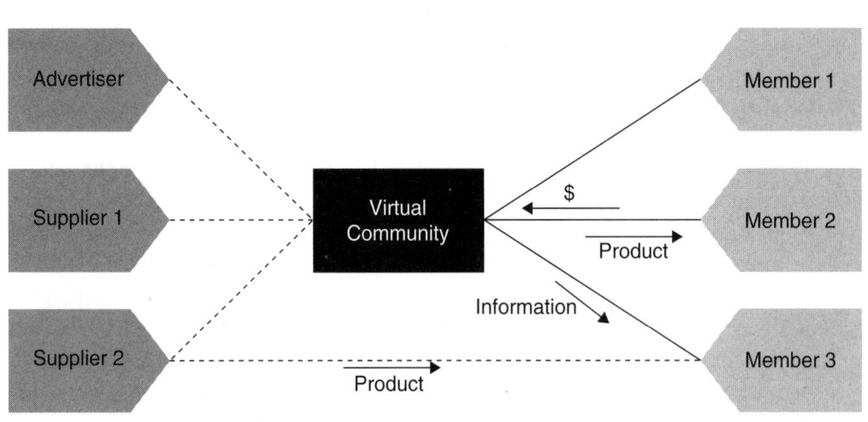

Figure 9.15 Virtual community business schematic

Semantic Technology Opportunities

- Although the critical success factors for virtual communities are less dependent on technical infrastructure concerns than other business model schematics, the reliance on partners and content suppliers presents opportunities to automate key content provisioning channels.
- Challenges with ASP (application service provision) integration for both back office infrastructure and community content provide opportunities to enable more flexible and on-demand type capabilities.

Generative Integration Infrastructure Pattern

Because the interconnectedness of highly complex software applications is not a core demand for the Virtual Community schematic, a less demanding infrastructure pattern is optimal. This type of infrastructure deployment would create highly automated connections between advertisers, suppliers, and the community—further enabling new revenue streams and maximizing efficiencies surrounding the maintenance of those interfaces.

The generative integration capability would enable Virtual Community technical administrators to generate integration scripts for content based on a taxonomy or model-driven paradigm. These generative capabilities would reduce the number of employees required to set up and configure interfaces that would otherwise be hard-coded.

Capability Case Solution Examples

Additional details about capability cases and the applicable solution stories may be found in Chapter 10. The capability cases that apply to this business model and architecture pattern are the following:

- Testcar Configurator
- US Army Knowledge Asset Management Network

Value Net Integrator

The Value Net Integrator schematic describes the capability to manage the value-added product and information resources from a series of firms that all participate in the same value net. A value net integrator takes advantage of the separation between physical and virtual goods and services by gathering, synthesizing, aggregating, and disseminating information to both upstream and downstream suppliers. Although many value net integrators will still operate in both the physical and virtual worlds, purely virtual integrators can maintain fewer physical assets. Because information is the core asset for the value net integrator, semantic interoperability can provide dramatically heightened value to those firms that leverage its benefits.

Example Companies

- Cisco
- Coles Meyer
- 7/11 (Seven Eleven) Japan

Who Owns What?

	Customer Relationship	Customer Data	Customer Transaction
Value Net Integrator		✓	

Business Model Schematic

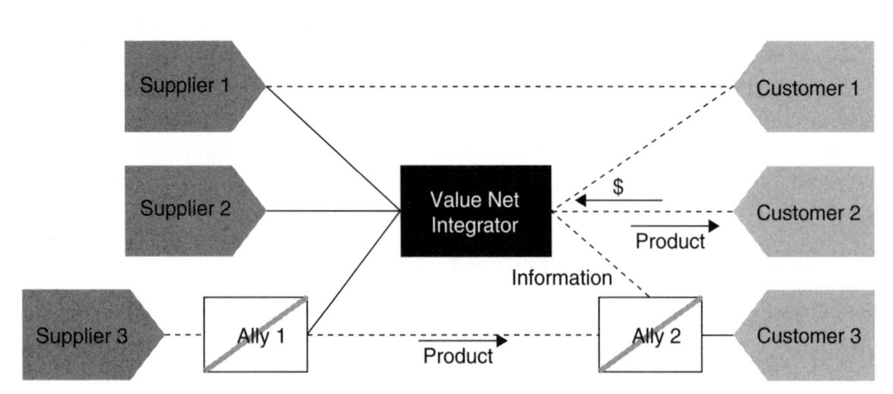

Figure 9.16 Value net integrator business schematic

Semantic Technology Opportunities

- Because the value net integrator strives to own the customer data, there is a great deal of opportunity to capture upstream and downstream customer information and synthesize it in innovative ways.
- The technology infrastructure required to gain access to the value chain is crucial if the integrator is to become a conduit for that customer data.
- Value net integrators earn their keep by adding value to information and then repurposing it to suppliers; this presents an opportunity to leverage a dynamic information infrastructure that can enable on-demand data configurations that are custom tailored to specific needs.
- The challenge of collecting, synthesizing, and distributing data from a diverse array of resources can challenge the IT infrastructure to keep pace, opening up opportunities to build in more efficient and less costly maintenance solutions.
- Discovering new facts about customers and inferring implicit buying trends and other relationships can be crucial to growing a customer's loyalty; the opportunity with semantic technologies is to provide advanced real-time capabilities for data analytics.

Dynamic Information Hub Infrastructure Pattern

A Dynamic Information Hub infrastructure enables value net integrators to achieve greater timeliness and quality for the information that is crucial for their success. Additionally, the information hub infrastructure has a low-cost maintenance model whereby the value net integrator can manage its information (local or distributed) in a model-driven environment that requires little hands-on code for repurposing information into new software systems.

In more complex value net scenarios, perhaps similar to Cisco's model, an Adaptive System of Systems infrastructure would be a better pattern choice. The high degree of collaboration required, coupled with the relative homogeneity of the Cisco extended supplier network, suggests that a centralized model-driven environment for both service connections and information could be quite successful at improving information access and driving down costs.

Capability Case Solution Examples

Additional details about capability cases and the applicable solution stories may be found in Chapter 10. The capability cases that apply to this business model and architecture pattern are the following:

- RightsCom Policy Engine
- FAA Passenger Threat Analyzer
- Cogito's Integrated Data Surveillance Solution
- Oceanic & Atmospheric Sciences Data Automation

Content Provider

Any business that generates and disseminates information, products, or services in digital formats to its customers via intermediaries is a content provider. Often the content provider will exploit channels similar to the Direct to Customer schematic, but the core business value proposition surrounds the content provider's ability to resell its content to third parties as value-added content. Content providers flourish in a number of domains including financial services, geographic and spatial data, weather, news, and advertising demographic data.

Example Companies

- Weather.com
- Morningstar
- Reuters
- MapQuest

Who Owns What?

	Customer Relationship	Customer Data	Customer Transaction
Content Provider			

Business Model Schematic

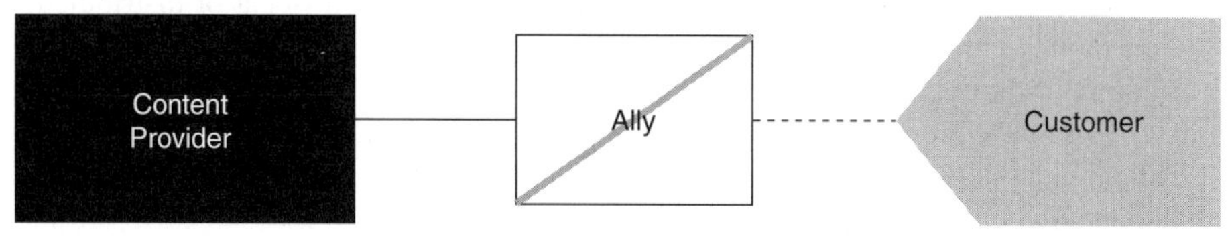

Figure 9.17 Content provider business schematic

Semantic Technology Opportunities

- Because the focus of the content provider is to deliver digital content in flexible and tailored ways to its intermediaries, the opportunity to restructure its information infrastructure in more dynamic ways is quite significant.
- Opportunities to impact digital information include:
  - Modularized information and content
  - Flexible storage and retrieval of digital content
  - Combination, aggregatation, and synthesis of digital information
  - Delivery of content in the proper context and views

Dynamic Information Hub

The content provider can use the Dynamic Information Hub pattern as a way to configure the delivery of content and information on an on-demand basis. By leveraging ontology and context-sensitive mappings to provide essential metadata to run time components, the infrastructure can tailor outbound content to the specific tastes and formats of the intermediaries. This capability will drive down the costs and lag associated with bringing on new partners, intermediaries, and customers.

It is clear that content management systems are insufficient for the growing demands placed on businesses that earn a living on digital content. The sheer volume of content coupled with the demand for flexible presentation and delivery channels require highly customized IT infrastructures, which can be costly to maintain. The shift toward semantic technologies and the Dynamic Information Hub infrastructure pattern will alleviate these pains and enable content-driven businesses to scale well beyond what has been possible in the recent past.

Capability Case Solution Examples

Additional details about capability cases and the applicable solution stories may be found in Chapter 10. The capability cases that apply to this business model and architecture pattern are the following:

- McDonald-Bradley Annotation of DoD Documents
- MIND SWAP RDF Editor
- Children's Hospital Lab Data Mediation

FINAL THOUGHTS ON INFRASTRUCTURE PATTERNS AND BUSINESS SCHEMATICS

You may or may not agree that these infrastructure patterns are indeed patterns. However, the intention was not to establish a new classification system to debate in academic circles. Rather, these patterns are intended to help architects make decisions about how to approach building an adaptive IT infrastructures and to offer some guidance on where to seek further design practices.

Because the semantic technology space is moving so rapidly, it should be obvious that a deep technical treatment would have a short-lived utility. The patterns presented here, and the E-Business schematics with which they are associated, are specifically held to be general—not specific—to allow for future adaptation to new technologies. By synthesizing the semantic infrastructure choices into four key archetypes (monitor policies, generate code, mediate conflict, and govern complex systems) the underlying technology can change without impacting the patterns. Likewise, the associations with the Place to Space E-Business schematics should provide some interesting decision criteria on into the future.

Part Three

Adopting Semantic Technologies

The third part of this book is intended to dispel any myths that these semantic interoperability technologies are only science projects. Chapter 10 is a comprehensive look at capabilities these technologies offer to businesses. Much attention was given to the case studies aspect of this chapter, ensuring that solution stories of actual implementations are provided. These capability cases and solution stories should provide a basis on which to analyze how companies are using the technology and what successes they have had thus far on the adoption path.

Chapter 11 is aimed toward the manager who wants to know how to go about implementing semantics-based technology. The chapter begins with an exploratory look at the solution-envisioning approach for identifying project goals and next goes through a range of management and staffing concerns such as skills assessments, staffing, and rollout planning. Finally, the chapter provides a review of available pragmatic methodologies for the ontology engineering life cycle and steps the reader through many adoption challenges faced at the design and implementation phases of a project.

Chapter 12 concludes the book by offering a synthesis of many of the main points made throughout the book. It also provides a forward-looking element to assist the reader in making informed judgments about what advances can be expected in the years following this book's publication. As part of this "futures" discussion, several leading-edge research programs will be discussed in light of the goals and timelines they have identified for themselves.

Adaptive Information, by Jeffrey T. Pollock and Ralph Hodgson
ISBN 0-471-48854-2

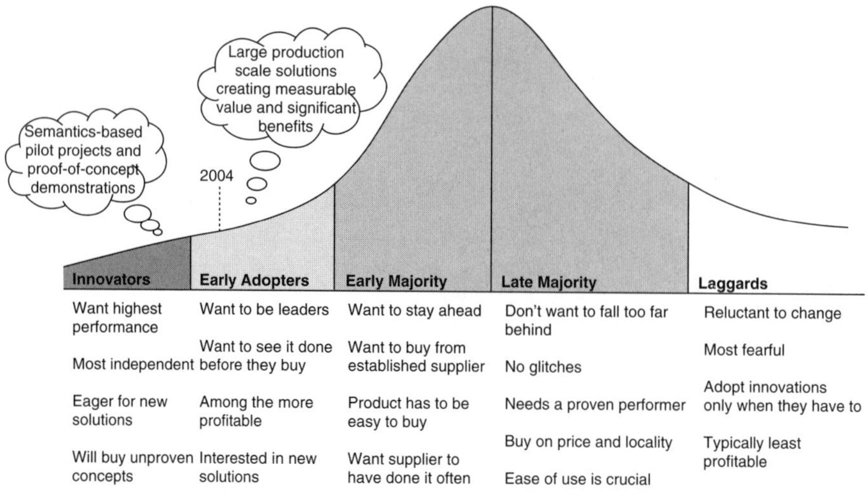

Large production scale solutions creating measurable value and significant benefits

Semantics-based pilot projects and proof-of-concept demonstrations

2004

Innovators	Early Adopters	Early Majority	Late Majority	Laggards
Want highest performance	Want to be leaders	Want to stay ahead	Don't want to fall too far behind	Reluctant to change
Most independent	Want to see it done before they buy	Want to buy from established supplier	No glitches	Most fearful
Eager for new solutions	Among the more profitable	Product has to be easy to buy	Needs a proven performer	Adopt innovations only when they have to
Will buy unproven concepts	Interested in new solutions	Want supplier to have done it often	Buy on price and locality	Typically least profitable
			Ease of use is crucial	

Chapter 10

Capability Case Studies

KEY TAKEAWAYS

- Interoperability is the most critical problem that the industry must solve to realize the benefits of adaptive enterprises that are capable of responding to changing market conditions and opportunities.
- Semantic Technology is showing great promise in a number of interoperability problem areas ranging from application, data, services, policy, and rules to workplace integration.
- Standards for ontology-based engineering are now ready for adoption.
- Web Services need semantics to realize their full potential of automatic service discovery and services composition.

This chapter gives an industry perspective on the growing use of Semantic Technologies in a wide range of applications, including Knowledge Management, Workplace Portals, Database Integration, Enterprise Application Integration (EAI), and B2B eCommerce. We review where semantic technology is being adopted and what is motivating this adoption.

Throughout the chapter case studies are used to illustrate applications of Semantic Technology. Capability Cases[1] are used to contextualize case studies and to explain solution ideas that either have been or could be implemented with Semantic Technologies.

INTRODUCING CAPABILITY CASES

A Capability Case is the *Case for a Capability*. By "capability" we mean the potential to deliver business functionality. Case is different from the Use Case meaning of the word "case." It is borrowed from two places. In the Case-Based Reasoning world the word "case" has a long-established meaning of defining a "problem-solution" pair for reuse and adaptation in new problem contexts.

[1] Irene Polikoff, Ralph Hodgson, and Robert F. Coyne, *Capability Cases*, to be published by Addison-Wesley. TopQuadrant's website also provides more information on Capability Cases, see http://www.topquadrant.com/diamond/tq_capability_case.htm.

Adaptive Information, by Jeffrey T. Pollock and Ralph Hodgson
ISBN 0-471-48854-2

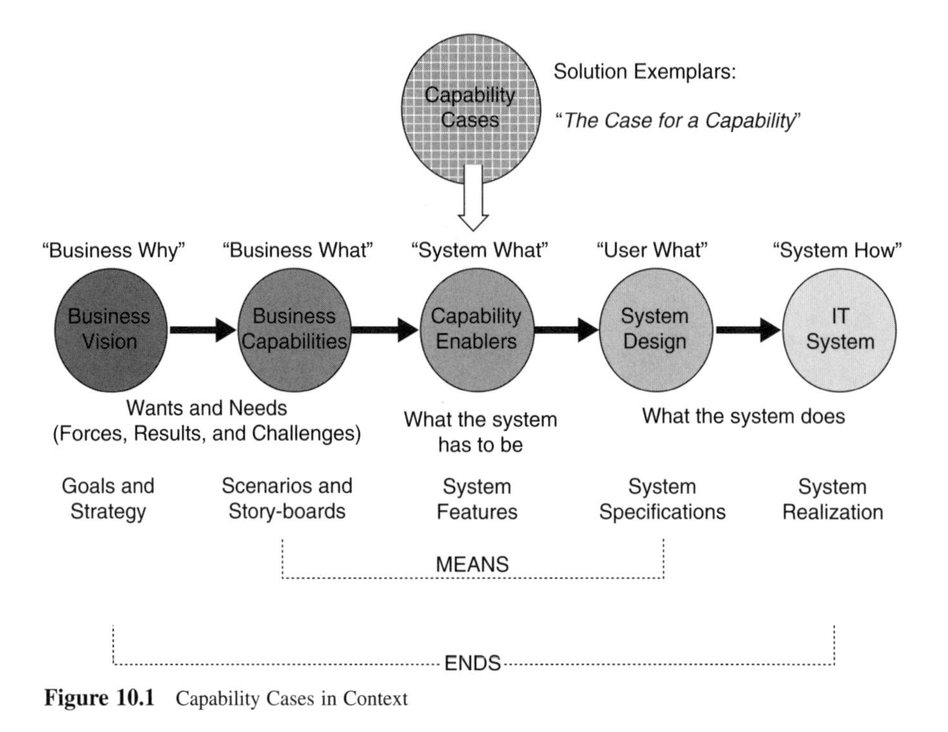

Figure 10.1 Capability Cases in Context

Capability Cases are reusable solution concepts. In the business world "case" has a well-established meaning of making a business justification for spending effort or money.

Why Capability Cases?

The communications gap between business and technical people accounts for some of the biggest failures that can occur in realizing IT solutions for business. This gap arises because of the different languages and vocabularies spoken by business strategists, solution users, IT, and technology evangelists. To bridge this gap, we introduce "Capability Cases." By using the languages of business as well as technical communities, a Capability Case provides a way to connect solution concepts to the business situation by using a *Business Problem—Technology Solution* pattern.

Even for the established, well-adopted technologies it is difficult for decision makers to determine the business value of a technology in a particular situation simply through an explanation of the technology. It is necessary to see the technology in a context in which it is providing business value. This need becomes even stronger with an emerging technology that introduces new capabilities and new paradigms. Because Capability Cases convey the business value of technology in a real work setting, they provide an excellent tool to satisfy this need. Each Capability Case demonstrates the value of Semantic Technology through one or more stories.

The Language of Capability Cases

> *A Capability Case expresses an IT solution pattern in a business context with stories of successful use, applicable technologies, and leading practices—all connected to business forces and results.*

A *force* is a business driver, internal or external to the business that requires a response. By force we mean a new or existing condition that is affecting the business. A force can be an issue, where the business is failing in some way, or an opportunity to realize new benefits. Forces can arise from a number of different sources that we categorize as Regulatory, Customers, Enterprise, Marketplace, and Technology. Forces can be constant or dynamic. For example, existing government regulation can be considered a stable force. On the other hand, *increasing availability of wireless infrastructure* represents a dynamic force that is on the increase. Such dynamic forces can be thought of as trends.

A *challenge* is a predicament that the business is experiencing. Challenges are often revealed by the pain that users and stakeholders are experiencing. We state a corresponding challenge in the form of a "How to" statement, for example, "How can we speed up order processing?" Sometimes it is hard to distinguish between a challenge and a force. Often forces could be expressed as challenges. The main difference between forces and challenges is that forces originate outside of a business system and challenges are their effects—what is inflicted upon the business system.

A *business result* is a statement of a desired outcome—something the business wants to accomplish. A result is a change in state of some aspect of the business or in the impact of the business in its environment. Capability Cases use Kaplan and Norton's Balance Scorecard framework[2] for distinguishing four categories of results: Financial, Customer, Internal, and Learning and Growth. Like Balanced Scorecard it places a strong emphasis on measurable results. In our work, however, we also adopt Neely, Adams, and Kennerley's "Performance Prism"[3] support for multiple stakeholder groups: end users, employees, suppliers, regulators, pressure groups, and local communities.

A Capability Case contextualizes technology-enabled capabilities by identifying forces, challenges, and business results associated with each capability.

What is the Relation Between Capability Cases and Use Cases?

As stated earlier, a Capability Case is the *Case for a Capability*, with "case" used in the sense of making a business case through the stories and business contexts they embody. A Use Case is a *Case of Use*.

[2] Robert S. Kaplan and David P. Norton, *The Balanced Scorecard: Translating Strategy into Action*, ISBN: 0875846513, Harvard Business School Press, 1996.

[3] Andy Neely, et al., *The Performance Prism—the Scorecard for Measuring and Managing Business Success*, FT, Prentice-Hall, Pearson Education, ISBN 0-27365334-2, 2002.

In software development Use Cases have become the standard way of expressing requirements for how a system should work. They explore the behavior of a system from the user's goal—what the user is intending to accomplish. Use Cases express the user's intent and the flow of interactions at a system boundary for both the successful cases of use and the failure cases.

In the early stages of the life cycle, Use Cases are simply listed with their intents. It is not until later in the life cycle, when the nature of the system is known, that more detailed elaboration of a Use Case, explaining how things work, can be described. What the "system should be" in order to respond to business forces and challenges and to deliver desired results is not a question that Use Cases are designed to answer. This question needs a concept-driven approach. Capability Cases address the need to present solution ideas earlier in the life cycle so that the concepts of the solution can be explored from a large space of possibilities. As such, they serve as "educators" and "catalysts" for sparking ideas.

The term "Use Case" has also been coopted by the technology vendor community to mean something different from its use in software requirements, design, and development. It is used to describe exemplar application areas for a technology. For example, a vendor may identify E-commerce as a "Use Case" for its technology. Sometimes, this is also called "usage scenario," which in our opinion is a more appropriate and less confusing name. Unlike "traditional" Use Cases that use structured templates specifying actors and normal and abnormal flows of events, usage scenarios are written in a narrative format.

Capability Cases have an intent similar to that of usage scenarios in that they convey the essential ideas of a system. Unlike free-form usage scenarios, they position the system in the business context by connecting the story to business drivers, challenges, and results. They also serve as a standard way to identify and describe capabilities or common building blocks for the solution. In this way, stories serve as a way to further explain and illustrate capabilities.

The Capability Case Template

A Capability Case is expressed with a template that captures the commonalities of one or more solution stories. The template for an elaborated Capability Case is detailed in Table 10.1.

When a Capability Case is first identified it is described, at a minimum, by an intent statement, one solution story, and a summary of benefits and technologies. If the capability is of interest to the business, it is elaborated to include more information on business situation and implementation specifics. It may also include additional solution stories. In the Case Studies that follow we use only the outline template.

Why Use Capability Cases for Case Studies?

Human beings understand the world through stories. We relate to one another's experiences through the themes and images that a good story conveys.

Table 10.1 Data Table—Capability Case Template

	Capability Case Template
Description	An explanation of how the capability works. Such explanations can range from a short statement to an extended overview.
Vintage	The maturity of the capability. Vintage can be "conceptual," "research prototype," "early commercialization," "mature commercialization," or "general industry adoption."
Challenges	The business challenges that could be overcome.
Forces	Business forces that indicate the need for the capability. By business forces we mean any new or existing condition that is affecting the business.
Results	Results business wants to accomplish. For each key result we identify measurements that could be used to assess effectiveness of the capability toward achieving it.
Best Practices and Lessons Learned	Typical use scenarios for the capability, implementation guidance, and some information on obstacles, technical or organizational, that an enterprise may need to overcome to successfully deploy the capability.
Applicable Technologies	A list of the technologies that can be used to realize the capability. For each technology, it may be necessary to list separate implementation considerations.
Implementation Effort	An order of magnitude estimate of implementation costs and complexity to help in prioritization and decision making.
Integration	Capabilities are building blocks for business solutions. Two perspectives of integration are of importance: mechanism and style.
Integration Mechanism	Mechanisms define ways in which a capability are invoked as a part of the overall solution. The following 4 mechanisms are distinguished: 1. **UI Integration**—The simplest form of integration. Examples include making a capability accessible through a link on the Web or as a "portlet" in a portal. 2. **Task-centric Integration**—One example of this mechanism is an Instant Helper screen with a "Can I help you?" message popping up when a user hesitates at a check out in the e-shop. 3. **Data-centric Integration**—When data is shared, aggregated, or exchanged. 4. **Process-centric Integration**—When a capability is triggered by events in a process or has to generate events for other capabilities or processes.
Integration Style	Styles represent different architectural strategies for integrating capabilities. The following distinctions are made: 1. **Proprietary Plug-ins**—Used when integrating additional capabilities into large-grained functional components that offer some degree of openness through proprietary APIs. The style could operate on multiple levels.

Table 10.1 Continued

	Capability Case Template
	2. Cooperative Applications—An example is a cooperation of MS Word and Groove, where each serves a clearly separate function and operates independently. At the same time, a MS Word document could be stored and version controlled by Groove. Another example is two custom business applications that serve separate business functions where one sends a weekly data extract to another. **3. Common Integration Framework**—Requires a set of agreed protocols used by different applications to communicate. Examples include various Enterprise Application Integration (EAI) platforms, Service-Oriented Architectures (SOA) for Web Services, COM/DCOM, J2EE, and CORBA platform and architecture frameworks. Interoperability is achieved through an application profile.

In an organization, stories catalyze change and enable the organization to reinvent itself. Steve Denning calls these kinds of stories "springboard stories"[4]:

> *A springboard story has an impact not so much through transferring large amounts of information, as through catalyzing understanding. It can enable listeners to visualize from a story in one context what is involved in a large scale transformation in an analogous context. . . . In effect, it invites them to see analogies from their own backgrounds, their own contexts, their own fields of expertise.*

A well-written Case Study has the same power. When a Capability Case is expressed with business forces and results, it makes the case for a solution idea in a compelling and substantive manner. The evidence is there—in the forces and measured results. The ability to communicate is there—in the independent stories of adoption.

The principle of looking for commonality across independent cases of solutions was well established by the Design Patterns community. When a number of independent stories line up and suggest a common solution, we have the basis for a Capability Case. The Capability Case is named so that it conveys the intent of the general idea.

In summary, a Capability Case:

- Facilitates a shared understanding of forces, challenges, and desired results
- Expands the space of solution possibilities through the use of Capability Cases as "innovation catalysts"
- Enables business and IT to do joint creative work toward a shared vision

[4] Stephen Denning, *The Springboard: How Storytelling Ignites Action in Knowledge-Era Organizations*, Butterworth-Heinemann, October 2000.

- Builds confidence and commitment toward implementing the shared vision
- Helps communicate a solution vision to other parties

Semantic Technology Capability Cases Summary

Across a number of industries some clear trends can be seen in the areas in which semantic technology is being used to solve interoperability problems. These are:

- Application interoperability
- Data interoperability
- Services interoperability
- Process interoperability
- Policy and rights interoperability

In all of these application areas there is a need to represent meaning (semantics) to allow information to be exchanged, reconciled, or, in general, "to be made sense of" for specific purposes. In the sections that follow examples of Capability Cases will be used to illustrate the role and business value of semantic technologies.

For many of these Capability Cases, there is a common set of forces and results. Examples of forces are listed in Table 10.2.

Table 10.2 Data Table—Semantic technology capability case driving forces

Perspective	Force
Customer	• Customers expect real-time access to business processes inside your organization and your supplier's organization. • Customers want to use multiple channels to interact with an organization including web, telephone, mail, and face to face.
Marketplace	• Expectation that responses can be immediate • Increasing glut of data and information • Multiple supply chain partners
Enterprise	• Conflicts in data within legacy systems • Growth and variety of departmental and related information • Incompatible technology infrastructure within a single organization • Knowledge in one or more systems, some external to the organization • Rapid pace of business change
Technology	• Growing number of disparate information repositories including unstructured (HTML pages, Word documents, etc.), semistructured (XML, RDF), structured (databases) • Point-to-point custom integration is hard to maintain • Traditional EAI solutions are proprietary and complex and require substantial upfront investment • Emergence of XML standards • Web has moved from a browsable media to a transaction processing platform

Table 10.3 Data Table—Semantic technology capability case business results

Balanced Scorecard Perspective	Business Result
Financial	• Lower operational costs
Customer	• Consistent interactions with customer through multiple channels • Faster response to customer inquiries
Internal	• Adherence to legal and regulatory statutes • Better-informed business decisions • Flexible IT infrastructure—ease of integration and modification • Improved information flow between individuals, departments, and systems • Increased efficiency and data integrity through automation • Single point of retrieval for information from multiple sources
Learning and Growth	• Increased confidence in the fidelity of metadata descriptions to actual databases and applications for data integrity, impact, and application dependency analysis • Increased sharing of knowledge across different parts of the organization

Examples of results, categorized according to the perspectives of the Balanced Scorecard, are listed in Table 10.3.

The Capability Cases and Solution Stories are summarized in Table 10.4.

APPLICATION INTEROPERABILITY

We know of a bank with over 40 different call center systems, a financial services company with more than 1000 databases, and a manufacturing company with over 2000 CAD/CAM systems. These systems contain valuable information and often are still good for supporting the specific tasks they were intended for. Unfortunately, the information they contain can not be leveraged by other systems without a considerable effort. When changes in business needs or available technology require modifications to these applications to provide additional capabilities and to streamline work flows, integration and extension become a very expensive undertaking. Simply tracking all the enterprise data sources and their relationship to each other is proving to be a challenge. In fact, many IT organizations spend up to 80% of their budgets maintaining legacy systems, leaving limited funds to support new business opportunities or to satisfy new regulatory requirements.

The next Capability Cases will show examples of new ways to integrate applications using semantic models. They describe the knowledge that the users of the systems need access to, rather than the data that implements that knowledge. We can envision future applications composed of very thin components that dynamically change their behavior based on the interactions with the business knowledge embedded in the model.

Table 10.4 Data Table—Summary of capability cases and solution stories

Category	Capability Case	Solution Story	Vendor/Technology
Application Interoperability	Semantic Application Integrator	Cargo Airlines System Integration	Network Inference/ OWL
		US Army Semantic Integration	Modulant
	Product Design Advisor	Test Car Configurator	Ontoprise/ F-Logic
Data Interoperability	Content Annotator	AeroSWARM Automated Markup	Lockheed Martin DAML
		MINDlab RDF Editor	RDF
		McDonald-Bradley Annotation of DoD Documents	RDF
	Semantic Data Integrator	FAA Air Passenger Threat Analyzer	Semagix
		Cogito Data Fusion	Cogito
		Major Electronics Manufacturer Data Interoperability	Unicorn Solutions
		Children's Hospital Lab Data Mediation	RDF
		Consulting Services Data Quality	Unicorn Solutions
		US Air Force Information Management	Modulant
	Semantic Data Acquirer	Oceanic and Atmospheric Sciences Data Automation	Thetus Publisher
	Semantic Multimedia Integrator	CoAKTinG Distributed Meetings Manager	AKT (UK)/ RDF
	Semantic Content Registry	European Environment Agency Content Registry	EEA/ RDF
Services Interoperability	Task Composer	STEER—a User-Centered Environment for Task Computing	Fujitsui Labs/ OWL-S
	Process-Aware Web Services Match-Maker	Service Discovery using Process Ontology Queries	MIT/ OWL-S
Process Interoperability	Process Aggregator	Ford e-Hub	XML
Policy/Rules Interoperability	Rights Mediator	RightsCom Policy Engine	Network Inference/ OWL

Capability Case: Semantic Application Integrator

To enable fast, effective, and flexible semantic integration of multiple systems using a separate semantic layer that abstracts the enterprise business model from the application and data layers. The underlying model is used to interpret meaning and transform content of messages exchanged by multiple legacy applications. It is directly reusable in a number of disparate application architectures and business contexts.

Key Benefits

- Substantially reduced project setup and maintenance costs
- High scalability for low-cost integration of additional applications
- Flexibility and control of changes in systems and interfaces

Solution Story: Cargo Airlines System Integration

A typical cargo airline operates a number of related systems and associated data stores, such as Capacity, Schedule and Inventory, Rating, Operations, and Finance and Accounting. Internal systems, partner, customer, and third-party systems need to be integrated. Typically, each uses differing terminology and interprets interactions differently.

With Network Inference's Cerebra[5] all of these interfaces are defined and driven by a central model that enables integration based on both the relationship between fields and field content. With inferencing, the number of links needed to add new interfaces is linear rather than exponential. Complex transformations are supported, where the content of multiple fields in one system is related to a single field in another. For example, different ULD typologies combine deck and type into a single typology; others express classification and deck separately.

At the field level, Cerebra Construct allows "ATA Container Type" to be mapped to "IATA Container Type." "IATA Container Type" can in turn be mapped to the composite concept comprising "Type" and "Deck." Through inference, "ATA Container Type" is now mapped to "Type" and "Deck." "ULD" can then be mapped to "IATA Container Type." Cerebra automatically creates the links between all four typologies.

Solution Story: US Army Knowledge Asset Management Network

After searching through a variety of systems and hours of manual work to gather data, staff members at the US Army Aviation and Missile Command (AMCOM) concluded that flying the UH-60 Blackhawk helicopter would be easier than pulling together information about it. Engineering drawings were kept on one proprietary system and configuration data on another, yet users throughout the organization frequently needed simultaneous views of both. Much of this data could not be accessed with PCs, and Web-based access was out of the question.

AMCOM required the ability to integrate multiple legacy systems and off-the-shelf software packages so that they could find, correlate, and use Blackhawk information across all systems. Using the webMethods integration platform[6] for the

[5] For more information, see: www.networkinference.com/Products/Cerebra_Server.html.

[6] For more information, see: www.webmethods.com/cmb/solutions/integration_platform/.

underlying integration backbone, AMCOM developed the Knowledge Asset Management Network (KAMNET), a system that provides access to engineering, technical, and programmatic data for the Blackhawk weapon system.[7] The solution encapsulates the business rules unique to access, viewing, and configuration management of Blackhawk engineering drawings. Future development plans included incorporating all available and pertinent information on each Blackhawk helicopter, including maintenance work orders, hours logged on major components, serial numbers, and avionics packages. As a result, AMCOM can determine the configuration of any aircraft, anywhere in the world, at any time, from a Web-based system.

Capability Case: Product Design Advisor

A 'Product Design Advisor' supports innovative product development and design by bringing engineering knowledge from many disparate sources to bear at appropriate points in the process. Possible enhancements to the design process that result include rapid evaluation, increased adherence to best practices, and more systematic treatment of design constraints.

Solution Story: Test Car Configurator

At a large German car company, manufacturing engineers design, build, and test new prototypes as part of the innovation process. The faster this cycle can be completed, the greater the number of innovations that can be brought to market, and the sooner. The company is evaluating semantic technologies provided by Ontoprise to represent complex design knowledge in electronic form.[8] Ontoprise technology brings together knowledge from many different sources and draws logical conclusions from the combined information. The capability provides a computational representation of complex dependencies between components of research test vehicles. These dependencies play a key role in the configuration and development of new vehicles. For example, for testing to proceed smoothly, the engineer must know whether a selected engine can be built into the chosen chassis, whether the brakes are sufficient for the engine performance, or whether correct electronics is present in the vehicle. "*We expect a shortening of the development cycle, while at the same time improving development quality*," said a representative of the company—"*The electronic advisor shall take care of routine tasks, allowing our engineers to concentrate on creative efforts*."

[7] For more information, see:
www.webmethods.com/content_cmb/1,2681,SuccessDetailcmb_82,FF.html.

[8] Andreas Maier, Hans-Peter Schnurr, and York Sure, *Ontology-Based Information Integration in the Automotive Industry*, The Semantic Web, ISWC 2003 Proceedings, D. Fensel et al (Eds), pp. 897–912, Springer-Verlag Berlin, Heidelberg, 2003.

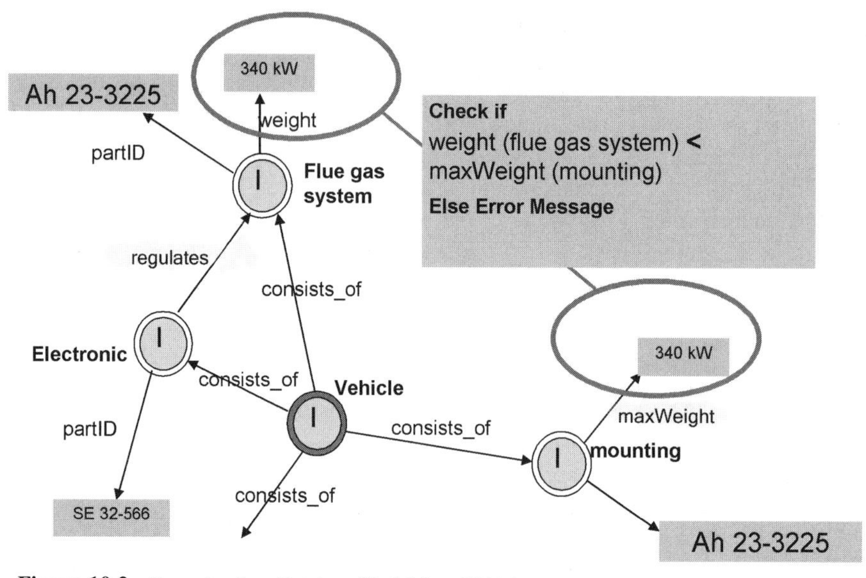

Figure 10.2 Example of an Ontology Model for a Vehicle

The knowledge model holds rules about constraints on how parts should be chosen for assembling particular engines. An example of the rules that can be supported is as follows:

> *The tolerated transmission power of a car's crank N_{crank} must be higher than the power of the motor N_{motor}, but must not exceed that of the gear N_{gear}.*

DATA INTEROPERABILITY

The most common solution to data integration and translation remains field-to-field mapping, where schemas from two data sources are imported and fields are mapped to each other. Rules can be defined to split or concatenate fields or to perform other simple transformations. Once this is done the integration tool can do data translations either directly at run time or by generating code that will perform the transformations. There are a number of tools on the market that support this approach. Major vendors include IBM and Microsoft. Some of the tools have been available for nearly a decade, but the adoption has been slow for a number of reasons:

- Field-to-field mapping works on a small scale. However, the number of maps grows exponentially with each new data source. Maintenance and evolution become a problem because any change in the schema of one data source will require you to redo multiple maps.
- Enterprises working with this technology often discover that creating correct maps is a challenge. It requires that the person responsible for each mapping

have an in-depth knowledge of both data sources, which is rarely possible. As a consequence, mapping mistakes are quite common.

- Mapping and translating between two schemas that are using different design paradigms (i.e., different degrees of normalization or nesting) can be very difficult. There is more then one way to design a schema. Performance considerations may result in denormalized database schemas. When schemas are expected to change, the designer may opt for a reflective design. Some XML schemas are deeply nested; others are shallow. Mapping between relational (RDBMS) and hierarchical (XML) stores can suffer from significant impedance mismatch of the models.
- Direct mapping may fail in situations requiring more conceptual and conditional transformations.

Many companies have been moving to XML to take advantage of standards-based integration. However, XML doesn't capture the contextual meaning (or semantics) of the data, and a growing number of "standard" XML dialects (currently over 400) intended to standardize business vocabularies make the need for a semantic translation layer even more apparent.

The next Capability Cases will show examples of new ways to integrate data. Typically, before making mappings, an ontology of a given business domain is defined. This can be "jump-started" by importing data schemas. The model contains business concepts, their relationships, and a set of rules, all expressed in a knowledge representation language.

The attraction of logic as a technology for supporting semantic integration stems from the capability of logical languages to express relationships in generic ways and the availability of sophisticated automated systems for finding combinations of related items that satisfy certain constraints. The variants of logic used for semantic integration [including Horn logic (prolog), frame logic, and description logic] differ primarily in the expressiveness of the logic and the tractability of the reasoning system.

Capability Case: Content Annotator

Considerable interest exists in creating metadata to assist organizing and retrieving information. The precision of search and context-based information retrieval and repurposing is greatly improved through accurate metadata. Content annotation is becoming increasingly important to assist the work of annotating information resources.

Content Annotator provides a way for people to add annotations to electronic content. Annotations can be comments, notes, explanations, and semantic tags. Three solution stories are provided: AeroText *S*emantic *W*eb *A*utomated *R*elation *M*arkup (AeroSWARM), MINDlab's SWAP RDF Editor, and McDonald-Bradley Annotation of DOD Documents.

Solution Story: AeroSWARM Automated Markup

The creation of markup from unstructured text sources such as web pages is tedious and time-consuming. Anyone who produces documents on a regular basis (e.g., intelligence analyst, commander) or has a large quantity of legacy documents needs some form of automated markup assistance. The Lockheed Martin DAML team has experimented with the application of information extraction technology to reduce the effort required for markup. They have built a tool called AeroSWARM, which automatically generates OWL markup for a number of common domain-independent classes and properties. The author can then manually do markup additions and corrections to the output of AeroSWARM. The processing of raw text is very difficult but sufficient levels of precision and recall are being attained to make this automated assistance approach worthwhile. AeroSWARM can also be customized for domain-specific markup generation.

A user can specify the set of web pages to markup, choose a target ontology and, then, AeroSWARM generates OWL markup like that shown in Figure 10.3. The sample markup includes entities (e.g., person, place, organization), relations (e.g., Pinochet persToLoc Santiago) and co-references (e.g., Pinochet sameIndividualAs Augusto Pinochet). A table on the AeroSWARM site describes all the entities and relations that are automatically identified and marked-up.

There are a number of advanced features that AeroSWARM supports. If the markup creator does not want to use the native AeroSWARM ontology as a target then she can choose a popular upper ontology (e.g., OpenCyc or IEEE SUMO which have predefined mappings) or AeroSWARM provides a drag-and-drop tool to create ontology mappings for user specified ontologies. The user can also semantically check the generated markup against constraints specified in an OWL ontology (e.g., only one person can be the biological mother of a person) by sending the output to

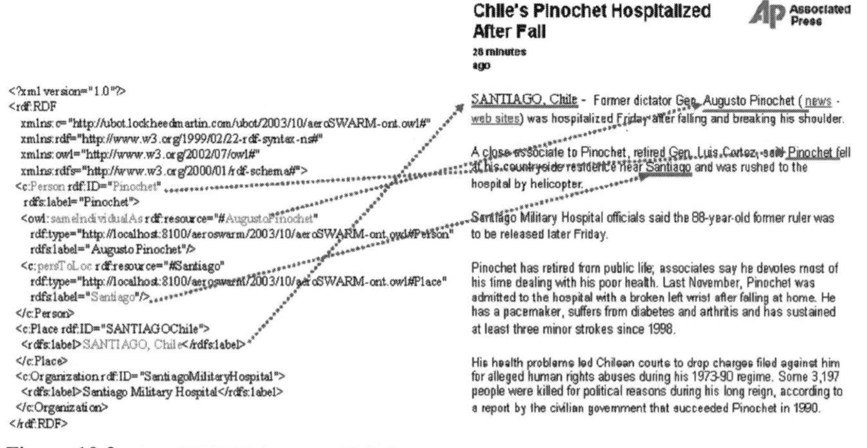

Figure 10.3 AeroSWARM Automated Markup

a "Consistency Reasoning Service" which uses the ConsVISor tool from Versatile Information Systems Inc. AeroSWARM is a Semantic Web service, which can be registered with a matchmaker and discovered and invoked by external tools or applications.

Although AeroDAML does not successfully tag all the subtle relationships in a text, it catches a large number of the simplest ones, reducing the workload for the remaining items to a level manageable by human effort.

Solution Story: MINDlab SWAP RDF Editor

The RDF editor from the University of Maryland's MINDlab[9] MINDSWP Group provides users with the ability to create Semantic Web markup, using information from multiple ontologies, while they simultaneously create HTML documents. The aims of this software are:

- Provide users with a flexible environment in which personal web pages can easily be created without markup hindrances.
- Allow users to semantically classify data sets for annotation and generate markup with minimal knowledge of RDF terms and syntax.
- Provide a reference to existing ontologies on the Internet in order to use more precise references in the user's own web page/text.
- Ensure accurate and complete RDF markup with the ability to make modifications easily.
- Allow users to extend ontological concepts.

To achieve these ends, the application has three functional parts: an HTML Editor, an Ontology Browser, and a Semantic Data Trees Classifier.

The HTML Editor with Preview Browser provides a standard WYSIWYG editor for creating and deploying web pages. The Ontology Browser helps users to work with multiple ontologies. Although many existing tools allow users to create their own ontologies for use in RDF documents, this tool encourages users to work with and extend preexisting ontologies, exploiting the distributed nature of the Semantic Web. The tool allows the user to browse through existing ontologies on the Internet with the aim of finding relevant terms and properties.

The default starting page is the DAML Ontology website, where a user can issue search queries using Class/Property names as keys. Once an appropriate ontology has been found, the user can add it to the local database. The properties of the ontology are automatically added to the Local Ontology Information.

The Semantic Data Trees part of the interface is what allows users to classify the data semantically into one of four basic elements: Class, Object, Property, and Value.

[9] MINDSWAP, the Semantic Web Research Group, is a group of people working with Semantic Web technology at the MIND LAB, the University of Maryland Institute for Advanced Computer Studies. MINDSWAP stands for "Maryland Information and Network Dynamics Lab Semantic Web Agents Project". More information on MINDSWAP can be found at http://www.mindswap.org/.

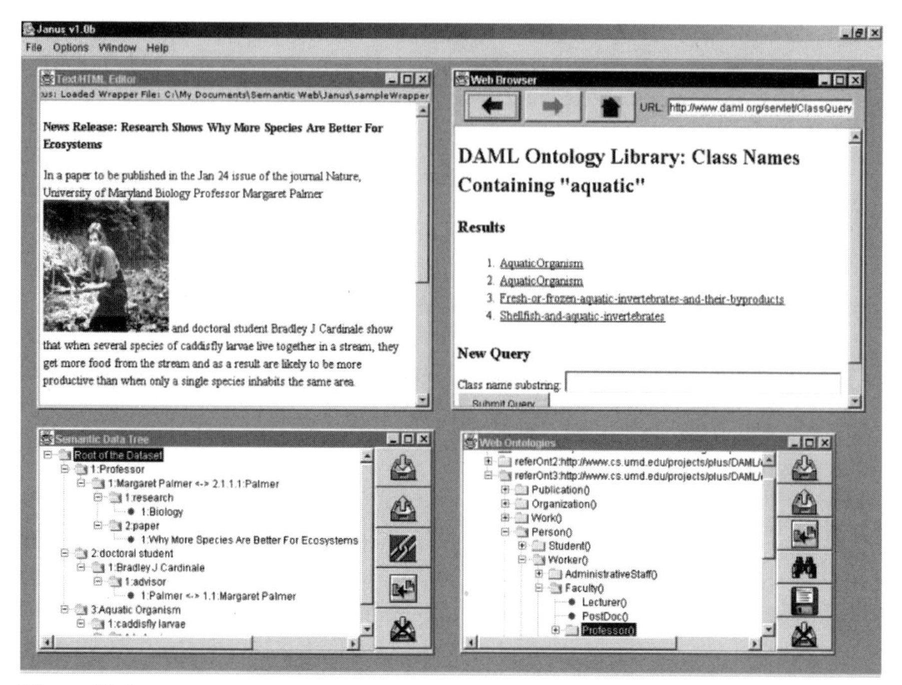

Figure 10.4 The MINDlab SWAP RDF Editor Interface

As users select classes from ontologies, the MINDlab Semantic Web Portal returns results in a separate window. The fetched data is immediately available for reference or incorporation to the current document. When a user publishes a document, the portal can include all of the new references in its knowledge base.

Solution Story: McDonald-Bradley Annotation of DoD Documents

McDonald-Bradley provides an RDF-based annotation for DoD documents. An example is shown in Figure 10.5. By clicking on the pencil icon, the user can bring up an annotation field and add comments to the text. Annotations expand, clarify, refute, or add a link to a portion of a document. This adds an additional dimension to the data source by allowing additional perspectives.

Capability Case: Semantic Data Integrator

Systems developed in different work practice settings have different semantic structures for their data. This makes time-critical data access difficult. Semantic Data Integration allows data to be shared and understood across a variety of settings.

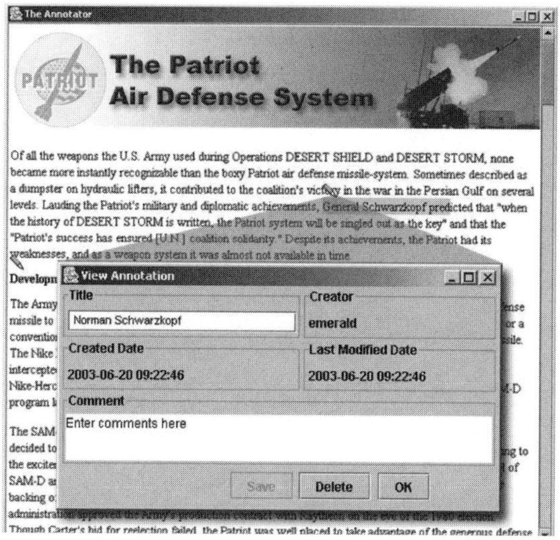

Figure 10.5 RDF Annotation of DoD Documents

A knowledge model is mapped to fields in databases, XML Schema elements, or operations, such as SQL queries or sets of screen interactions. Once the data sources are mapped to the model it can be used as an enterprise data management tool and to transform and validate data at design or run time. This approach solves many maintenance, evolution, and schema compatibility problems.

Solution Story: FAA Air Passenger Threat Analyzer

Data for passenger threat analysis comes from a wide range of heterogeneous, structured, and unstructured sources, including the FBI's Most Wanted List, flight details, news, public records, and biometrics. Airline authorities need to quickly aggregate and co-relate this information for important decisions such as adding a marshal to a flight, searching luggage, or requiring the brief interrogation of a particular passenger before letting him or her advance to the gate.

A robust solution to aviation security must address the following types of requirements:

- Analysis of government watch lists containing publicly declared "bad" persons and organizations
- Security applications for the sequence of kiosks at the airport departure location
- Aggregation and intelligent analysis/inference of valuable information from multiple sources to provide valuable and actionable insight into identifying high-risk passengers

Figure 10.6 Actionable information related to passenger profile[12]

- A scalable and near-real-time system that can co-relate multiple pieces of information to detect the overall risk factor for the flight before departure

The solution, built with Semagix Freedom,[10] allows security personnel to assess passenger threats while maintaining a high rate of passenger flow.[11] Semagix Freedom interfaces with diverse information sources, automatically extracts metadata from heterogeneous content, organizes, normalizes and disambiguates metadata based on the ontology. It co-relates the information from different sources to determine possible threats. Ontology-based link analysis/inferencing identifies high-risk passengers by discovering hidden relationships between seemingly unrelated pieces of information.

[10] For more information, see: www.semagix.com/technology_architecture.html.

[11] A. Sheth, et al., "Semantic Association Identification and Knowledge Discovery for National Security Applications", Journal of Database Management, 2004.

[12] Technical Memorandum # 03-009, LSDIS Lab, Computer Science, the University of Georgia, August 15, 2003. Prepared for Special Issue of JOURNAL OF DATABASE MANAGEMENT on Database Technology for Enhancing National Security, Ed. Lina Zhou.

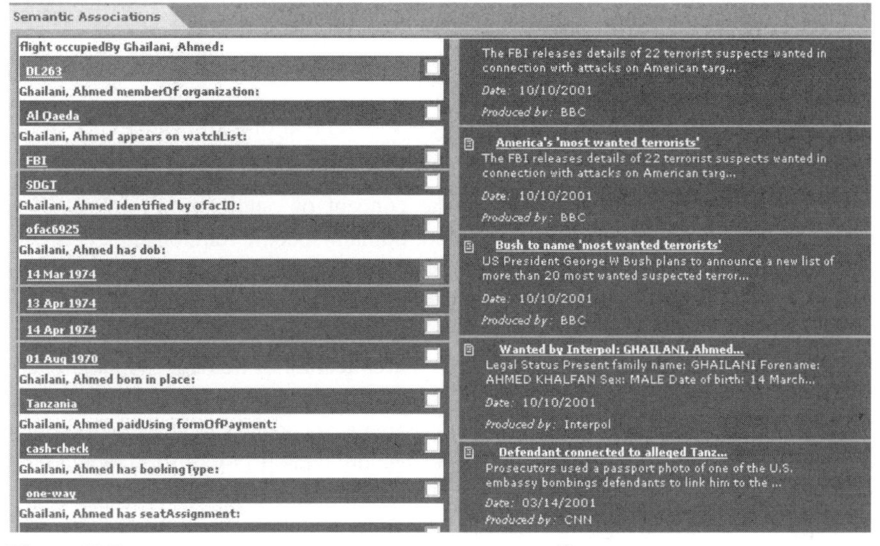

Figure 10.7 Semantic Associations and Semantically Relevant[12]

On the basis of a threat score for each passenger, the passenger either will be allowed to proceed from one checkpoint to another in a normal manner, or will be flagged for further interrogation along concrete directions as indicated by the semantic associations in the application. Figure 10.6 shows a "passenger profile" screen that provides a 360-degree view of information related to a passenger.

Section 1 in Figure 10.6 presents a listing of the semantic associations of the passenger to numerous other entities in the ontology. It provides a passenger (entity)-centric view of the ontology that reveals a number of semantic associations, both direct and indirect (hidden). Only those relationships that are regarded as relevant to the given context are displayed. Such semantic associations form the basis of identifying connections between two or more seemingly unrelated entities. Section 2 presents a listing of all the content that is contextually relevant to the passenger, but not necessarily mentioning the name of the passenger. Once again, this approach exploits semantic associations in the ontology to decide relevance of content.

The main idea behind the strategy of the application is to automatically attach a threat score to every passenger who boards any flight from any national airport, so that flights and airports could be assigned corresponding threat levels. This threat is based extensively on semantic associations of passenger entities with other entities in the ontology such as terrorist organizations, watch lists, and travel agents. The following semantic associations are considered in the generation of a passenger's threat score:

- Appearance of the passenger on any government-released watchlist of bad persons or bad organizations
- Relationship of the passenger to anyone on any government-released watchlist of bad persons or bad organizations

- Deviation from normal methods of ticketing, flight scheduling, use of a travel agent in reservation of tickets
- Origin of the passenger and the flight
- Association of the name of the passenger with that of a known bad person in any public content, etc.

Semagix Freedom is built around the concept of ontology-driven metadata extraction, allowing modeling of fact-based, domain-specific relationships between entities. It provides tools that enable automation in every step in the content chain—specifically, ontology design, content aggregation, knowledge aggregation and creation, metadata extraction, content tagging and querying of content and knowledge. Figure 10.8 shows the domain model-driven architecture of Semagix Freedom.

Semagix Freedom operates on top of a domain-specific ontology that has classes, entities, attributes, relationships, a domain vocabulary, and factual knowledge, all connected via a semantic network. The domain-specific information architecture is dynamically updated to reflect changes in the environment.

The Freedom ontology maintains knowledge, which is any factual, realworld information about a domain in the form of entities, attributes and relationships. The ontology forms the basis of semantic processing, including automated categorization, conceptualization, cataloging, and enhancement of content. Freedom enhances

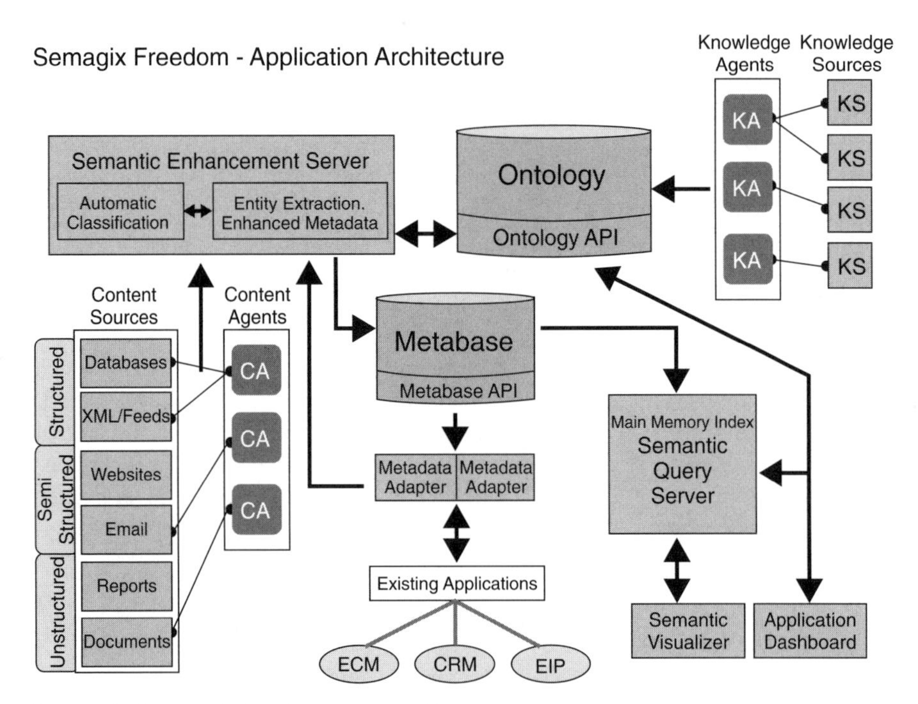

Figure 10.8 Semagix Freedom Architecture[12]

a content item by organizing it into a structured format and associating it with instances within the ontology.

The ontology is automatically maintained by Knowledge Agents. These are software agents created without programming that traverse trusted knowledge sources and exploit structure to extract useful entities and relationships for populating the ontology automatically. Once created, they can be scheduled to perform knowledge extraction automatically at any desired interval, thus keeping the ontology up-to-date.

Freedom also aggregates structured, semi structured, and unstructured content from any source and format, by extracting syntactic and contextually relevant semantic metadata. Custom meta-tags, driven by business requirements, can be defined at a schema level. Much like Knowledge Agents, Content Agents are software agents created without programming by using extraction infrastructure tools that extract useful syntactic and semantic metadata information from content and tag it automatically with predefined meta-tags. Incoming content is further "enhanced" by passing it through the Semantic Enhancement Server module.

The Semantic Enhancement Server tools classify aggregated content into the appropriate topic/category (if not already pre-classified) and subsequently perform entity identification and extraction, and content enhancement with semantic metadata from the ontology. Semantic associations in the ontology are leveraged to derive tag values if such metadata is not explicitly mentioned in the content. The Semantic Enhancement Server can further identify relevant document features such as currencies, dates, etc., perform entity disambiguation, and produce XML-tagged output.

The Metabase stores both semantic and syntactic metadata related to content in either custom formats or one or more defined multiple metadata formats such as RDF, PRISM,[13] Dublin Core, and SCORM.[14] Content is stored in a relational database as well as a main memory checkpoint. At any point in time, a snapshot of the Metabase (index) resides in main memory (RAM), so that retrieval of assets is accelerated with a patented Semantic Query Server.

The ontology for the national security domain was developed at the University of Georgia's LSDIS lab (http://lsdis.cs.uga.edu) in a project called PISTA[12]. A sample of this ontology is shown in Figure 10.9.

The PISTA ontology provides a conceptualization of organizations, countries, people, terrorists, terrorist acts, etc. that are all interrelated by named relationships

[13] PRISM, Publishing Standards for Industry Standard Metadata, is a specification for exchanging "descriptive metadata" for a document's author(s), title, subject, rights and permissions, relations, and time-stamps. It is an open standard developed by publishers and vendors. Originally developed to meet the production, repurposing, aggregation, syndication, and rights clearance needs of magazines, it applies more broadly to publishing-like operations within other types of organizations. Version 1.0 was released in April 2001. PRISM uses RDF and builds on the Dublin Core standard. See: www.prismstandard.org for more details.

[14] SCORM, the Sharable Content Object Reference Model, defines a Web-based learning "Content Aggregation Model" and "Run-Time Environment" for learning objects. The SCORM is a collection of specifications adapted from multiple sources to provide a comprehensive suite of e-learning capabilities that enable interoperability, accessibility, and reusability of Web-based learning content. See http://www.adlnet.org/index.cfm?fuseaction=scormabtn for more details.

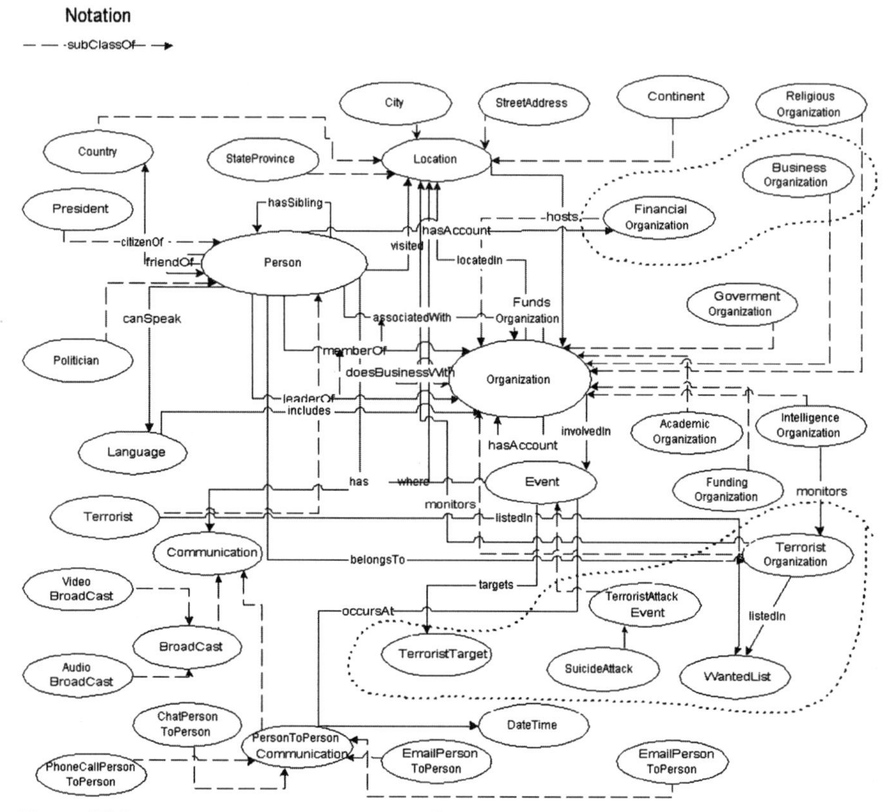

Figure 10.9 A Sample of the PISTA Ontology[12]

to reflect realworld knowledge about the domain (i.e., "terrorist" "belongs to" "terrorist organization"). This ontology is populated using Semagix's Freedom product as part of a non-public portion of a very large populated ontology, called SWETO, having about 1 million entities and over 1.5 million relationship instances.

More information on the use of Semagix's semantic technology for homelands security is available at the website of Semagix (http://www.semagix.com) and the University of Georgia's LSDIS (Large Scale Distributed Information Systems) laboratory website (http://lsdis.cs.uga.edu/Projects/SemDis/).

Solution Story: Cogito's Integrated Data Surveillance Solution

The market for military and other types of intelligence technologies has taken on an acute urgency in recent years. The problem lies less with capturing intelligence data than with sorting it quickly and massaging it into meaningful formats while it is still

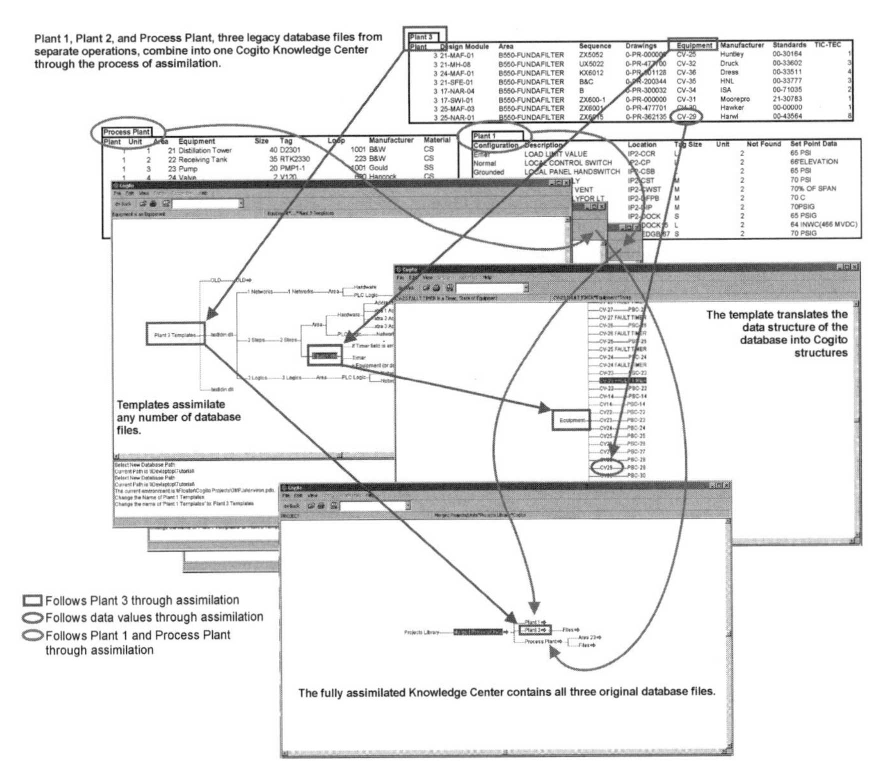

Figure 10.10 Cogito Knowledge Center

fresh. The problem is exponentially exacerbated by the complexity of interfacing multiple data sources in a manner that enables creation of multiple meaningful views.

As a demonstration project for a client in the intelligence community, Cogito[15] integrated a number of enterprise databases from a nuclear power plant, a petrochemical plant, and another process plant into a single Knowledge Center™. Templates allow data from different data sources and different data formats to be integrated together. The Cogito Assimilation Module solved the problem for a government agency by eliminating the interfaces altogether, fusing the data into a single Cogito Knowledge Center™.

Templates were written for each legacy data source (a simple task for someone who understands the logic of the legacy source), describing how each one relates to a uniform semantics. The Cogito Assimilation Module uses these templates to fuse the data from all the sources into a single Knowledge Center™. The result is a seamless information source, which fuses all the original data so that it becomes seamlessly interusable for security surveillance.

[15] For more information, see: http://www.cogitoinc.com.

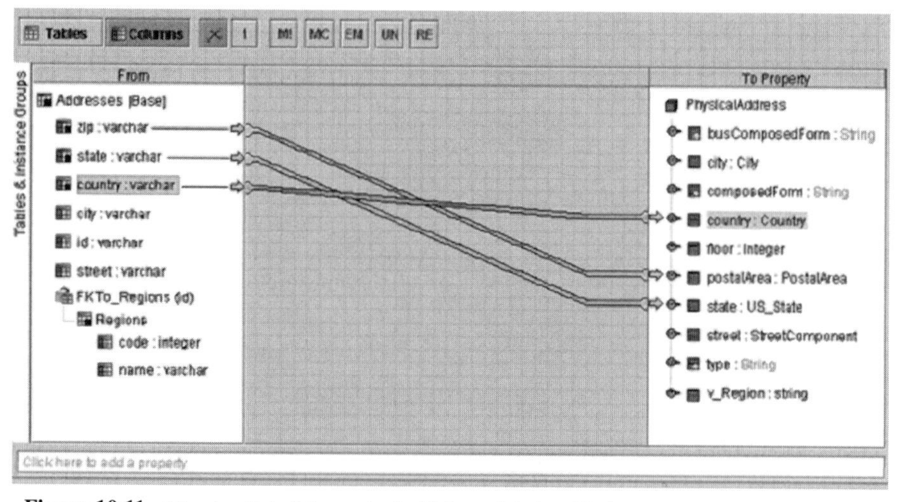

Figure 10.11 Mapping Data Schemas in the Unicorn Solutions Tool

Solution Story: Major Electronics Manufacturer Data Integration

This case study concerns a Fortune 500 electronics manufacturer who owns and operates over 20 factories in which components are constructed, assembled, and tested around the world. The solution described uses technology from Unicorn Solutions.[16]

Multiple applications support materials transactions and the processing of customer orders. Significant semantic differences between them made interoperability difficult. The company had reasons to believe that unknown discrepancies among the various systems were inflicting financial damage on its operations.

During the course of a 2-week project, the following steps were taken:

- A business domain and problem were scoped.
- An Ontology model was constructed.
- Relevant relational database schemas were semantically mapped to the ontology model.
- Incompatible databases were compared and differences identified.

At the end of the project, all discrepancies were identified and known. Additional benefits of a Semantic Mapping Approach were higher quality of the data transformation scripts and fewer mappings and code to maintain.

Benefits included cleaner data, increased scalability of data integration, fewer mappings between data sources, higher code reusability, easier code maintenance, and higher quality of transformation code.

[16] For more information, see: www.unicorn.com.

Solution Story: Children's Hospital Lab Data Mediation

We describe a solution that was built for the Boston Children's Hospital by Yao Sun as a research project toward his Ph.D. at MIT.[17] The solution uses the MEDIATE system to integrate semantically diverse lab test data from different clinical sources. Labs in different clinics (or even within the same hospital) have different standards for recording lab data. During clinical consultation, it is necessary to exchange this data quickly and accurately. No useful standards exist to ensure that the data that is exchanged is semantically relevant.

We describe an example for a 2-year-old boy with multiple congenital anomalies including structural heart disease, tracheoesophageal fistula, vertebral anomalies, and renal problems (i.e., VATER complex), who has an appointment to be seen by his new pediatrician. To familiarize herself with the patient's problems and past treatment, the pediatrician uses a single interface to view the medical history from several sources: the cardiology foundation computer, the pediatric hospital main computer, and the previous pediatrician's office.

The pediatrician locates and reviews the last progress note from each of the systems. The information taken from these sources is presented in a way that is consistent with the semantics of the new clinic, enabling the pediatrician to establish an efficient agenda for the initial visit without duplicating the semantics of the evaluations that have been performed at the other facilities. A clinician selects the lab tests that she wants to examine, irrespective of the source of the data, by using the ontology of her own home laboratory (left side of Fig. 10.12). The results are taken from the remote database, according to the elements in its ontology that are semantic matches for the selected elements. These results are displayed according to the display form constructed in the right side of Figure 10.12.

Solution Story: Consulting Services Company Data Quality

An international services company wanted to see side-by-side information from its American and European divisions. Different divisions had their own definitions of key business indicators such as utilization rates.

Because of differences in business rules, converting the information into a compatible format required a lot of manual coding and comparison. The complex manual process was impacting the company's ability to make decisions in a timely fashion.

An Information Model was constructed with the Unicorn System; this Model was then published via a Web browser. Because the Information Model was built after extensive interviews with employees from each division, it enabled data consistency across geographically disparate units.

[17] Yao Sun, *Information Exchange Between Medical Databases Through Automated Identification of Concept Equivalence*, Ph.D. Thesis in Computer Science, Department of Electrical Engineering and Computer Science, Massachusetts Institute Of Technology, February 2002.

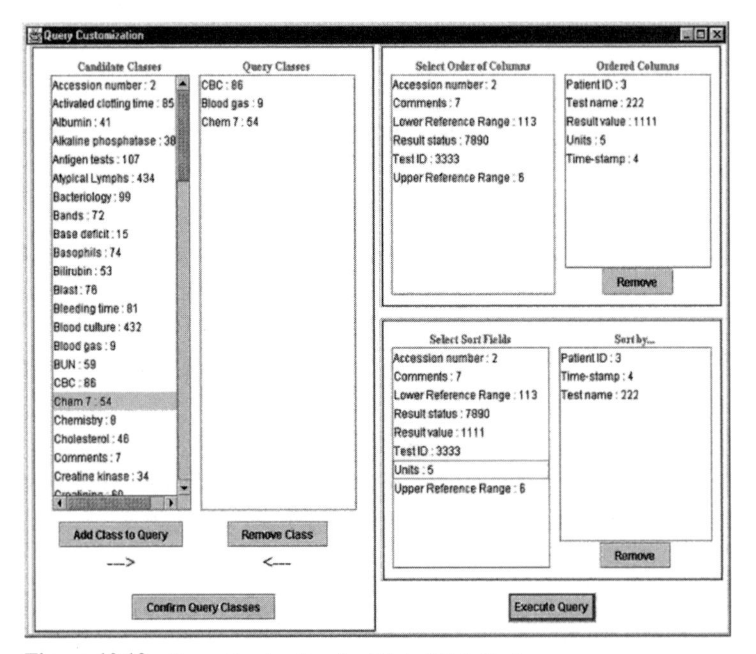

Figure 10.12 Query Construction for Clinical Data Exchange

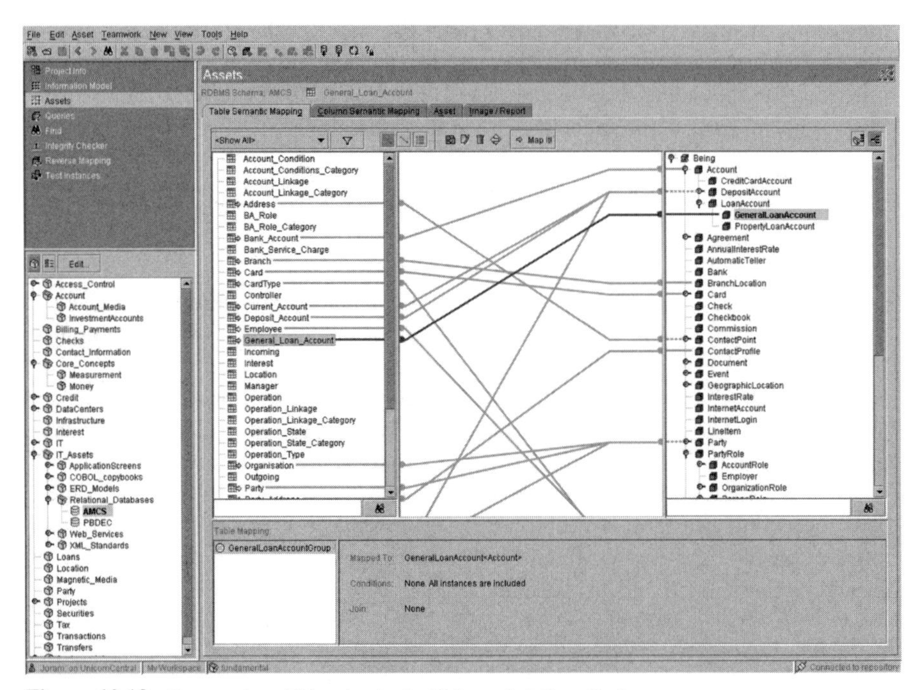

Figure 10.13 Browser-based Mapping in the Unicorn Solutions Tool

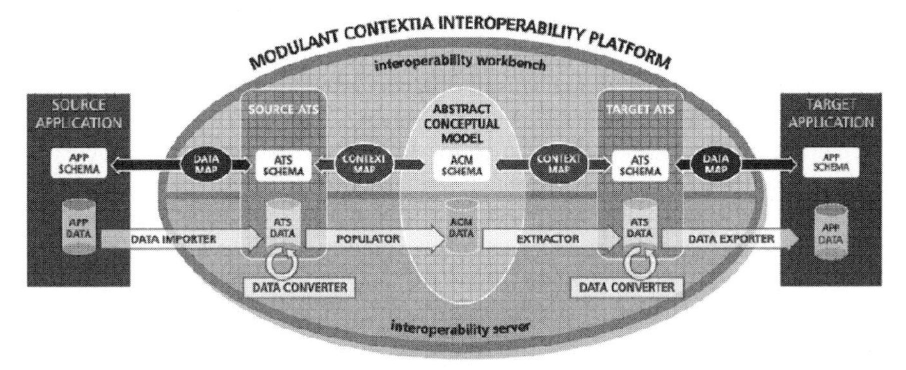

Figure 10.14 Contextia™ Interoperability Workbench

The company now has an accurate and accessible list of its overall utilization statistics for comparing departmental results. Each subsidiary continues to work in the way it always has. Mapping different data elements into a common model enabled management to easily evaluate business results and understand company-wide issues.

Solution Story: US Air Force Enhanced Information Management

The US Air Force's Technical Order management system is a series of pieced-together legacy ties. The Air Force decided to consolidate them into one aggregated system. The Air Force did not have a good understanding of the information used in its many existing systems and did not have documentation for some of the systems.

Modulant™[18] used the Contextia™ Information Interoperability Methodology (CIIM) to build a semantic model of the entire Technical Order information system and to identify which information was inconsistent and missing.

Modulant was able to identify over 175 specific questions for domain experts that enabled complete documentation of the logical model.

Once a single logical model for the program was created, the Contextia process was used for data cleansing to ensure the accuracy of the information. To achieve similar results would have required three person-years' worth of effort; Contextia enabled the process to be complete within three-quarters of a person-year.

Capability Case: Semantic Data Acquirer

A "Semantic Data Acquirer' uses ontologies to move intelligence close to sensors in industrial and scientific data acquisition systems. These "Semantic Data Acquisi-

[18] For more information, see: www.modulant.com.

tion Systems" are capable of extracting more meaningful information and relations from data.

Although many communities have collaborated to agree on standard community knowledge models, other communities have not. For many it is difficult to have a common language—their data is used for different purposes, often unanticipated. Take for example satellite data. Examples of its uses are weather forecasting, transportation logistics planning, and military simulations. A common challenge for these communities is that they are faced with new data sets with unknown relationships between the data and its situations of use.

For these application areas it is difficult to define a rigid data description. A semantic extraction approach is needed. Semantic Data Acquisition Systems define semantics and the structure of knowledge for a community organically. They build a description based on the properties and relationships that are ascribed to data. They evolve as additional properties and relationships are added.

This approach of building the description of the data from the data itself is needed in many communities where needs are specialized, where knowledge is emerging, and where data descriptions are diverse.

Solution Story: Oceanic and Atmospheric Sciences Data Automation

Oregon State University's College of Oceanic and Atmospheric Sciences (COAS) is one of many institutions that collect and processes data from multiple satellite data sources and blend that data with data from other sensors and devices to do multiscale science.

COAS needed to automate the collection, processing, analysis, and distribution of its satellite data and to track complete data lineage for the purposes of data reuse and scientific defensibility. Raw data from satellite instruments needed to be processed into levels of products that could be used and annotated by scientists. Users needed to be able to search on both data and metadata, perform analysis on the data and continue to use it in different applications.

COAS[19] uses Thetus Publisher, from Thetus[20] Corporation, to classify data as it is collected, and to keep an evolving, rich metadata description of the data. The metadata complies with WC3 standards for knowledge description and discovery: XML Schema (XML), Resource Description Framework (RDF), and Web Ontology Language (OWL). The Publisher Server is developed in Java and runs on a Linux system using a MySQL database. The Publisher is storage system independent and can be used with a wide variety of storage systems including direct

[19] For more information, see: www.thetuscorp.com/Thetuscorp/clients/casestudies/COAS/page1.html.

[20] For more information, see: www.thetuscorp.com.

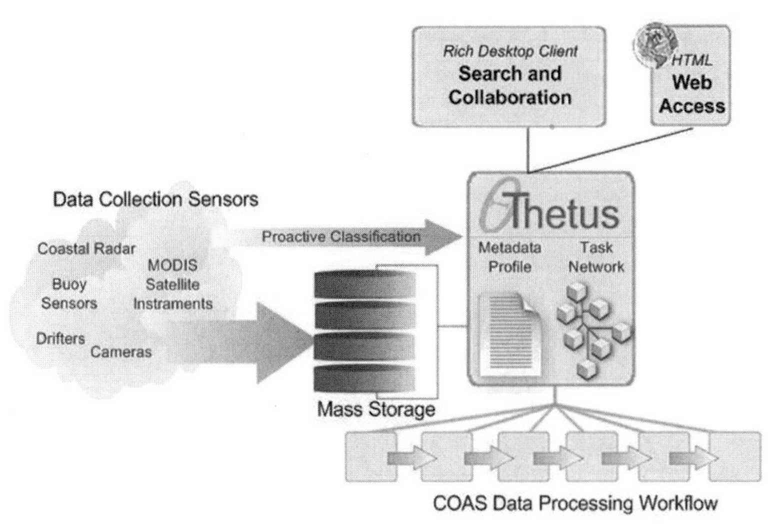

Figure 10.15 COAS System Context Diagram

attached storage, network attached storage (NAS), and storage area networks (SANs).

COAS did not have any data description to leverage—there was no satellite data ontology. As users of the data-added properties, the ontology was built by the Thetus Publisher. By using the properties to build the ontology, it was guaranteed to be relevant but it also could have conflicting and redundant terms and relationships. It was essential that properties could be changed, but it was also important that the software let individual data users use their own ontological jargon. This organic approach to building an ontology by extracting semantics worked well for satellite data and allowed COAS to build a data ontology that can be used (and modified) by other satellite data users.

The system automated the collection and processing of data, accelerating the "time to science," and dramatically reduced the IT infrastructure support needed to support the collection of this massive stream of data. Central to its operation is a satellite data ontology that is standards based and exportable.

The solution uses a rich client- and Web-based search capability that allows users to easily access selective slices of data based on a wide range of data properties or to be notified of the availability of new data. A data reporting system allows data users to view a comprehensive yet evolving data history. Another capability is the automated publishing of selective slices of data to the COAS website and collaborative partners.

A portion of the COAS satellite data ontology that allows users to easily view large amounts of data sorted by ever-changing data properties is shown in Figure 10.16.

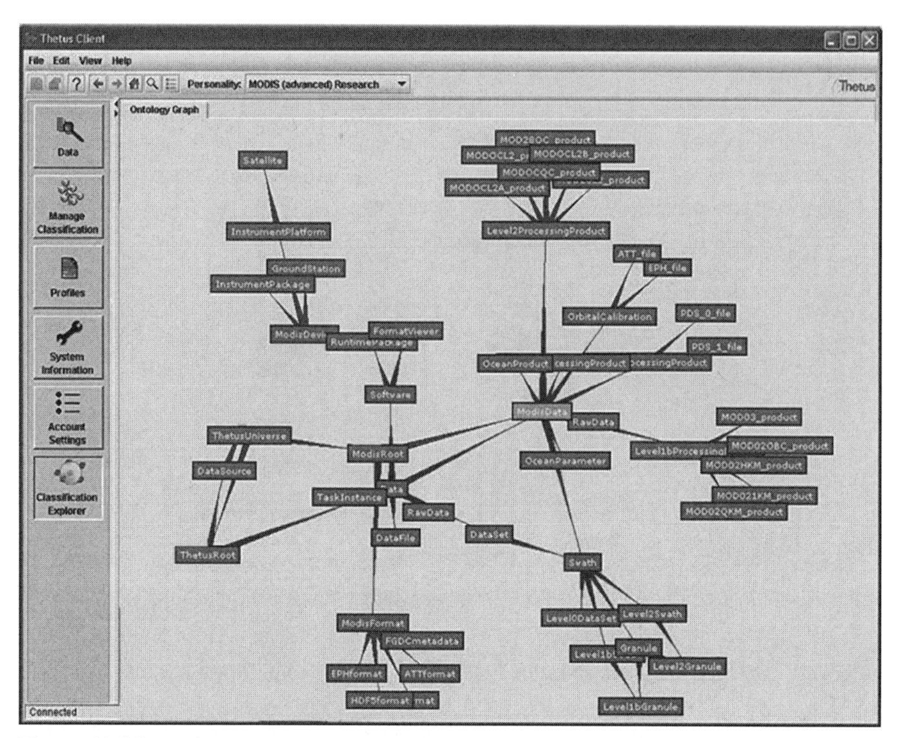

Figure 10.16 COAS Satellite Data Processing

Capability Case: Semantically Enriched Multimedia Integrator

A Semantically Enriched Multimedia Integrator' uses knowledge models to control how audio-visual streams can be integrated to provide an improved user experience of both real-time information and the retrieval of libraries of multi-media resources.

Solution Story: CoAKTinG Distributed Meetings Manager

The CoAKTinG test bed (Collaborative Advanced Knowledge Technologies in the Grid[21]) is extending and integrating technologies from the UK AKT (Advanced Knowledge Technologies) consortium to support distributed scientific collaboration—both synchronous and asynchronous—over the Grid and standard internet. The project has integrated technologies from the UK's Open University (OU) for instant messaging/presence awareness, real-time decision rationale, and group memory capture. Issue handling and coordination support uses technologies from Edinburgh University. The semantically annotated audio-visual streams use technologies from Southampton University.

[21] For more information, see: www.aktors.org/coakting.

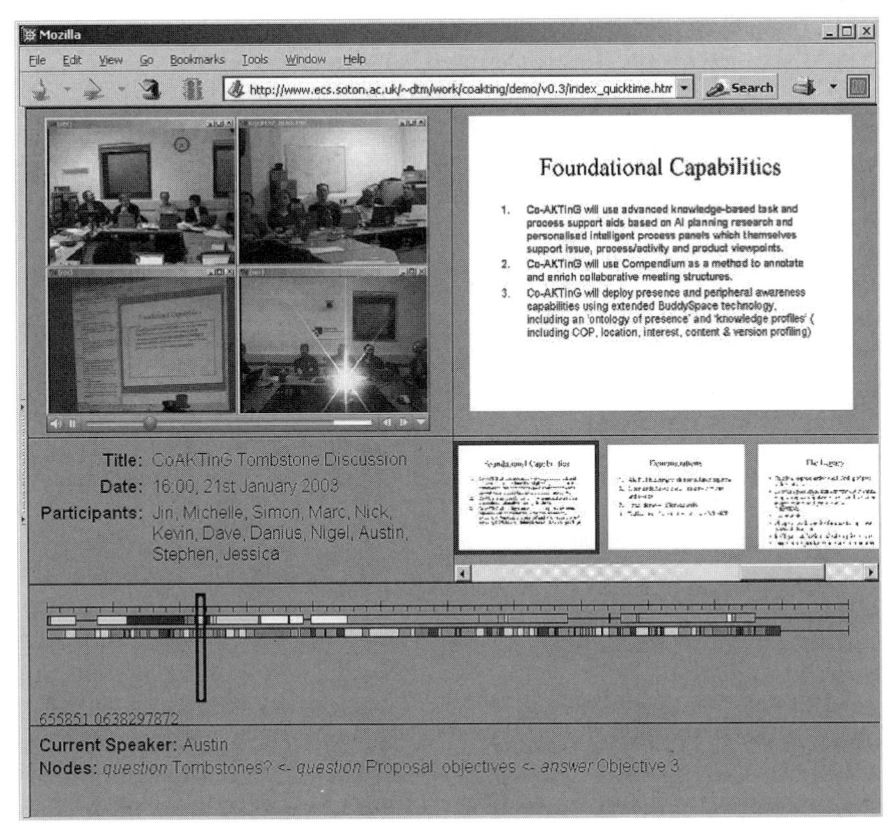

Figure 10.17 CoAKTinG proof of concept web interface

An example of multiway integration is the meeting navigation and replay tool illustrated in Figure 10.17.[22] This integrates metadata grounded in a meeting ontology for scientific collaboration, with time-based metadata indicating current slide, speaker, and issue under discussion. The two colored bars indicate slide transitions (top) and speaker (bottom). The current speaker at the time selected is indicated at the foot, plus an indication of the current issue under discussion, extracted from the Compendium[23] record.

Ontologically annotated audio/video streams enable meeting navigation and replay. Few researchers have the time to sit and watch videos of meetings. An AV record of an on-line meeting is only as useful as the quality of its indexing. Prior work has developed ways to embed "continuous metadata" in media streams. Within

[22] Bachler, M.S., Buckingham Shum, S., De Roure, D., Michaelides D. and Page, K.: "Ontological Mediation of Meeting Structure: Argumentation, Annotation, and Navigation", 1st International Workshop on Hypermedia and the Semantic Web, ACM Hypertext, Nottingham, 2003.

[23] Compendium (http://www.CompendiumInstitute.org) is an approach to collaborative support that provides "lightweight" discussion structuring and mediation plus idea capture, with import and export to other document types. Dialogue maps can be created on the fly in meetings, providing a visual trace of issues, ideas, arguments, and decisions.

CoAKTinG, this work forms the platform for integrating multiple metadata streams with AV meeting records. This is illustrated in the meeting replay tool.

Issue handling, tasking, planning, and coordination are provided in applications using I-X Intelligent Process Panels and their underlying <I-N-C-A> (Issues, Nodes, Constraints and Annotations) constraint-based ontology for processes and products. The process panels provide a simple interface that acts as an intelligent "to do" list that is based on the handling of issues, the performance of activity, or the addition of constraints. It also supports semantically task-directed "augmented" messaging and reporting between panel users. A common ontology of processes and process or collaboration products based on constraints on the collaborative activity or on the alternative products being created via the collaboration supports the applications.

The project features enhanced presence management and visualization. This moves the concept of presence beyond the "online/offline/away/busy/do-not-disturb" set of simple state indicators to a rich blend of attributes that can be used to characterize an individual's physical and/or spatial location, work trajectory, time frame of reference, mood, goals, and intentions.

The challenges addressed are:

- How best to characterize presence in collaborative environments
- How to make it easy to manage and easy to visualize collaborative environments
- How to remain consistent with the user's expectations, work habits, and existing habits of using Instant Messaging and other communication tools

Capability Case: Semantic Content Registry

A "Semantic Content Registry" is a directory of information resources with metadata that expresses the provenance of each resource. The system registers metainformation about content stored in diverse Data Repositories. It allows searching by metainformation fields, including the data definitions, and locates the registered resources, offering a link to the registered resources whenever possible. It also stores information about the delivery dates so that supervising the processes can be supported with simple work flow functions.

Solution Story: European Environment Agency Content Registry

A huge amount of environmental data and information is reported annually by countries in Europe to the European Community and international organizations, involving networks such as the EEA/EIONET, the European Commission, and the OECD. To address the reporting challenges, the European Environment Agency has developed a reporting system called ReportNet[24] with components for

[24] For more information, see: www.eionet.eu.int/reportnet.html.

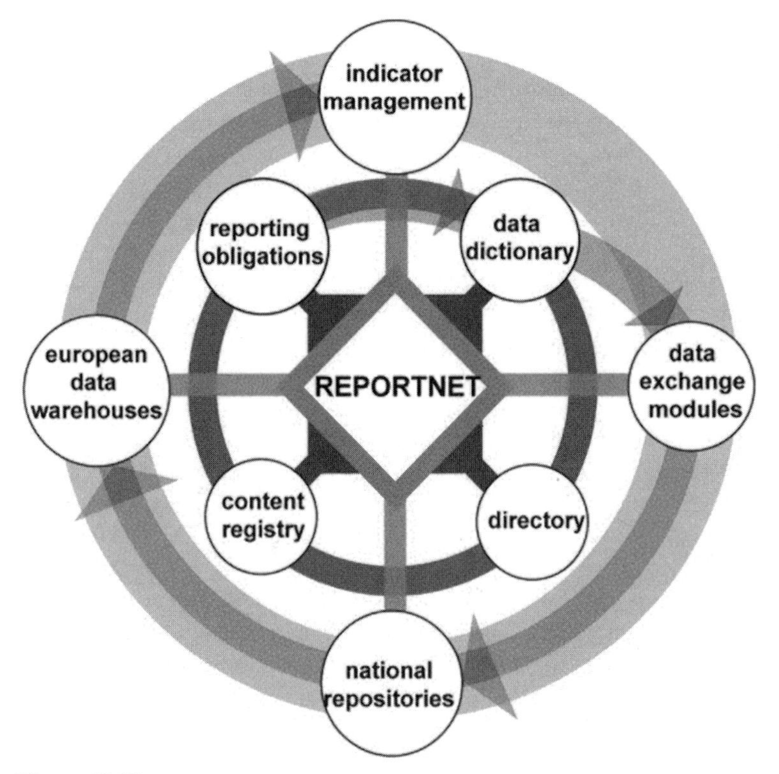

Figure 10.18 Components of the EEA Reportnet

reporting obligations, metadata, directory services, data repositories, and process monitoring.

ReportNet is an implementation of a data collection network in a situation where data volumes vary and the frequency of reporting is typically once a year, where users are widely distributed but committed to a network solution to data flows and dissemination, and where, unlike in the case of economic data, there are few, if any, examples where access to data is classified by commercial in-confidence restrictions. The components of ReportNet are illustrated in Figure 10.18. The inner circle illustrates the normative metadata components. The outer ring indicates the main "real" data flow.

The Network Directory is the database of all persons, organizations, and their roles that are active with or of interest to the EEIS. The Directory provides functions that allow user authentication, security services, getting contact information, and routing of work flow processes.

The Reporting Obligations Database (ROD)[25] provides descriptions of the requirements of data and information, including the legal basis for why such content will have to be or is desirable to be provided, and periodicity.

[25] For more information, see: rod.eionet.eu.int/index.html.

The Data Dictionary is a central registry where other applications and projects that need data definitions can share their definitions. It stores definitions of data elements, their attributes, their data types, allowable values, and relationships between other elements. It also manages compound elements and provides the functions of a XML Schemas registry.

The Data Exchange Modules (DEMs) are smart electronic questionnaires used to gather data from various sources to agreed formats. Optionally, such data can be delivered to agreed places. DEMs provide functions for importing data in popular formats or, alternatively, allow direct input, format the data technically correct, package the data, and upload or otherwise automatically deliver it.

The Data Repository is the location where the deliveries of data, information, and knowledge to the international reporting system are stored. This is a content management system providing functions for uploads, downloads, versioning, approvals and sealing of official datasets, packaging in virtual envelopes for delivery, and tagging with metainformation. The content in DR is in document format. A Data Repository is accessed by the DEMs when they upload content. It makes metainformation available for the Content Registry in RDF Dublin Core format.

The Data Warehousing (DW) component provides query access to the data reported on particular topics. The Data Warehouses can be queried by using SQL, intelligent agents, or Web-based forms. Data to the Warehouses is processed from the documents contained in Data Repositories. The Data Warehouses are accessed by Indicator Management tools for the purpose of producing assessments and scenarios.

The Indicator Management (IM) component denotes the various models and other systems that are used to process the delivered data into indicators describing state, trends, and scenarios. The functions include those of decision support systems, model management and operation, multimedia, and Web access to results.

The Content Registry's (CR) objective is to provide a Web-based reference center and its communication services for resources at distributed national, topic-specific, and other information systems around EIONET. The Semantic Content Registry gets its information from the multiple Data Repositories by harvesting them for metadata (pull) or through notifications after upload events (push). The CR uses RDF to keep track of the deliveries of data sets to the international reporting system, and also other content, as applicable. It requires that the harvested services provide Dublin Core descriptions of their content in RDF format.

Users that supervise the data reporting activities use the Content Registry to search delivered data sets by data flow, country, and date. It can also search documents by the Dublin Core elements. Contents are created automatically but controlled by the EIONET Network Management Centre. There is one central Content Registry on the ReportNet. The Content Registry is built with the Java language, MySQL relational database, and an XML Server. The CR Harvester uses an RDF parser called ARP (Another RDF Parser) The parser is part of the HP Jena toolkit.

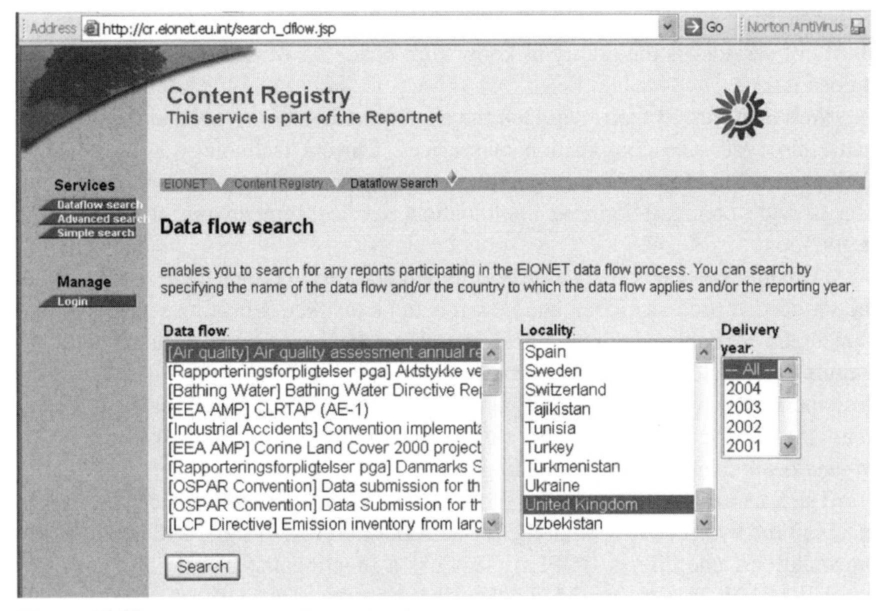

Figure 10.19 EEA ReportNet Content Registry

SERVICES INTEROPERABILITY

Web Services promise a new level of functionality of executable services over the Internet. To employ their full potential, appropriate description means for Web Services need to be developed. The current technology around UDDI, WSDL, and SOAP provides only limited support in facilitating service recognition and comparison, service configuration, and service composition. As a result, it is hard to realize complex work flows and business logics with current Web Services. These needs are addressed by the Semantic Web Services work that led to the OWL-S (formally DAML-S) specification.

In a business environment, the promise of flexible and autonomous Web Services is automatic cooperation between enterprise services. An enterprise requiring a business interaction with another enterprise should be able to automatically discover and select the appropriate optimal Web Services relying on selection policies. Services are invoked automatically, and payment processes are initiated. Any mediation would be applied based on data, process, and policy ontologies, with semantic interoperation providing the run time orchestration. An example would be supply chain management where an enterprise manufacturing limited-lifetime goods must frequently seek suppliers as well as buyers dynamically. Instead of employees constantly searching for suppliers and buyers, the Web Services infrastructure does it automatically within defined constraints.

The Web is moving beyond a network of unstructured pages of information to a collection of services that can be executed over the Internet. Web Services promise a new level of service on top of the current Web. A key requirement for interoperability of services is the ability to know with precision what a service is and what it does.

Web Services on their own lack the expressive constructs needed for the automatic discovery and composition of services. Current technology around UDDI, WSDL, and SOAP provides only limited support in mechanizing service recognition, service configuration and combination, service comparison, and automated negotiation.

Dynamic service composition and interoperability requires a precise profile of the service, its process model, and how it is to be invoked. Locating services is fundamentally a problem of semantics—matching the meaning of requests with the meaning associated with a service in its intent, behavior, and results. In OWL-S the description of a web service is organized into three specifications: *what a service does* (the service profile), *how the service works* (the process model), and *how the service is implemented* (the grounding).

The Service Profile allows the service to be discovered and evaluated at matchmaking time by specifying what the service does in terms of input and output types, preconditions, and effects (IOPEs). This has a functionality similar to the yellow pages in UDDI. The Process Model describes how the service works in terms of the internal processes with their control and data flows. This information can be used during service discovery, but it is also intended for execution monitoring. The Process Model can be likened to the business process model in BPEL4WS.[26] Each service is either an *Atomic Process* that is executed directly or a *Composite Process* that is a combination of other subprocesses. A process is described with a minimal set of control constructs consisting of *Sequence*, *Split*, *Split + Join*, *Choice*, *Unordered*, *Condition*, *If-Then-Else*, *Iterate*, *Repeat-While*, and *Repeat Until*.

The Grounding specifies the details of how the service is invoked by specifying a communications protocol, parameters to be used in the protocol, and the serialization techniques to be employed for the communication. It is simply a mapping from OWL-S to WSDL.

The role of the Profile, Process, and Grounding Models of Semantic Web Services across the life cycle activities of Web Services-based development is shown pictorially in Figure 10.20, taken from an article[27] written by Sheila A. McIlraith, David L. Martin, and James Hendler (ed.) and published in *IEEE Intelligent Systems*.

Table 10.5 compares Web Services and Semantic Web Services.

The potential of Semantic Web Services for realizing the promise of the Web as a backplane for discovering and composing services is starting to be evident in some interesting application prototypes. Currently these are research projects, and

[26] BPEL4WS (Business Process Execution Language for Web Services) is a specification co-authored by IBM, Microsoft, BEA, SAP and Siebel Systems. Influenced by Microsoft's XLANG and IBM's WSFL, it is a work flow language with simple control flow constructs such as *if*, *then*, *else*, and *while-loop*.

[27] Sheila A. McIlraith, David L. Martin, James Hendler (ed.) "Bringing Semantics to Web Services", IEEE Intelligent Systems, 2003.

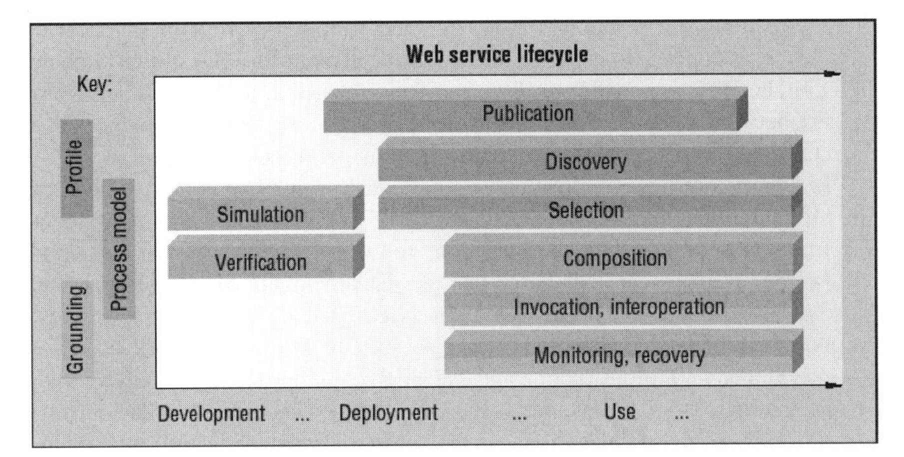

Figure 10.20 Semantic Web Services and the Web Service Lifecycle

Table 10.5 Data Table—Comparison of Web Services and Semantic Web Services

	Web Services	Semantic Web Services
Capabilities	Service discovery and execution	Discovery, Execution and Interactions and reasoning at the agent or application level
Specification Intent	Protocol-level description of *how a service is invoked*	Semantic-level description of: • *what a service can do* (service profile) • *what happens when the service is invoked* (service model) • *how a service is invoked* (the service grounding).
Realization	Messages using XML	XML + Web service ontology models
Composition of Services	No direct support for composite applications	Automated Web services including service composition and interoperation
Standards	UDDI, SOAP, WSDL, WS-Security, WS-Transaction, WS-ReliableMessaging	SOAP, WSDL, WSFL, XLANG and BPEL4WS using RDF[S], OWL, and OWL-S

the Capability Cases we will describe illustrate some emerging capabilities for process aggregation, task computing, service discovery, and service mediation.

Capability Case: Task Composer

Increasingly, nonexpert users are obliged to use feature-rich applications and complicated digital devices to accomplish everyday tasks. We look up the address of a store or directions to a restaurant on the Web and print the information on paper to be used once, as opposed to downloading it to our cell phone or PDA. When booking a hotel on the Web, we cannot readily send the information to our calendar. Copy and Paste between applications can be avoided through Semantic Web Services that connect one service to another to accomplish an overall task.

"Task Composer" is a realization of a new field of interoperability, defined by Fujitsu Laboratories and the MINDlab as "Task Computing"[28]—a shift of focus away from the details of how tasks are done to the intent of what a user wants to do. Task Composer presents to the user the tasks that are possible in the user's current context as an abstract view of devices and services. The user need not need know how and where the actual constituent tasks are executed. For more involved tasks, the user is guided in the construction and execution of composite tasks using simpler tasks as building blocks. Once complex tasks are defined by users, they can become part of a library of reusable tasks. Tasks may span multiple applications and multiple computing platforms.

Solution Story: STEER—a User-Centered Environment for Task Computing

The STEER (Semantic Task Execution EditoR) is a tool for discovering and composing Web Services built by Fujitsu Laboratories. An illustrative user task is the scenario of a user wanting to make an impromptu presentation in a conference room equipped with Semantic Web-enabled infrastructure. To make the presentation, the presenter wirelessly connects her PDA to the conference room network. She uses the composition of "Local File" service on her PDA and "View on Projector" service in the conference room [Local File + View on Projector]. As a result, her presentation is projected on the screen and the control page shows up on her PDA. Using the control page, she can control her presentation wirelessly. If someone wants a copy of her presentation, the [Local File + Bank (File)] service makes the presentation file available through the "Bank (File)" service in the conference room.

Figure 10.21 shows STEER's compose page. Service compositions are categorized based on their semantic inputs and outputs. For each pair, any service on the right-side drop-down menu can be combined with any service in the left-side drop-down menu. A drop-down menu is used to select a service, which is then executed through the "Execute" button. The user can click the "Construct" button to create a more complex composition starting from the two-service composition in this Compose page.

The swirl in the upper right corner is known as the "White Hole." A user can drag and drop objects such as files, URLs, contacts, and schedules from PIM to create services that provide semantic versions of the objects dropped into it. After the drag and drop, it swirls for a while to let the user know the object is successfully dropped. In the lower right corner a management pane is shown through which

[28] Simos, M. et al., *Organization Domain Modeling (ODM) Guidebook*, 1995.

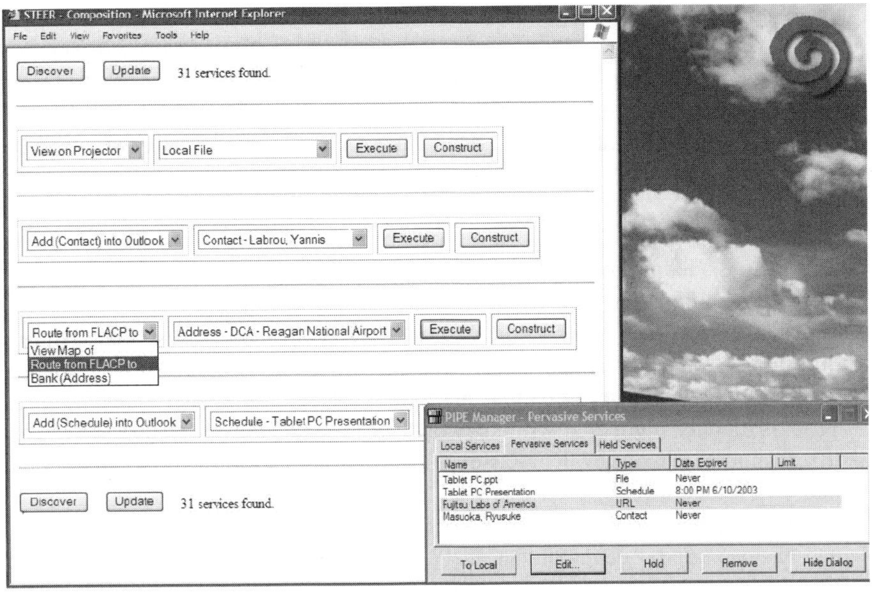

Figure 10.21 Task Computing with STEER

the user can manage the services. A user can make the services local or pervasive or temporarily hold those services for possible future use.

Capability Case: Process-Aware Web Services Match-Maker

Process-Aware Web Services Match-Maker uses a process ontology to describe the semantics of a Web Service profile. By having a well-defined semantics, the process ontology allows more precise matching than a simple keyword search against a UDDI directory.

Solution Story: Service Discovery Using Process Ontology Queries

Abraham Bernstein and Mark Klein[29] in their paper "Discovering Services: Towards High-Precision Service Retrieval," describe a promising approach to process ontology-based Web Services directory that overcomes the imprecision of simple keyword-based Web Services discovery. A motivating example of how simple keyword search breaks is the case of someone looking for a particular kind of mort-

[29] Abraham Bernstein and Mark Klein, "Discovering Services: Towards High-Precision Service Retrieval", in Proceedings of the CaiSE 2002 International Workshop, WES 2002, on Web Services, e-Business, and the Semantic Web, Springer, pp. 260–275, 2002.

gage. A keyword-based query to find mortgage services that deal with "*payment defaults*" would also match descriptions like "*the payment defaults to $100/month.*" A further example is when we may be looking for a mortgage service where insurance is provided for payment defaults. A keyword search would not distinguish this from a service that provides insurance for the home itself. Improving the precision of retrievals is a matter of ensuring that important roles and relationships are made explicit in the query and the service model, thereby eliminating unintended meanings.

Process modeling languages allow the salient behavior of a service to be captured in an unambiguous way. The process models, as well as their component subtasks and resources, are described in the process ontology. Queries are then defined, with a process query language called PQL, to find all the services whose process models include a given set of entities and relationships. The greater expressiveness of process models, compared with keywords or tables, offers the potential for substantively increased retrieval precision, at the cost of requiring that services be modeled in a more formal way.

The system builds on the MIT Process Handbook project, a process ontology, which has been under development at the MIT's Center for Coordination Science (CCS) for over ten years.[30] The growing Handbook database currently includes approximately 5000 process descriptions ranging over areas as diverse as supply chain logistics and staff hiring.

PROCESS INTEROPERABILITY

Process interoperability is the ability to orchestrate and/or compose processes both within and across working settings or organizational boundaries. An interest in process orchestration has been well established in the B2B communities for supply chain management in virtual enterprises. Process composition is being driven by the increasing interest in Web Services and Service-Oriented Architectures for building more flexible systems from both existing legacy functions and services that can be found dynamically through directories. The area of interoperability is currently an emerging field and in the authors' opinion will benefit from the developments that are occurring in Semantic Web Services and Context Mediation.

Capability Case: Process Aggregator

A Process Aggregator provides a single point of management for processes that require coordination across a variety of working contexts. A Process Aggregator collects and assesses information from disparate data sources and information services, resolving semantic differences. Through coordination mechanisms, it manages

[30] T. W. Malone, K. Crowston, J. Lee, B. Pentland, C. Dellarocas, G. Wyner, J. Quimby, C. Osborn, A. Bernstein, G. Herman, M. Klein, and E. O'Donnell, Tools for inventing organizations: Toward a handbook of organizational processes, Management Science, vol. 45, pp. 425–443, 1999.

processes that comprise multiple work flows, resolving any semantic or contextual differences between work settings. Through context mediation, the Process Aggregator can resolve potential semantic conflicts between the users and providers of a Web Service.

Solution Story: Ford e-Hub—a Process Aggregation Architecture

We provide only a short mention of Ford's e-Hub which is described more fully in a paper by Mark Hansen et al. entitled "Process Aggregation Using Web Services" published in the WES 2002 CAiSE workshop proceedings.[31] Although this is not currently a semantic technology implementation, it is typical of the applications that semantic interoperability can address to overcome the challenges of semantic alignment and context mediation. Designed to provide Ford with EAI and B2B integration for its dealer network and its suppliers, this solution is very representative of what we expect Semantic Web Services to be doing more of in the future. The current solution uses Microsoft BizTalk as the platform for the Process Aggregation Architecture.

POLICY AND RULES INTEROPERABILITY

As the Web becomes a place for people to do work, there is a growing need to integrate process, application, and knowledge with the support for collaboration. Distributed working environments can share knowledge and work settings if there are knowledge models that govern policies concerning access rights in the environment. A representative capability in this area is a "Rights Mediator."

Capability Case: Rights Mediator

A rights Mediator' provides a single point of mediation for authorizing access to services and resources that need controlled access policies. Rights Mediators are traditionally found as access authorization engines for systems that manage access to critical data and services. In Digital Rights Management (DRM), Rights Mediators sanction access to diverse types of multimedia objects, against complex models of ownership, royalties, and access permissions.

Solution Story: RightsCom Policy Engine

Using OWL and semantic technology from Network Inference, RightsCom has built an integrated solution for rights management in the media and entertainment

[31] Mark Hansen, Stuart Madnick and Michael Siegel, "Process Aggregation Using Web Services", Web Services, E-Business and the Semantic Web, CAiSE International Workshop, WES 2002, Toronto, Canada, pp. 12–27, 2002.

industry. The "Digital Rights Management (DRM)" problem is a complex semantic challenge:

- Complex agreements cover exploitation of intellectual property rights—individual and "blanket"
- Overlapping, nesting, and overriding global requirements, territorial variations
- Different legal frameworks and understanding, frequently inconsistent semantics
- Agreements control very large financial flows, and "Approximation" is not good enough

Effective rights management requires metadata for content and rights to be aggregated from many different sources. Services need to be provided both for "Right Users", people and organizations that want to use content, and for "Rights Controllers," people and organizations that want to benefit from rights, as creators, owners, or administrators. Consider the scenario of having to manage rights for a TV program that uses complex multimedia objects. The context model for such a scenario is illustrated in Figure 10.22.

The model illustrates how:

- Composers "assign" their rights to music rights societies and music publishers.
- Recording artists "assign" rights to record companies and sound recording rights societies.
- Music publishers are members of (or assign rights to) music rights societies, enter into subpublishing agreements with other music publishers.
- Record companies are members of (or assign rights to) sound recording rights societies, enter into subdistributions agreements.
- Rights societies enter into reciprocal agreements with each other.

Consider the following motivating scenario. A producer at an Australian TV company is compiling a multimedia documentary for broadcast and resale anywhere in the world. The documentary draws together content of many different media types, with metadata from different sources and structures. The producer needs to discover and capture information about the rights that apply for the resources used in the documentary, to secure the necessary licences and to support his own resale or licensing of the documentary when it is made.

RightsCom has developed a mapping tool set that provides the basis for translating between one data set and another, thus enabling semantic interoperability. This enables disparate DRM systems—internal or external—to work together in a consistent, automated way. The solution uses a DRM ontology, from Ontologyx, supporting the mapping and integration of complex metadata with semantics based on a powerful "context model." Cerebra Server, a software platform from Network Inference, provides querying and reasoning services for dynamic, policy-

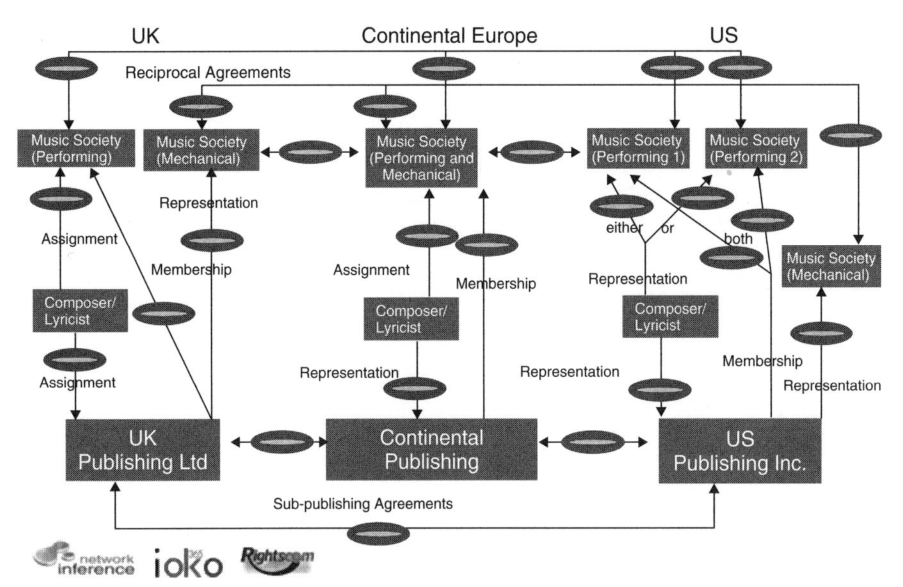

Figure 10.22 Media Rights Metadata Context Diagram

driven applications using the OWL ontologies. Systems development for the solution demonstrator was done by ioko. The DRM Ontology is illustrated in Figure 10.23.

Ontologyx recognizes a group of seven different "context types" that make up the basis of rights management:

1. **Assignment**—how rights are assigned and delegated
2. **Rights Statement**—claims to control particular rights
3. **Permission**—allowing someone to use something
4. **Prohibition**—stopping someone from using something
5. **Requirement**—obligations that must happen
6. **Agreement**—assertions about how rights can be used
7. **Proposal**—offers and requests for using rights

The user is able to search for *Rights Statements* (or "claims") for selected Resources. Ontologyx and Cerebra Server interact to identify candidate *Rights Statements* and enforce precedence and override policies to determine the valid, applicable *Statements*.

These contexts can extend and interlink to any required degree of complexity using the Context Model. A user is able to search for *Offers* for selected resources according to a number of dimensions. Ontologyx and Cerebra Server interact to identify candidate *Offers* and enforce precedence and override policies to determine valid *Offers*.

The benefits of the solution that Cerebra Server with OntologyX provides are:

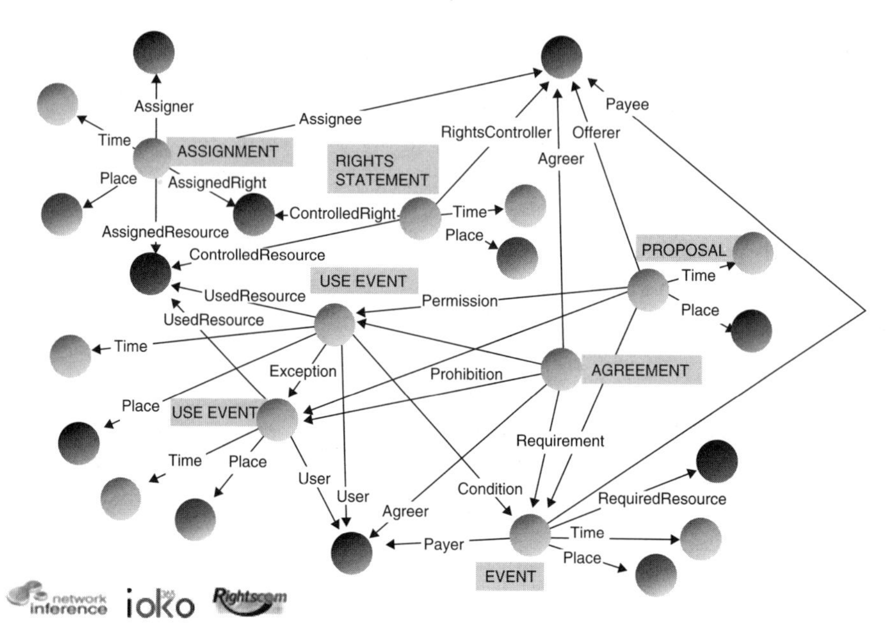

Figure 10.23 Ontologyx DRM Ontology Model

Manageability

- Individual offers are described in relation to direct content only.
- Changes are local to Ontologyx and applied by Cerebra Server at run time, maintaining the single virtual view of federated data repositories and disparate underlying formats.
- Automates find-and-acquire processes for complex multimedia rights and offer management within an example domain.
- Provides a single system point of access for multiple disparate repositories, syntaxes, and standards, underpinned by a consistent, robust semantic model.
- Automates execution of business "policies" within the query cycle.

Smart Query

- Information is maintained in multiple sources that are queryable with single "data-source and syntax agnostic" queries.
- Queries retrieve offers for complete works and subworks.
- Provides fast, deterministic query resolution for direct and indirect business queries

Accuracy

- Answers are deterministic, not probabilistic, overcoming the problem that data in this domain is especially prone to ambiguity.

- Policies are "executed" through the overlaps and overrides to provide only provable and relevant results.

Efficiency

- Digital Rights Offers are retrieved according to complex nested hierarchies, overrides, and properties.

Extensibility

- The System can be dynamically updated to reflect pace of change within the industry.
- Incomplete knowledge can be extended over time to improve quality and determinism.
- Enables simple change to be propagated dynamically for all logical consequences, without application and legacy data impact.

Interoperability

- Ontologyx mapping enables integration of different semantics structures and names from any domain.
- Integrates a data description (syntax), domain structure (how things work in the real world), and "policies" (what overrides what, at what level and dimension).
- Allows disparate data structures to be managed in a single place.
- Answers from multiple locations are maintained in disparate structured and unstructured formats and syntaxes. The underlying semantic model accurately expresses the relationship between data elements.

SUMMARY

Integration is arguably the most pressing and expensive IT problem faced by companies today. A typical enterprise has a multitude of legacy databases and corresponding applications. The disconnected systems problem is the result of mergers, acquisitions, abundance of "departmental" solutions, and simply implementation of many silo applications created for a specific purpose.

Semantic differences remain the primary roadblock to interoperability and smooth application integration, one that Web Services alone won't overcome. This point was well made by Larry Ellison, the Chairman and CEO of Oracle Corporation, when he stated:

> *Semantic differences remain the primary roadblock to smooth application integration, one which Web Services alone won't overcome. Until someone finds a way for applications to understand each other, the effect of Web Services technology will be fairly limited. When I pass customer data across [the Web] in a certain format using a Web Services interface, the receiving program has to know what that format is. You have to agree on what the business objects look like. And no one has come up with a feasible way to work that out yet—not Oracle, and not its competitors. . . .*

Semantic technology is moving into mainstream adoption, driven by the need to solve complex interoperability problems in a business ecology that is increasingly dynamic. We have shown that there are many companies at work on implementing semantic solutions. Today these are typically pilots that are showing great promise. In the next two to three years they will evolve into scalable operational infrastructure that will connect applications, data, and business processes in a secure and flexible way. The pilot experiences will give these companies a real competitive advantage over the coming years.

We hope that Capability Cases and Solution Stories discussed in this chapter have conveyed how ontology-based solutions are solving a number of interoperability problems.

Chapter 11

Adoption Strategies

KEY TAKEAWAYS

- Developing an organizational competency in modeling should be a priority.
- Be ready to change the engineering lifecycle.
- Be ready to change architecture approaches.
- Adopt a formal modeling methodology for guidelines.
- Avoid modeling myths and follow best practices.

Every few years a paradigm-shifting technology comes around that challenges IT departments to change the way they do things. We experienced such shift with the move from the mainframe to the client server architectures; we experienced it again with the move to the Internet. In each case, adopting the new technology meant acquiring brand new skills and new tools and establishing new organizational roles, new methods, and processes. Semantic technology represents such a shift.

Semantic, ontology-based technologies show great promise for the next generation of more capable information technology solutions because they can solve interoperability problems much more simply than before and make it possible to provide certain capabilities that have otherwise been very difficult to support. The way to fulfillment of the vision will be paved by pioneers in large enterprises, in organizations that can reap the fastest bottom-line returns from large-scale semantic integration. But despite the clear advantages of this technology, there are still major obstacles to be overcome before the potential of semantic interoperability will become a reality. Technology managers wanting to take advantage of semantic interoperability solutions will need to get their organizations fit for the task.

At the heart of semantic solutions is the engineering of shared knowledge structure(s) that embrace the awareness and concerns of multiple stakeholders and the commonality and variability among their needs. These structures are called "Ontologies," and their creation is done via "Ontology Engineering." Like other software engineering practices, there are resources available: tools, standards, methodologies, reusable components. However, as an emerging technology area, these resources are not yet mature or well-coordinated.

Adaptive Information, by Jeffrey T. Pollock and Ralph Hodgson
ISBN 0-471-48854-2

Gartner[1] says that the greatest challenge for the ontology domain is the relative scarcity of skills. It adds that organizations can draw to some extent on the established integration competencies, but semantic modeling and the unification of information models will severely task most enterprises.

Challenges notwithstanding, as we have shown in Chapter 10, ontologies have been deployed (often originally in a research setting) in several business areas, including health care, government, and commerce. Not only do these projects provide "proof of concept" for semantic applications in varying contexts, they also provide resources for future ontology projects.

In this chapter we give advice on how to realize semantic interoperability solutions. We cover methods and techniques across the lifecycle and issues relating to adoption by:

- Identifying skills needed for design and implementation of semantic interoperability solutions
- Examining methods and processes for ontology engineering
- Outlining a practical approach for the application of ontology-enabled solutions

The approach focuses on understanding the evolving capabilities of the technology well enough to execute an effective solution, together with recommending an essential ingredient—forging a shared understanding among key stakeholders in a suitable initial project.

ACQUIRING NEW SKILLS AND COMPETENCIES

New technology requires new skills, new tools, and new ways of working. Like the introduction of databases two decades ago and the introduction of Web technology more recently, introducing semantic models into IT departments will have a significant impact. Table 11.1 compares and contrasts the impact of introducing semantic modeling to organizations with the changes that accompanied some earlier technology transitions.

Although ontology design is in some ways similar to database design, there are also important differences. For example:

- Operational databases are intended for one application; conversely, ontology models are intended for many applications across a wide domain.
- Databases do not typically anticipate future design considerations; ontology models need to be designed for extensibility and be abstract enough to accommodate future changes to the domain.

Semantic modeling can benefit from insights and best practices that were formulated for object modeling. Unlike entity relationship modeling for relational databases, tools used for ontology development do not enforce a formal approach for

[1] Gartner Report, "Semantic Web Technologies Take Middleware to the Next Level", 8/2002.

Table 11.1 Data Table—Semantic technologies skill matrix

	Databases	Web Technology	Semantic Models
Organizational Roles	Database designers, database administrators	Web masters, Web content managers	Ontology modelers, information resource managers
Methods and Techniques	Data models, ERA models, normalization	Information architecture, content models	Logic programming, domain modeling
Languages and Skills	SQL, data modeling, product specific skills	HTML, Web Scripting Languages such as Java Script, product-specific skills	RDF, DAML, OWL, product-specific skills
Tools and Technologies	Database servers, data modeling tools	Web servers, Web application servers, Web development tools	Ontology servers, inference engines, ontology engineering tools

creating semantic models in RDF and OWL. This leaves it to the individual modeler to address questions such as:

- When it is appropriate to structure properties by grouping related properties together? When should they be kept flat?
- When should an object refer to its constituent parts ("hasPart")? When should the constituent parts refer to the object ("isPartOf")?
- When is it necessary to explicitly type the objects at creation? When is it better to leave them untyped and infer type dynamically?

The flexibility of the tools makes it critical for organizations to establish formal ontology design guidelines.

How does an organization successfully take new methods and tools on board and use them to the best advantage? A solution that has worked well during the previous waves of new technology adoption is to combine the technical adoption with training in new development techniques. A serious investment in new technology demands an equally serious investment in training. In fact, we recommend this as the first step in the adoption journey. In addition to reading books like this one, technology managers will benefit from attending briefings on semantic technologies that feature examples and case studies and discuss lessons learned. We also recommend sending your key architects to technical training on RDF, OWL, and modeling techniques. Not only will you begin to grow required technical skills, but the learning experience can also help to identify the right first projects and areas of application of semantic solutions in your organization. A growing number of conferences showcase semantic interoperability solutions.

The International Semantic Web Conference (ISWC) is the place to learn about new developments in the field of semantic technology. Although it still has a strong research orientation, it also features a rapidly expanding industrial applications track. The Metadata Registries Open Forum is another international conference with a strong focus on semantic technologies. It concentrates on the following topics:

- ISO/IEC 11179 Metadata Registries
- Model Driven Architecture
- ebXML
- MetaModel Facilities
- Case studies
- Terminology/Ontology/Concept Structures

The XML Conference and Exhibition from IDEAlliance offers interoperability demonstrations as well as talks and tutorials on Topic Maps. Some of the recent presentations at this conference covered the following topics:

- Information Architecture with XML: From *Lingua Ubiquita* to *Lingua Franca*
- Interoperability—The Undelivered Promise?
- Towards Semantic Interoperability of XML Vocabularies
- Semantic Integration at the IRS: The Tax Map
- Semantic Web Servers—Engineering the Semantic Web
- Know What Your Schemas Mean: Semantic Information Management for XML Assets

Semantic interoperability solutions are also discussed at Wilshire Meta-Data Conference and Delphi Group events.

Outside of conferences, training programs on semantic technology are now becoming publicly available. In collaboration with Jim Hendler's[2] MINDLab, TopQuadrant offers a regular program of tutorial and workshops. The program includes training for managers as well as technical personnel. The classes cover business cases for semantic technology, case studies, best practices for ontology development, semantic web services, Topic Maps, RDF, and OWL education. Custom seminars are also offered.

Several companies offer Topic Maps education. Isogen has a two-day training seminar. e-Topicality and Infolum in the United States and Techquila in the United Kingdom offer private seminars and tutorials.

Once you are ready to start your first semantic interoperability project, a well-designed training program can help the entire team to learn the methodology, techniques, language, and tools needed. We also find that bringing expert consultants in

[2] Jim Hendler is a professor at the University of Maryland and the Director of Semantic Web and Agent Technology at the Maryland Information and Network Dynamics Laboratory. As Chief Scientist and Program Manager at DARPA for the DAML program, he has been one of the major drivers in the creation of the Semantic Web and continues to be a prominent player in the W3C's Semantic Web Activity.

to advise and mentor the team can be valuable in jump-starting a project that is using a new technology.

MANAGING IMPACT ON THE ENGINEERING LIFECYCLE

Lifecycle Activities of Ontology Engineering

Gartner notes that "beyond the challenges of initial development, many ontology proponents underplay the needs for ongoing information-management processes at the enterprise level." Regardless of the details of the overall methodology for ontology construction and deployments, there are five phases that an ontology goes through. The first four phases deal with the initial development and deployment of the semantic models; the fifth phase concerns the ongoing maintenance. These phases are:

- Creating an ontology
- Populating an ontology
- Validating an ontology
- Deploying an ontology
- Evolving and maintaining an ontology

A complete ontology lifecycle is shown in Figure 11.1. The circle indicates the ongoing nature of processes needed to support semantic solutions.

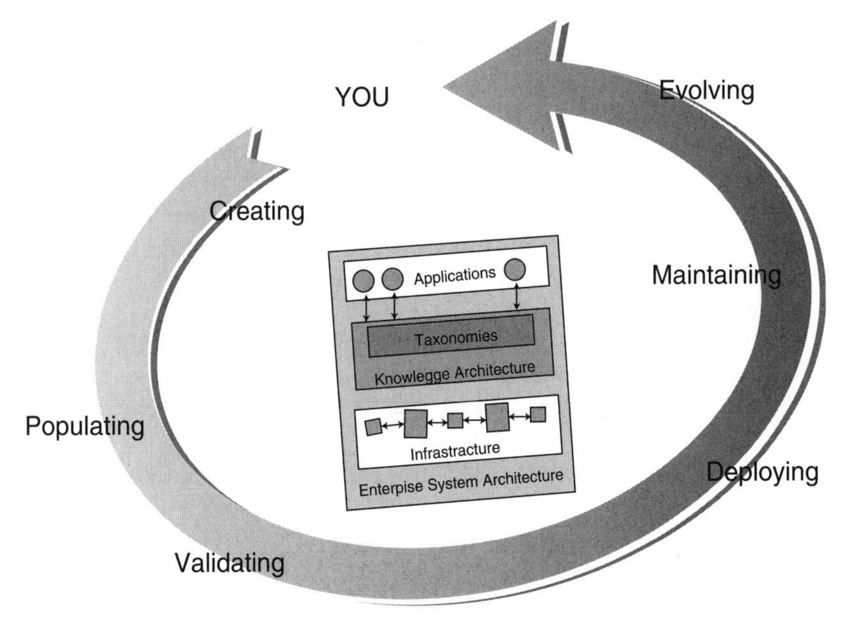

Figure 11.1 Lifecycle of ontology engineering

Several methodologies described later in this chapter offer processes for creation of ontologies. Typically they separate the tasks of capturing ontology concepts from the tasks of encoding them into knowledge representation languages such as RDF and OWL. As ontology editing tools mature and enter the mainstream, the need for a separation between these tasks diminishes. Most organizations have ontology information available in legacy forms, such as database schemas, product catalogs, and yellow pages listings. Many of the recently released ontology editors import database schemas and other legacy formats, for example, Cobol copybooks. Often, it is also possible to reuse, in whole or in part, ontologies that have already been developed in the creation of a new ontology. This brings the advantage of being able to leverage detailed work that has already been done by another ontology engineer. Finally, text mining can be used to extract terminology from texts, providing a starting point for ontology creation. Machine-learning algorithms may also be useful on database instance data when attempting to derive ontology from large quantities of denormalized database (or any semistructured source) data.

Ontology population refers to the process of creating instances of the concepts in an ontology and linking them to external sources such as databases, part lists, product catalogs, and web pages. Population can be done manually or be semiautomated. Semiautomation is highly recommended when a large number of knowledge sources exist.

Deploying ontology requires integrating it with the applications that will use the ontology. Today, the proprietary methods, those developed by individual vendors to support deployment of their tools, offer the most comprehensive guidance on ontology deployment. It includes integration with other software in the enterprise environment such as databases, application servers, middleware and popular software packages, technology integration (i.e., using ontologies with J2EE and .NET programming frameworks), and run time environment considerations.

An ontology must bridge a gap between different groups of human experts and their terminology and/or between viewpoints represented in different systems. Once an ontology is created it must be examined and critiqued to ensure its ability to bridge this gap effectively. There are many ways in which an ontology can be critiqued; the most effective critiques are based on strict formal semantics of what the class structure means.

Ontologies, like any other component of a complex system, will need to change as their environment changes. Some changes might be simple responses to errors or omissions in the original ontology; others might be in response to a change in the environment (perhaps even caused by the knowledge system of which the ontology itself is a part!). As additional models are developed, ontology integration becomes important. Some of these models will be developed separately for overlapping domains, creating integration challenges. Fortunately, there are a number of proven technical approaches to solving these challenges, including:

- Enhancing two ontologies to produce a single, larger ontology. This strategy has proven useful in many Topic Map and RDF-based solutions

- Point-to-point mapping. Two ontologies are shown side by side. A user explicitly maps concepts and relations from one side to the other.[3] These connections are translated into logical constraints on the ontologies. A number of vendors provide tools to support this kind of mapping.
- Merging ontologies into a single master ontology[4]

As more examples of situations in which ontologies need to be merged in enterprise settings are encountered, these and additional methods will be refined and enhanced. In addition to the technical issues in merging ontologies, there are also methodological issues surrounding this process. An ontology that is designed for a specific small application must be able to participate in an enterprise-wide data federation with other ontologies. This means that the original design of the ontology must take into account, in some way, applications beyond the single application for which it is initially designed. This is a classic problem[5] of Domain Analysis.[6] Domain Analysis methods, such as the Organization Domain Modeling method described later in this chapter, suggest that a solution to this design goal is inclusion of an analysis of stakeholder groups in the design of the ontology.

Project Management and ROI Considerations

One of the signs of a mature technology is the development of a clear value proposition for its application in particular settings.[7] The boundaries between various technologies are clear, as are the trade-offs of using one technology vs. another. Yet even in such a situation, it can be difficult to determine the right niche for a new project that can deliver a clear return on investment. Even though the capabilities of a mature technology are well-known, there can be many ways to apply them, in both small-scale projects and large-scale ones. The relationship of the value proposition of the technology to the enterprise's bottom line can vary dramatically from one project to another. The process of settling on project requirements that set realistic goals with a reasonable outlay of resources is a difficult one.

It is no wonder that, even in the case of mature technologies, a common source of failure for software projects stems from a failure of business managers to be able to match their needs with the capabilities of technology solutions.[8]

[3] Maier, J. Aguado, A. Bernaras, I. Laresgoiti, C. Pedinaci, N. Pena, T. Smithers, Integration with Ontologies, WM 2003, Lucerne, Switzerland.

[4] N.F. Noy and M.A. Musen. PROMPT: Algorithm and tool for automated ontology merging and alignment. In *Proceedings of the National Conference on Artificial Intelligence* (*AAAI*), 2000.

[5] Simos, M. et al., *Organization Domain Modeling* (*ODM*) *Guidebook*, 1995.

[6] Simos, M.A.: Organization domain modeling (ODM): Formalizing the core domain modeling life cycle. In: *Proceedings of the 1995 Symposium on Software Reusability*, Seattle, Washington (1995).

[7] Moore, G.: *Crossing the Chasm*. Harper Business, revised edition, 2002.

[8] There is a common consensus that the business/IT enablement gap is responsible for many IT projects failures: cancellations, cost overruns, underdeliveries. It is also the reason for somewhat different, but much-related failures—the failures to realize business value and the full potential of technology investments. See, for example, surveys from *Standish Group International, Inc. 2000*, the *Gartner Group, 9/00*, and the *IBM Chantilly Study, 1998* that concluded that over 70% of failed projects fail in the "Requirements Phase."

For emerging technologies, the situation is further complicated by the fact that the value proposition of the technology is not clear, nor the distinction between the capabilities of distinct offerings. Innovators and early adopters face the challenge of determining needs in their organization that can be reflected in some measurable value, without the benefit of either a clear statement of capabilities for a technology or a body of referencable success stories.[9]

There are also some unique challenges that a business unit faces when deploying semantic technologies. To bring value to an organization, semantic technologies entail several stakeholders interacting in many roles. Providing value for all these stakeholders requires executives to think enterprise-wide, but adopting a new technology requires them to act locally. In this sense, the adoption path of semantic technologies is similar to a business plan.

Semantic Technology promises to allow an organization to bring all its data together, to find what it needs to know. This is an enterprise-wide application of technology. Whenever a new technology is brought into an enterprise, there are challenges that change agents and technology innovators face to make such a project successful.

The W3C is outlining a grand vision of a semantic web[10] that is as extensive as the WWW of today. In the Semantic Web the meaning of content will be confected using the semantic capabilities of ontologies. This will allow semantics-based web spiders to collect information from web pages and assemble it in new and informative ways that are unimaginable with today's text and layout-based information sources.

If you believe in the Semantic Web vision; Just as circa 1998 it was time to "wake up and smell the XML," the same wake-up call is true today for Ontologies. As was the case with the Web in the mid-1990s, early adopters will be the ones who will disseminate their information to a wider audience earlier and will reap the most benefits from the new Web.

The grand vision offers a way to inspire work on semantics, but the immediate business value will come from smaller efforts within enterprises: semantic integration of multiple data sources, resolution of merger and acquisition units, and rationalization of information silos within large organizations. What are some specific areas that can be used as a launch pad on the way to the grand vision? With today's increasing focus on ROI, semantic solutions face the challenge of finding areas where they can relatively quickly outperform alternative, more mature technology.

Some prominent ontology methods discussed in this chapter, such as TOVE and On-To-Knowledge, stress that beyond assaying the domain and selecting appropriate representations (standards), tools, methods, and so on for a project,[11] you should understand the application area first and then build the ontology. We agree with this perspective but take it a step further by observing that to select and understand the

[9] Moore, G.: *Crossing the Chasm.* Harper Business, revised edition, 2002.

[10] "The Semantic Web", Tim Berners-Lee.

[11] Grueninger and Fox, The Role of Competency Questions in Enterprise Engineering, *Proceedings of the IFIP WG5.7 Workshop on Benchmarking—Theory and Practice*, Trondheim, 1994.

application area you must first envision the value proposition and business case based on what is currently possible with the technology.

This poses a problem for the adoption of ontology-based technologies. If a domain has to have a certain technical complexity, and relate to a large number of stakeholders, before the value of semantic technology can be shown, then how can an early adopter "establish a beachhead" with a simple example success?

The way out of this dilemma is to look carefully at the value propositions for semantic technologies, and find ways to provide value without having to commit to a large, enterprise-wide (high-investment, high-risk) project. Fortunately, while each of these value propositions increases with the scope of the application, there are possibilities for some value to be found in smaller projects. For example:

- **Semantic integration**—Although the value of semantic integration increases with the number of systems integrated, and indeed the "big win" only comes when integration happens at an enterprise level,[12] it is still possible to reap substantial gains in a situation with fewer stakeholders. The complexity of adoption for a semantic integration solution increases not only with the number of legacy applications but also with the number of stakeholders. Situations in which a small number (best: one!) stakeholders control a few legacy data sources are ideal.
- **Concept-based search**—The value of concept-based search increases with the technical complexity of the domain of search. In the case of many legacy help desk applications, all the data is controlled by a single stakeholder and much of the semantic data is already represented—syntactically—in some database. A fairly modest migration project can provide substantial gains.
- **Semantic "publish and subscribe"**—This opportunity area is compelling but involves two significant but different stakeholder groups—the content creation group and the content usage group. The secret to establishing a beachhead in this case is to find a situation in which the content providers will be empowered by the capability to make their knowledge known.

Insider Insight—Jeff Dirks on the Financial Validation for Semantic Interoperability

Projects that enable semantic interoperability contribute both easily measurable and hard-to-quantify financial benefits. While one is tempted to emphasize critical success factors such as improved data quality, interoperability, simpler access to distributed information, and enhanced business agility, since these are hard to quantify, they may not be incentive enough for a skeptical CFO.

However, by focusing on hard ROI, one can create a defensible economic justification, which must illustrate superior payback versus other investment alternatives. Examples of readily quantified ROI include:

Continued

[12] "The Semantic Web", Tim Berners-Lee.

1) Reduced costs associated with reuse, measured by the number of times a database schema, XML Schema, shared taxonomy, or larger ontology is accessed from an enterprise-accessible catalog and copied or used as a starting point, replacing the effort and likely incompatibilities of reinvention.
2) Reduced cost in days required (or size of team) for information integration, measured by the swiftness of interconnected systems being mapped to a common semantic model, often replacing fragile and unscalable point-to-point integrations. Since most integration projects are expensive, greater efficiency will generate substantial savings.
3) Reduced system administration and change management expenses, by streamlining the process of proposing, analyzing, and approving changes to the semantic model, schema, or ontology, with time to update each system reduced from hours to seconds.

In addition to the hard-to-measure (but important) justifications and the quantifiable cost reductions indicated above, there is also a new—often irresistible—justification based on the mandate for greater governance and transparency. Every public company has an urgent need for a secure, visible, and thorough process to control schema, metadata, and semantic models that drive financial reporting systems. Since, for example, XML by itself does not capture contextual meaning (semantics) of data, C-level executives should endorse the urgency of semantic interoperability, since misunderstanding has serious personal consequences.

Jeff Dirks, CEO
SchemaLogic

ENVISIONING THE SOLUTION

As mentioned previously. Ontology Engineering is challenging on three levels:

Ontology Engineering Faces the Usual Challenges of Any Enterprise Project

- The difficulty of specifying technology requirements increases with the number of stakeholders.

Ontology Engineering Is an Emerging Technology

- There are a number of different business value propositions for ontologies; many of these are not clear, and some may even be in conflict with others.
- It is necessary to establish one or more small "beachhead" applications before moving on to large-scale applications.

Ontology Engineering Applies to Integration Across Contexts and Stakeholders

- Much of the value appears to be available only in enterprise-wide applications.

- Ontology applications involve several stakeholders.
- Ontologies must be specific to their context to be useful.

Many of these constraints are contradictory and pose unique challenges for an ontology engineering project. For example, we need a large number of stakeholders to achieve the value of an ontology engineering project, but large stakeholder sets increase the difficulty of system requirements specification. We need an enterprise-wide solution, but for a beachhead application we need something small. We need specific requirements to pursue a project, but the value propositions provided by the technology are themselves vague and conflicting.

Resolving these contradictions requires us to find a way to:

- Bridge the gap between a diverse set of stakeholders and a large number of indistinct value propositions.
- Provide integration value even in a small-scale project.
- Create clear plans for extending the value for an enterprise-scale project.
- Exploit currently available technology, while being extensible for technology that is to come.
- Develop ontologies that are specific to the work practice of the organization.

In the following sections, we outline an approach for designing solutions based on semantic technologies that has proven to be effective in realizing these goals. This method uses Capability Cases, business solution patterns we introduced in Chapter 10, and is called Solution Envisioning with Capability Cases.

Solution Envisioning with Capability Cases

New business solutions start in a "fuzzy," creative space. Solution success depends on a number of people with diverse interests and different ways of seeing and speaking about the world coming together to form a shared vision. These solution stakeholders include businessmen, technology vendors, solution users, and IT. Each stakeholder may relate to and speak about a future solution in a different way. Clearly, to forge shared understanding and commitment an effective process for working together is needed.

What are the qualities or requirements that such an envisioning process must satisfy? It must:

- Be both exploratory and convergent.
- Provide a common ground and common language for business people and developers.
- Provide a rapid means to fully explore the solution space.
- Create a framework for navigating today's jungle of technology options.
- Communicate best practices in the way that makes them into "catalysts" for innovation and fosters a shared vision.

- Provide a traceable decision flow from business objectives to solution capabilities.

A series of workshops on "System Envisioning" was conducted at the annual OOPSLA Conference[13] from 1996 to 2000 in an attempt to formulate a repeatable and practical set of methods that would satisfy these requirements. The workshop ideas provided important input to Solution Envisioning with Capability Cases. One of the key principles of Solution Envisioning is an acknowledgement that it is often impossible to know what a solution will look like in advance of knowing what is possible. Even with a mature technology, it is difficult to know what is possible. In an emerging technology, the situation is even worse. To address this problem, Solution Envisioning uses Capability Cases to allow stakeholders to get an idea of what is possible.

As explained in Chapter 10, Capability Case for some technology has two essential parts:

- A story of how the technology has been applied, indexed by
- A particular business value that the story demonstrates.

The business value provides a handle for the story, a mnemonic summary of why the story is relevant in its context. The Capability Case begins with a brief description of the Intent of the capability—the business value that the story provides. The Capability Case continues with a brief outline of a story where a technology was used to achieve this value.

Solution Envisioning is the process by which a number of stakeholders are brought together with a prepared set of Capability Cases. The Capability Cases are presented as a sort of "gallery" through which participants can browse.

The gallery of Solution Stories serves two major purposes in the Solution Envisioning process. First, it serves as an "existence proof" that progress can be made in certain situations with the emerging technology. Second, it provides a context of use that can be compared to current stakeholder interests. The Capabilities become a shared vocabulary for the session, allowing stakeholders to express their needs and concerns in a way that is understandable to the other stakeholders in the session. The result is a vision of a possible solution, made by taking pieces of Solution Stories and combining them into a vision that satisfies the needs of the stakeholders in the session.

How does Solution Envisioning with Capability Cases address the challenges listed above? Solution Envisioning directly facilitates "bridging the gap" between business need and technological promise by allowing stakeholders to see technologies in action.

It is difficult to determine the business value of a technology simply by understanding the technology itself; it is necessary to see it in a context in which it provides business value. This is exactly what the gallery of Solution Stories does; it provides a business value for each technology, in a real work setting. Some technologies may appear more than once, offering multiple business values. It also

[13] ACM's Annual International Conference on Object Oriented Programming, Systems and Languages.

addresses the specificity of ontology engineering to an organization's work practice, by bringing in stakeholders at an early stage of system development, even earlier than requirements gathering. The impact of a particular stakeholder group on a project is realized right at the beginning. In this way, Solution Envisioning with Capability Cases provides a setting for and a way to manage the strategic conversations on how an enterprise can achieve innovation given an increasingly diverse choice of enabling technologies and solution alternatives.

Solution Envisioning's Three Key Phases

How does Solution Envisioning with Capability Cases work? As Figure 11.2 illustrates, there are three major phases at the core of Solution Envisioning:

- **Business Capability Exploration** that results in a Business Case
- **Capability Envisioning** that creates the Solution Concept
- **Capability Design** that delivers the Solution Blueprint

Business Capability Exploration focuses on creating a structured model of the situation by expressing it in terms of:

- Forces that are influencing the business situation
- Challenges that are faced by solution stakeholders
- Business results that would satisfy them

Once Forces, challenges, and Results are confirmed, they will be used to link to the best possible set of capabilities that might prove useful to the organization. This sets the stage for the next phase of Solution Envisioning—Capability Envi-

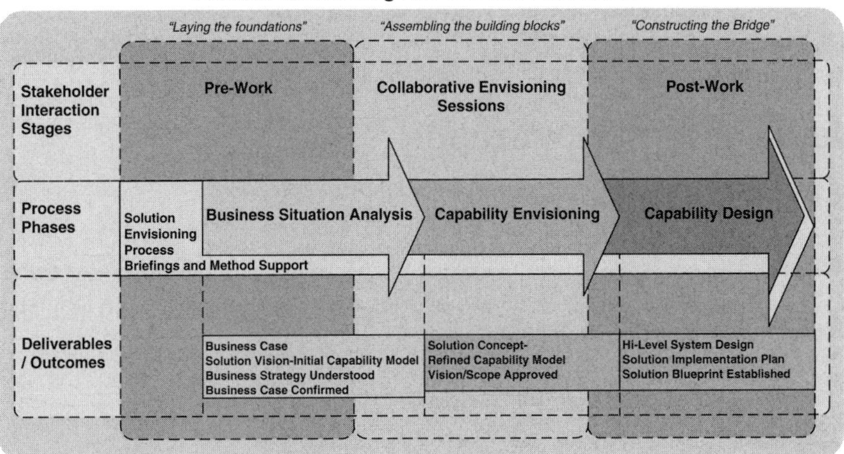

Figure 11.2 High-level process structure of solution envisioning

sioning—and progresses the process in a structured way. The gallery of selected Capability Cases is the centerpiece of the Capability Envisioning step, enabling the stake-holders to interact with one another in the context of actual Solution Stories.

Finally, the consensus from the Capability Envisioning step is made concrete through a process of Capability Design, in which capabilities are combined and architected to provide a conceptual architecture for a solution that addresses the common understanding of stakeholder needs.

We have applied Solution Envisioning with Capability Cases successfully in a number of semantic technology projects, including:

- Planning a system to manage engineering designs
- Determining requirements for long-term archives
- Knowledge management for a services company
- Designing a help desk for an electronics manufacturer
- Providing citizens improved access to government services

The applicability of the process is limited only by the availability of Solution Stories. As the field of Semantic Engineering matures, more and more successful Solution Stories will become available, providing a larger range of material for the Solution Envisioning process to draw on.

CHOOSING THE RIGHT ONTOLOGY ENGINEERING METHODOLOGY

Ontological Engineering is still a relatively immature discipline. The first ontology engineering methodologies were proposed in 1995 on the basis of the experience gathered in developing the Enterprise Ontology[14] and the TOVE (TOronto Virtual Enterprise) project ontology[15] (both in the domain of enterprise modeling). Today, there is no single commonly accepted comprehensive methodology; instead, there is a variety of methods. Some of the methods are generic and tool independent (we call them nonproprietary methods), whereas others are more specific to the technology used. What follows is an overview of the key methods.

NONPROPRIETARY METHODS

Uschold and King (1995)

This methodology is based on the experience of developing the *Enterprise Ontology*, an ontology for enterprise modeling processes. It recommends the following process for developing ontologies:

[14] Uschold and Grueninger, "Ontologies: Principles, Methods and Applications", *Knowledge Sharing and Review*, 1996.

[15] Grueninger and Fox, The Role of Competency Questions in Enterprise Engineering, *Proceedings of the IFIP WG5.7 Workshop on Benchmarking—Theory and Practice*, Trondheim, 1994.

Table 11.2 Data Table—Ontology engineering methodology summary

Methodology	Level of Detail	Strategy for Application Integration	Strategy for Identifying Concepts	Life Cycle Coverage	Tools
Uschold and King	Does not describe details of techniques and activities.	The process is totally independent from the application of ontology.	The most relevant concepts are identified first, then the rest of the concepts are identified using generalization and specialization.	Focused on ontology creation, it does not cover deployment, maintenance, or the work associated with requirements.	No specific tools are suggested or required.
TOVE	Provides specific guiding questions for *activity* and *organization* ontologies.	Ontology use scenarios are identified in the specification phase.	The most relevant concepts are identified first, then the rest of the concepts are identified.	Covers the requirements phase, ontology creation; also offers some guidance on deployment and evolution.	No specific tools are suggested or required.
ODM	Elaborate "plug-in" architecture to cover detailed methods; many are built in.	Ontology is structured as part of a "reuse" activity. Stakeholder analysis is used to anticipate diverse application needs.	Strong emphasis on legacy artifacts to determine basic distinctions. Stakeholder commonality & variability drive further modeling.	Covers life cycle phases of reuse library, from stakeholder analysis (prerequirements) to asset base deployment.	No specific tools are suggested or required.
Methontology	Well detailed	The process is totally independent from the application of ontology.	The most relevant concepts are identified first, then the rest of the concepts are identified.	Offers a good coverage of the life cycle processes.	No specific tools are suggested or required. ODE has been used for ontology creation.

Table 11.2 Continued

Methodology	Level of Detail	Strategy for Application Integration	Strategy for Identifying Concepts	Life Cycle Coverage	Tools
CommonKADS	Well detailed	Ontology use scenarios are identified in the specification phase.	The most relevant concepts are identified first, then the rest of the concepts are identified.	Covers knowledge aspects of life cycle through the implementation phase.	Well tooled up for ontology creation, intended to work with any of the ontology deployment tools.
On-To-Knowledge	Well detailed	Ontology use scenarios are identified in the specification phase.	The most relevant concepts are identified first, then the rest of the concepts are identified.	Offers a good coverage of the life cycle processes.	Tool support for the ontology creation as plug-ins to OntoEdit
Modulant	Well detailed	Ontology is constructed to support a specific interoperability environment.	Uses bottom-up approach: first identify known concepts, then cluster to find higher-level concepts.	Primarily focused on the analysis and design phases.	Has been created for use with Modulant's tools.
Unicorn	Does not describe details of techniques and activities.	Ontology represents a common interlingua for multiple data sources. Integration is done by "star-pattern" translation.	Makes considerable use of legacy data structures.	Offers a good coverage of the life cycle processes.	Has been created for use with Unicorn's tools.

1. **Identify purpose.** It is important to be clear why the ontology is being built and what its intended uses are.
2. **Build the ontology**, a three-step process consisting of:
   a. **Ontology capture**, which means:
   - Identification of the key concepts and relationships in the domain of interest, that is, scoping. It is important to center on the concepts as such, rather than the words representing them.
   - Production of precise unambiguous text definitions for such concepts and relationships
   - Identification of terms to refer to such concepts and relationships
   - Agreeing on all of the above

   The authors use a middle-out (as opposed to top-down or bottom-up) approach to perform this step and recommend that rather than looking for the most general or the most particular concepts as key concepts, the most important concepts be identified, which will then be used to obtain the remainder of the hierarchy by generalization and specialization.

   b. **Coding.** Involves explicitly representing the knowledge acquired in step 2.a in a formal language.

   c. **Integrating existing ontologies.** During either or both of the capture and coding processes, there is the question of how and whether to use ontologies that already exist.
3. **Evaluate the ontology**, where the authors adopt the following definition: "to make a technical judgment of the ontologies, their associated software environment, and documentation with respect to a frame of reference. The frame of reference may be requirements specifications, competency questions, and/or the real world."
4. **Document the ontology.** The authors recommend that guidelines be established for documenting ontologies, possibly differing according to the type and purpose of the ontology

This methodology outlines a solid process for building ontologies. It has been successfully used on a number of complex projects. However, it does not cover a number of areas key for organizations getting started with semantic solutions:

- It does not precisely describe techniques to be used or details of the activities.
- It does not cover application-related concerns.
- It does not offer advice or recommendation on lifecycle activities.

Grüninger and Fox (TOVE)

This methodology is based on the experience in developing the TOVE project ontology[16] within the domain of business processes and activities modeling. Essentially,

[16] Grueninger and Fox, The Role of Competency Questions in Enterprise Engineering, *Proceedings of the IFIP WG5.7 Workshop on Benchmarking—Theory and Practice*, Trondheim, 1994.

it involves building a logical model of the knowledge that is to be specified by means of the ontology. This model is not constructed directly. First an informal description is made of the specifications to be met by the ontology, and then this description is formalized. The methodology recommends that ontological design decisions be grounded in an awareness of the applications in which an ontology will be used. The applications are specified with use cases, which in turn determine the most important concepts and relations for an ontology to cover. Design begins with these concepts and moves "outward," making it a "middle-out" methodology. The steps proposed are as follows:

1. **Capture of motivating scenarios.** According to Grüninger and Fox, the development of ontologies is motivated by scenarios that arise in the application. The motivating scenarios are story problems or examples that are not adequately addressed by existing ontologies. A motivating scenario also provides a set of intuitively possible solutions to the scenario problems. These solutions provide an informal intended semantics for the objects and relations that will later be included in the ontology. Any proposal for a new ontology or extension to an ontology should describe one or more motivating scenarios and the set of intended solutions of problems presented in the scenarios.
2. **Formulation of informal competency questions.**
   - These are based on the scenarios obtained in the preceding step and can be considered as expression of requirements in form of questions. An ontology must be able to represent these questions using its terminology and be able to characterize the answers to these questions using the axioms and definitions. These are the informal competency questions, because they are not yet expressed in the formal language of the ontology.
   - The competency questions are stratified, and the response to one question can be used to answer more general questions from the same or another ontology by means of composition and decomposition operations. This is a means of identifying knowledge already represented for reuse and integrating ontologies.
   - The questions serve as constraints on what the ontology can be, rather than determining a particular design with its corresponding ontological commitments. There is no single ontology associated with a set of competency questions. Instead, the competency questions are used to evaluate the ontological commitments that have been made to see whether the ontology meets the requirements.
3. **Specification of the terminology of the ontology within a formal language.** The following steps are to be taken:

   3.1. Get informal terminology. Once the informal competency questions are available, the set of terms used can be extracted from the questions. These terms will serve as a basis for specifying the terminology in a formal language.

3.2. **Specify formal terminology.** Once informal competency questions have been posed for the proposed new or extended ontology, the terminology of the ontology is specified using a formalism such as KIF. These terms will allow the definitions and constraints to be later (step 5) expressed by means of axioms.

4. **Formulate formal competency questions using the terminology of the ontology.** Once the competency questions have been posed informally and the terminology of the ontology has been defined, the competency questions are defined formally using tools such as situation calculus.

5. **Specify axioms and definitions for the terms in the ontology within the formal language.** The axioms in the ontology specify the definitions of terms in the ontology and constraints on their interpretation; they are defined as first-order sentences using axioms to define the terms and constraints for objects in the ontology. Simply proposing a set of objects alone, or proposing a set of ground terms in first-order logic, does not constitute an ontology.

   Axioms must be provided to define the semantics, or meaning, of these terms. If the proposed axioms are insufficient to represent the formal competency questions and characterize the solutions to the questions, then additional objects or axioms must be added to the ontology until it is sufficient. This development of axioms for the ontology with respect to the competency questions is therefore an iterative process.

6. **Establish conditions for characterizing the completeness of the ontology.** Once the competency questions have been formally stated, we must define the conditions under which the solutions to the questions are complete.

TOVE is an important methodology because of its attention to the application use scenarios and the use of competency questions as a technique for bounding and focusing ontology models. It also offers some provisions for extending previously developed ontologies. However, it does not sufficiently cover the lifecycle processes and does not detail activities and techniques to be used.

Organization Domain Modeling (ODM)

Because the purpose of an ontology is to share information, the process of designing an ontology is fundamentally a process of designing for reuse. Therefore, a methodology for ontology design must include a way for the context of use of the ontology to influence the design of the ontology itself. The details of ontology design must reflect certain aspects of the work flow or organizational context in which the ontology is to be used.

The Organization Domain Modeling (ODM) methodology, developed as part of the DARPA STARS program, is a multiphased process by which the various parties who are involved in the reuse activities around an ontology (the "stakeholders") and the existing knowledge sources are carefully inventoried before any ontology modeling takes place. This methodology, although not unique to ontology development (it is applicable to any software project interested in reuse), offers well-documented process for designing reusable models.

Stakeholder interests are evaluated to determine the scope of the ontology. Once this is determined, an ontology is constructed that models the commonalities and variations among the scoped entities. Differences of opinion or requirements between two or more stakeholders are resolved by representing their common concerns at one level of the ontology and the distinctions at another. This allows any knowledge sharing to take advantage of the common interests of the stakeholders while respecting their differences. ODM departs from other application-driven methodologies by recognizing that in a reuse situation, there are multiple applications and many of them might not have been clearly foreseen at the beginning of the ontology modeling process.

METHONTOLOGY

This methodology was developed within the Laboratory of Artificial Intelligence at the Polytechnic University of Madrid. The METHONTOLOGY framework[17] enables the construction of ontologies at the knowledge level and includes the identification of the ontology development process, a life cycle based on evolving prototypes, and particular techniques for carrying out each activity. The METHONTOLOGY framework is supported by ontology development environment from the University of Madrid called ODE.

This is one of the more complete methodologies, which covers project management (planning, control, and quality assurance), development, and support activities. Development-oriented activities include specification, conceptualization, formalization, implementation, and maintenance:

1. **Specification** states why the ontology is being built, what its intended uses are, and who the end users are.
2. **Conceptualization** structures the domain knowledge as meaningful models at the knowledge level.
3. **Formalization** transforms the conceptual model into a formal or semicomputable model.
4. **Implementation** builds computable models in a computational language.
5. **Maintenance** updates and corrects the ontology.

METHONTOLOGY also outlines support activities performed at the same time as development-oriented activities, without which the ontology could not be built. These include knowledge acquisition, evaluation, integration, documentation, and configuration management.

[17] Fernández-López, Gómez-Pérez, Pazos-Sierra, Pazos-Sierra, Building a Chemical Ontology Using METHONTOLOGY and the Ontology Design Environment, IEEE Intelligent Systems & their applications, January/February 1999.

METHONTOLOGY has been used to build numerous large ontologies and corresponding applications. It has good coverage of lifecycle activities and is described at a fairly detailed level.

CommonKADs

No discussion of ontology engineering methodology would be complete without CommonKADS, a methodology for development of knowledge-based systems, which is the result of the Esprit-II project (P5248) KADS-II. CommonKADS supports most aspects of a KBS development project, including project management, organizational analysis, knowledge acquisition, conceptual modeling, user interaction, system integration, and design. The methodology is result oriented rather than process oriented.

It describes KBS development from two perspectives:

- Result perspective: a set of models of different aspects of the KBS and its environment that are continuously improved during a project lifecycle
- Project management perspective: a risk-driven generic spiral lifecycle model that can be configured into a process adapted to the particular project

Recognizing that many factors other than technology determine success or failure of IT systems, CommonKADS recommends starting by identifying problem/opportunity areas and potential solutions and putting them into a wider organizational perspective. This analysis is accomplished by a feasibility study that serves as a decision support for economical, technical, and project feasibility, in order to select the most promising focus area and target solution.

On-To-Knowledge

The On-To-Knowledge methodology was developed and applied in the EU IST-1999-10132 project On-To-Knowledge for introducing and maintaining ontology-based knowledge management applications in enterprises. The methodology covers so-called Knowledge Meta Process and Knowledge Process. The first process addresses aspects of introducing a new KM solution into an enterprise as well as maintaining it; the second process addresses the handling of the already set up KM solution.

On-To-Knowledge takes after CommonKADS in putting emphasis on an early feasibility study and on constructing several models that capture different kinds of knowledge needed for realizing a KM solution. It also borrows from the KACTUS methodology in putting an emphasis on the application to be deployed early in the ontology design process. The Knowledge Meta Process consists of five main phases. The phases are "Feasibility Study," "Kickoff," "Refinement," "Evaluation," and "Application & Evolution." Each phase has numerous activities, requires a main decision to be taken at the end, and results in a specific outcome. The phases "Refinement—Evaluation—Application & Evolution" typically need to be performed in iterative cycles.

Once a KM application is fully implemented in an organization, Knowledge Processes essentially cycle around the following steps:

Knowledge Creation

- For example, contents need to be created or converted such that they link to the conventions of the company, e.g., to the knowledge management infrastructure of the organization. This step could also include import of documents and metadata from elsewhere in the enterprise.

Knowledge Capture

- Knowledge items must be captured in order to elucidate importance or interlinkage, e.g., the linkage to conventionalized vocabulary of the company by the creation of relational metadata.

Knowledge Retrieval and Access to Knowledge

- This includes the "simple" requests for knowledge by the knowledge worker. Typically, however, the knowledge worker is accepted not only to recall knowledge items but also to process it for further use in her context.

PROPRIETARY METHODS

Most vendors develop repeatable best practices for implementing solutions built with their products. In this section we describe two vendor-specific methods.

Modulant™[18]

Modulant™ is a software company that offers a semantic mediation approach for es-tablishing the interoperability of information among applications. Modulant was founded in 2000, but its core mapping methodology comes from a merger with Product Data Integration Technologies (PDIT), a company that has been in operation since 1989.

Modulant has developed Contextia™ Information Interoperability Methodology (CIIM), a comprehensive method for implementing semantic interoperability solution using the company's Contextia tool set. The methodology defines a framework for describing the data of many applications in a consistent, computer-processable manner using a so-called abstract conceptual model (ACM).

The ACM enables resolution of conflicts and accommodation of different perspectives by mapping each application viewpoint onto the common model. The mapping specifications are created and validated with the Modulant Contextia Inter-

[18] For more information, see: www.modulant.com.

operability Workbench. They are then used at run time to derive instructions to the Modulant Contextia Interoperability Server for information-preserving transformations among applications.

Each application's data is mapped to the abstract model according to a common mapping strategy for the interoperability environment. The mapping specifications are therefore mutually independent, so that new applications can be added to the environment, or existing applications can be updated, without any impact on other parts of the environment.

In a run time environment, the mapping specifications, which consist of data and context maps for each application, control:

- Population of nonpersistent data storage, within the Modulant Contextia Interoperability Server, with application data values and metadata that represents the meaning or context of the application data
- Resolution of conflicts in format, structure, and semantics among data from different applications
- Extraction of relevant application data for presentation to a target application
- Transformation of data formats and structures

Modulant has defined procedures and practices for development and validation of the mapping specifications that support an interoperability environment. These procedures and practices partition the mapping task into a number of iterative steps with clearly defined inputs and outputs. The process consists of the following steps:

- Identify interoperability environment
- Describe interoperability environment
- Identify common concepts
- Select ACM
- Develop/extend ACM
- Develop high-level mapping strategy
- Document/maintain mapping strategy
- Analyze application data structures
- Develop ATS schema (Application Transaction Set)
- Develop data map
- Test data map
- Define data conversions
- Map data elements
- Create structure mappings
- Define population rules
- Validate context map
- Test context map

Application of CIIM and deployment of its results involves three groups of people:

- Domain/Subject Matter Experts: the people who need to share information with others, and from whom knowledge of the meaning of application data can be acquired
- Interoperability Architects: experts who develop and validate mapping specifications based on the requirements specified by disparate groups of domain experts
- IT support staff who configure and administer the Modulant Contextia Interoperability Server run time environment

Unicorn Solutions[19]

Unicorn Solutions is a new software company established in 2001 and headquartered in New York City. Unicorn calls itself a data management company and defines its mission as enabling large enterprises to manage thousands of disparate information resources based on business meaning—or semantics. Unicorn's product combines metadata repository, information modeling, hub-and-spoke mapping, and automated data transformation script generation capabilities in a single platform.

To facilitate the creation of agreed-upon information models that describe businesses and their component parts Unicorn has developed a six-step Semantic Information Management (SIM) Methodology. The methodology is structured such that each stage adds value in its own right, while simultaneously progressing the enterprise toward the overall benefits of a comprehensive SIM architecture. SIM consists of the following stages:

Gather Requirements

- Establish the scope of the project, survey the data sources relevant to the project, and capture the organization's information requirements.

Collect Metadata

- Catalog data assets and collect metadata relevant to the organization and its use of data.

Construct Information Model

- Capture the desired business worldview, a comprehensive vocabulary, and business rules.

Rationalize (Semantics)

- Capture the meaning of data by mapping to the Information Model.

[19] For more information, see: www.unicorn.com

Publish/Deploy

- Share the Information Model, metadata, and semantics with relevant stakeholders; customize it to their specialized needs.

Utilize

- Create processes to ensure utilization of architecture in achieving data management, data integration, and data quality.

OVERCOMING POPULAR MYTHS

A key recommendation is to select and adopt a defined process for ontology engineering. The process will be most effective if it is customized for an organization, aligning with and taking into account its existing software development practices.

There are some traps that a semantic engineer can fall into, which, although they might seem to solve some problems easily, in fact short cut some stakeholder issues, and therefore are more costly by the time the solution is deployed. Often, projects that fall into one of these traps fail to reach deployment at all. The ontology engineering methodology you select for the use in your organization must help you to avoid these pitfalls.

Some common traps (each with a short mnemonic name) are:

- **"Modeling is its own reward"**—You make a lovely model of the area you want to share. But you haven't deployed it in a way that embeds it in the work practice of any stakeholder. The model doesn't actually represent any stakeholders at all, and none of them has any stake in adopting it.
- **"They just have a different word for everything"**—A common approach to stakeholder modeling is to believe that if you got all the stakeholders in a room together, they could agree on a set of terms for their domain. But different stakeholders have solid work context reasons for making different distinctions—often different stakeholders don't even share the same concepts.
- **"It ain't got that swing, if you leave out a thing"**—It is easy to feel that there is no value to a model, until it encompasses everything. Good modeling has to be able to provide value at a point that is small enough to be adopted as a beachhead by some real stakeholders.
- **"One size fits all"**—You can go to the libraries at Stanford or DAML and find ready-made ontologies. But these will not represent the stakeholder situation of your work context and hence cannot make up the complete semantic model for a specific enterprise solution.
- **"I'll do it my way"**—On the other hand, there are parts of the modeling work that can be reused. The available "starter sets" and ontology modules are being developed all the time.

- **"The truth is out there"**—There is no absolute "correct" truth that can be modeled; we are modeling the concepts that help stakeholder groups get work done in their own context.

IMPLEMENTING BEST PRACTICES

Adopting a methodology can also help you to reinforce the following good practices:

- **Understand the business problem**—Conventional ontology wisdom suggests that you must understand your application before beginning a solution. But because semantic solutions mediate between stakeholders and applications, a broader approach is to understand the business situation and what business value the ontology is to provide.
- **Develop and deploy iteratively**—Make sure that the first ontology deployment you propose provides some value and can be extended to provide more value as it is extended. Planning this trajectory can be as important as any other modeling decision in your ontology.
- **Leverage available infrastructure**—Your fastest road to realizing results may be from layering semantics on top of existing solutions. Web architectures, search engines, and controlled vocabularies are a few examples of structures that already contain some semantic information that has already proven its value to some stakeholder groups.
- **Define model architecture**—Your ontology doesn't have to be one model. Often it is imperative that you have different types of models working together, in a "model architecture." This can be a way to represent the relationships of multiple stakeholder views.

Insider Insight—Curtis Olson With A Pragmatist's Viewpoint on Adoption

In practice, ensuring that data is moved from one system to another without any meaning loss is a very difficult and complex task. The key is to understand where meaning loss may happen and to prepare and react accordingly. A truly valuable Semantic Interoperability solution would translate, inform you of possible meaning loss, and help reconcile those differences. Easy to market, hard to implement.

Most of the ROI justification for Semantic Interoperability technologies comes from introducing loose coupling, abstraction, and replacing human coding with vendor-implemented algorithms. However, these are exactly the characteristics that make it more difficult to trace from Point A to Point B. When evaluating Semantic Interoperability technology, pay special attention to the processes that are "automatically" done, for you. Make sure that you are able to query, monitor, and fine-tune these components—as any black box adds significant effort and risk to new projects. Here are a few questions to answer during any evaluation:

- How will I know when data is being transformed correctly?
- When are changes to loosely coupled systems identified?
  - At design time? At run time?
- Can data transformation rules be easily understood?
- How are data definition mismatches identified?
- Can data collisions be reconciled or transformed?
- How robust is the data error recovery/feedback loop to help reduce data collisions?
- Can I trace data from one source to another?
  - From one to many? From many to one?

Consider a similar market shift from hierarchical to relational databases. The promise was to release developers from having to understand the physical structure of the data (e.g., Get Next Child, symbolic links) and let them worry about the logical data structures driven by RDBMS optimizers. The reality was that developers had to rethink how they developed applications and, in many cases, how to work around new technology inadequacies. The promise of RDBMS was eventually achieved, and those early adopters were able to use that newfound expertise to create competitive advantage.

The new approaches created by Semantic Interoperability software will create new roles, initiate new methodologies, and require you to challenge long-standing assumptions. However, these technologies can be mastered using tried and true development principals and sound judgment. As Thomas Edison said, “The three great essentials to achieve anything worthwhile are first, hard work; second, stick-to-it-iveness; and third, common sense.” A pragmatic approach for technology innovation that has stood the test of time.

Curtis Olson
Founder, DataTier Technology

Chapter 12

Tomorrow's Adaptive and Dynamic Systems

KEY TAKEAWAYS

- Adoption is increasing at a rapid pace.
- Semantic interoperability is the natural evolution of computing.
- Research from universities and corporations continues to pave the way.
- Key technologies are already in place to enable more adaptive future.
- Semantic technologies have not yet "crossed the chasm."

A lot of ground has been covered so far. But for the final chapter in this book the authors will look toward the future. First, big picture market trends will be identified and shown to indicate that semantic technologies are poised to dramatically impact the entire software industry. Major forces, such as the ongoing information explosion and the speed of change barrier, continue to drive the industry toward change—adopt semantic technologies or die. Second, a brief survey of the current adoption rates and momentum for semantic technology will be evaluated for various commercial industries including manufacturing, financial services, and life sciences. Third, major research programs and key technologies to watch will be chronicled for the reader to evaluate. Finally, the most significant barriers to adoption will be considered and the authors will offer advice to companies interested in "crossing the chasm" with semantic technology.

BIG PICTURE MARKET TRENDS

The market for semantic interoperability is expected to grow fairly quickly, fueled by the needs of enterprises and by the growing maturity of the AI (artificial intelligence) technologies that underlie these solutions. According to a report released by the Business Communications Company,[1] it expects that AI technologies that assist

[1] For more information, see: www.bccresearch.com.

Adaptive Information, by Jeffrey T. Pollock and Ralph Hodgson
ISBN 0-471-48854-2

existing applications to handle more complex data analysis, addressing the potential variability in a situation via a set of rules, will see strong growth and implementation across sectors. Its estimate is that this technology will reach $4.8 billion in sales and an AAGR of 14.5% through 2007.

Sector growth numbers and predictions about semantic technologies cannot tell the whole story. Much depends on assumptions made by the analyst creating the numbers—Is the sector an artificial intelligence sector or a semantic web sector? Do you measure the value of underlying OEM semantic platforms in third-party tools or just the top-shelf semantic applications for application interoperability? No doubt there are many ways to slice the sector and draw boundaries around the technology. Perhaps the best evidence for future growth is the number of start-ups in the semantics sector, nearly two dozen between 2000 and 2003 alone. More significantly, most major software companies, including IBM, Microsoft, HP, and Sun, have had active research programs in the field of semantics since 2002 or earlier.

These indicators of future market growth are with good cause. Persistent challenges to software manufacturers in the form of information management and change management are reaching critical proportions. The sheer volume of information that large businesses must maintain and leverage is staggering; the quantity of data has outstripped software's ability to make sense of it. Likewise, with the speed of business changes increasing year after year, software is now incapable of keeping pace and adapting to new demands placed on it from business.

Responding to Information Explosion

As discussed in Chapter 1, new digital information is being created faster than any time in history. In fact, this exponential rise in information volume is expected to continue indefinitely into the future. Without new software capabilities to better manage all this new digital data—with its meaning intact—businesses will drown in data they can't use and struggle to achieve competitive advantage without knowing their own business as well as they could. This rate of digital overload is an uncontestable fact.

The data storage industry is responding to this trend with enterprise data center software and new hardware platforms with ever larger and faster disk storage technology. The infrastructure to write data to disk and manage the physical distribution of that data is sophisticated and well developed. But without semantic technologies, the infrastructure to enable the widespread *understanding* of all that new data is nonexistent. No other software technology on the horizon has the possibility to improve the usefulness of data to the same degree that semantics-aware software will.

Confronting the Speed of Change Barrier

The speed of change barrier is the minimum amount of time it takes to reconfigure and reconnect business systems (supply chain, product management, customer management, etc.) in order to adapt to new business demands like mergers, acquisitions, partnerships, and business strategy changes. This speed barrier has two key factors: human culture and technology rigidity. Semantic interoperability tackles both problem areas squarely with innovative approaches.

Human culture differences lead to conflicts of meaning among business information resources and processes. These culture differences are a natural and inevitable reality of large organizations; in many cases, culture differences are actually encouraged to improve morale and competitiveness. Likewise, the rigidity of technologies in business today is related to both the tight coupling required by most middleware solutions and the vast amounts of custom-written code in nearly every layer of an enterprise infrastructure.

Semantic technologies address the culture problems by providing a mediating force in these differences of meaning. Semantic technologies address the rigidity of enterprise infrastructure by enabling loose coupling of data and the use of models to drive granular software configurations. In summary, semantic interoperability is a capability earned by overcoming the most pressing performance barriers to change in modern business.

Harnessing Technology Waves

To start this book, the concept of the Third Wave was introduced to show how the forces seen in the industry today (information overload, speed of change barriers, etc.) are quite natural in this era of transition between industrial and information production modes. This is just the beginning of a long period of information-driven value creation. Within this new period of information-led growth, there will be several smaller waves of innovation and improved capabilities.

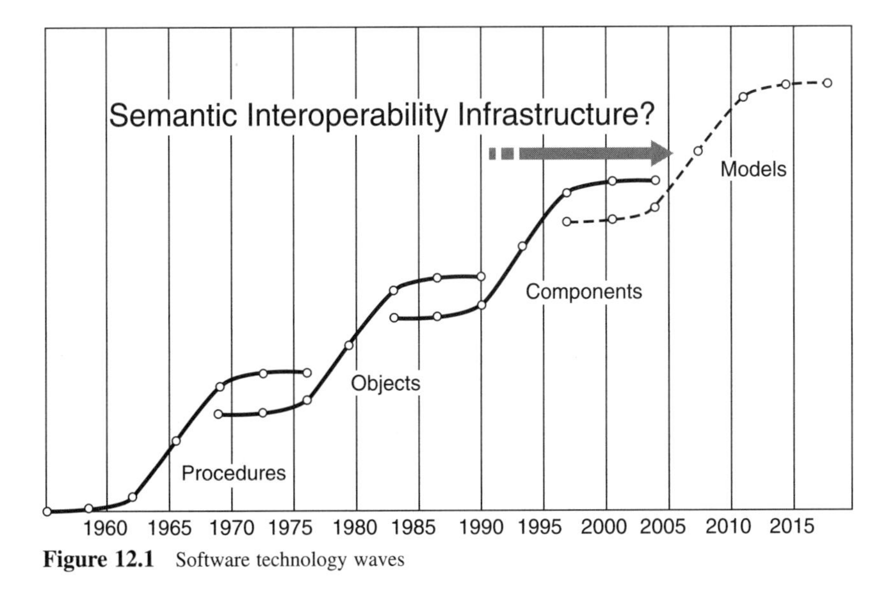

Figure 12.1 Software technology waves

Table 12.1 Data Table—Evolution of software technology

	Style	Focal Point	Data Layer
1955–1975	Procedural Programming	Syntax	Hierarchical
1975–1995	Object-Oriented Programming	Structures	Relational (RDBMS)
1990–	Component Programming	Services	Document Markup (XML)
2000–	Model-Driven Programming	Semantics	Data & Ontology (RDF/OWL)

Each of these technology waves has produced dramatic changes in the way engineers view software development, but also in the tools that they have at their disposal. Everything from the data persistence structures to programming style has changed during these 15 to 20-year cycles.

Today these shifting technology waves offer opportunity for innovators and early adopters. Whereas laggards are still shifting to object technology and the late majority is just starting to evaluate Web Services component technology, model-driven technology has yet to appear on the radar of many companies. Companies who can harness this shift to model-based, semantic interoperability technology will enjoy a lead on their competition.

INDUSTRY ADOPTION OF SEMANTIC TECHNOLOGY

As we have seen, a number of forces are driving the need for more embedded semantics in software systems. Many forces arise from how the Web is evolving from a fairly static world to being a dynamic place to do work. Other areas driving the need for more semantics include application integration, data consolidation, and better decision making across disparate knowledge sources both within and across extended enterprises. In these situations, a number of challenges confront enterprises wishing to use their information resources to obtain business value and gain competitive advantage:

- How to manage the full spectrum of information sources, from structured data to unstructured content
- How to deal with the growth in the variety of types and formats of information and an increasing use of richer media types
- How to access and reconcile content and data in decentralized and diverse locations across the enterprise in the presence of multiple-point solutions to discrete business problems at each and every business unit
- How to make knowledge actionable through integration in business processes where it can realize business value
- How to deal with changing business rules arising from dynamic partnerships and changing regulations

Traditional development platforms were designed for creating new software (i.e., writing code)—not for making sense of and reusing the tens of thousands of application services that already exist in the enterprise. As a result, developers and business analysts spend much of their time on mundane tasks—determining what applications, Web Services, and other information sources are "out there" and how they can be used to solve business problems. To further the problem, much of the intelligence gained from these tasks resides in the heads of IT personnel, preventing sharing and reuse. Middleware and "EAI" solutions such as integration brokers and adapters provide critical "plumbing" infrastructure, but these technologies do not reduce the labor required to integrate disparate information and services, nor do they answer the fundamental questions of any integration project.

In e-Commerce, the diversity of information and the desire to make well-informed decisions drive the need for new capabilities. In Supply Chain Management, matchmakers, mediators, and semantics-based negotiators are helping to create the next generation of B2B value nets and market hubs. Some reasons that motivate the use of semantic technologies in B2B e-Commerce include:

- Discovery of applicable products and services is greatly facilitated by advanced search and knowledge-based navigation.
- Profiles of buyers and sellers can be well represented in semantic markup languages.
- Context awareness, based on knowledge models of working settings that provide role-based personalization, can improve the user experience.
- Translating product and service information between different sellers and buyers is well supported by ontology-based mappings optionally with rules to handle complex matching and transformations.

Semantic Technology Application Areas

In the finance sector, semantic technologies are being used for data consolidation and information exchange across databases with different schemas. In one example, a mutual fund company with customer information found in hundreds of databases uses semantic technology to do conversions, real-time data integrity checking, data reconciliation, and inferencing of new knowledge about customer behavior.

As the Web becomes more of a place for doing work, support for effective collaboration is a key capability. An environment for collaboration provides an enriched user experience when there is awareness of personal interests and goals. Advanced personalization is driving new capabilities for context-aware retrieval of knowledge and the more accurate delivery of information from real-time sources. Business social networks are also becoming enabled by semantic technologies that allow common interests and working relationships between people to be explored.

With the adoption of Web Services, application integration is becoming more dynamic and offers the possibility of software reuse through the discovery and composition of capabilities over internets and intranets. Although Web Services have the basic mechanisms for executing software functions over the Web, they lack sufficient semantics for effective discovery and for ensuring the integrity of services composition. Semantic Web Services, based on emerging semantic standards, promises to provide solutions for efficient service discovery and reliable services composition.

An important role for semantic technologies also exists in Enterprise Architecture and Enterprise Technology Management. There is an increasing need for today's enterprises to have agile ways of managing policies and portfolios of enterprise applications. Alignment with business goals, strategies, and other stakeholder interests can be ensured when an adaptive policy management system is in place. Knowledge about enterprise architectures can be represented in semantic models, which can then be consulted to determine what things do, why they are important, and what impacts will occur if they are changed or removed. Adaptive ontology-centric systems like this can answer questions such as:

- What applications, Web Services, and other information sources exist in the IT environment?
- How can I combine these information resources to solve business problems?
- What happens when the IT environment changes?
- Which business goals are being addressed by which IT capabilities?
- What measures of effectiveness are the IT systems able to provide me?
- How do these measures relate to other business goals?

With real-time knowledge about business activities enabled through semantics-aware tools, we get closer to realizing the vision of the "Adaptive Enterprise." The adaptive enterprise is one where decisions and changes are made with quality real-time information from throughout an organization.

Insider Insight—Dr. Rod Heisterberg on Why Semantics Matter for Decision Process Reengineering and Are Crucial for Value Chain Management

Today's key business trends of globalized trade, virtual enterprise integration, and corporate transparency are driving the deployment of Collaborative Commerce (c-commerce) strategies. Furthermore, as the market forces of downsizing spawned Business Process Reengineering (BPR) to streamline process alignment and promote Enterprise Resource Planning solutions in the 1990s, so these trends are promoting Business Intelligence platforms and applications to support Decision Process Reengineering (DPR) initiatives in this first decade of the twenty-first century. Forward-looking enterprises that have already reaped the benefits of BPR inside their four walls are now seeking to achieve the benefits of c-commerce by leveraging semantic interoperability as an enabling technology for DPR in order to transform their core business management decision making practices. These value propositions for real-time enterprise performance management are based on managing information rather than inventory. The DPR business case for c-commerce is predicated on IT investments to facilitate collaborative business practices with shared demand/supply data and virtual enterprise visibility information in real time that will reduce uncertainty to improve decision effectiveness.

This concept of DPR utilizes a three-architecture approach as a framework for executing c-commerce strategies that provide real-time decision making for value chain management. The value propositions for the Technical, Application, and Business Architectures are summarized in terms of their Return on Investment (ROI) characteristics for enterprise information integration.

The Technical Architecture leverages Internet technologies for first internal and then external data sharing by providing loosely coupled application integration and semantic interoperability throughout the virtual enterprise. It consists of the infrastructure elements, such as the networks, hardware platforms, databases, data warehouses, operational data stores, and middleware that enable the interoperability between these elements to gain visibility of trading partner inventory across the value chain. The ROI for building this IT infrastructure is based on measures of enterprise efficiency.

The Application Architecture focuses on the enterprise's core competencies for engaging in collaborative business practices and integrating value chain management processes using CPC and CPFR business patterns. It consists of the value chain applications and associated process integration for Demand Chain Management, Customer/Partner Relationship Management, Supplier Relationship Management, and Supply Chain Management activities. It also provides real-time Business Activity Monitoring for event notification as appropriate to the value chain scenario business rules. The ROI for building virtual enterprises on trusted value chains to redefine competitive advantage is measured in terms of virtual enterprise effectiveness.

The Business Architecture is where the benefits of DPR are realized by generating virtual enterprise performance measurement information throughout the value chain in real time. This reengineering of value chain management decision making processes involves both planning and control functions for strategic, tactical, as well as operational activities. This adaptive real-time enterprise performance measurement system enables value chain managers to make strategy a continual process by monitoring performance against enterprise and line-of-business plans, updating the performance measures on the Balanced Scorecards using actual operations data in real time, collaborating as virtual teams with an appropriate management dashboard to interpret the performance metrics, developing valuable insights with knowledge management tools in order to formulate new competitive directions, and then reallocating enterprise resources to continuously reflect the dynamics of the business environment. This ROI is associated with value chain adaptability in response to either new business opportunities or threats.

Clearly, value chain operations that involve actions outside the enterprise's four walls extend the DPR scope. The management of these operations requires monitoring and measuring trading partner performance as never before because of virtual enterprise integration of the core competencies across the value chain. In order to realize this new capability, enterprises must reengineer their decision making processes in such a manner so that they are able to adapt to the dynamics needed to synchronize supply with demand. Decision makers must be able to evaluate performance across the virtual enterprise using common metrics articulated in collaborative service-level agreements in order to make effective decisions for guiding value chain strategies, directing logistics tactics, and managing operational activities in a real time manner. As c-commerce evolves as the mainstream e-business strategy for effective value chain management, DPR becomes the critical success factor for enterprise profitability and growth in the twenty-first century.

Dr. Rod Heisterberg
Managing Partner
Rod Heisterberg Associates

Table 12.2 Data Table—Industry and early adopter momentum

	Some Early Adopters
Aerospace	NASA, Boeing, Lockheed Martin, BAE Systems, Rolls-Royce
Automotive	Ford, Audi, Daimler-Chrysler
Bioinformatics	AstraZeneca, GlaxoSmithKline, US National Library of Medicine
Financial Services	Metlife, CIT Group
Government	Environmental Protection Agency, European Environment Agency, San Diego Super-Computer Laboratory
Media and Entertainment	RightsCom
National Security and Defense	Department of Defense, Department of Homeland Security
Health Care	Boston Children's Hospital
Professional Services	Cap Gemini Ernst & Young
Publishing	Adobe, Interwoven
Retail and Consumer-Packaged Goods	Coca-Cola
Science and Engineering	AKT, MINDlab, MIT, Oregon State University College of Oceanic and Atmospheric Sciences
Telecommunications	British Telecom, Nokia

In many ways the knowledge-based systems and expert systems of the 1970s, 1980s, and 1990s were semantic technology implementations—they had embedded models of aspects of the world they were part of. Impressive systems were built for a wide spectrum of applications including advisory and diagnostic systems for industries as wide ranging as medicine and engineering. However, these early knowledge systems were highly localized—whereas today the "intelligent" part of "artificial intelligence" is distributed. Intelligence is in distributed data, distributed processes, and distributed rules. Artificial intelligence has moved to middleware.

But today, the widespread need for more smarts in application networks is a pain point felt in all vertical industries. Industries taking early steps to implement semantic interoperability capabilities include those listed in Table 12.2.

RESEARCH PROGRAMS TO WATCH

With all the promising activity and adoption of semantic approaches by industry it may seem easy to conclude that all the important research has been done. That couldn't be farther from the truth. Many academic and commercial research programs continue to dedicate significant efforts toward advancing the readiness of semantics-based approaches. A short survey of key research programs follows.

Advanced Knowledge Technologies

Advanced Knowledge Technologies[2] (AKT) is an interdisciplinary research collaboration between research groups at the Universities of Aberdeen, Edinburgh, Sheffield, and Southampton and the Open University. AKT's key focus is in the area of representation and management for the knowledge life cycle. Key objectives include:

- **Knowledge Acquisition**—how to acquire knowledge from different sources
- **Knowledge Modeling**—how to model knowledge structures in ontology
- **Knowledge Reuse**—how to reuse knowledge patterns across contexts
- **Knowledge Retrieval**—how to automate knowledge extraction and retrieval
- **Knowledge Publishing**—how to present and visualize knowledge to users
- **Knowledge Maintenance**—how to keep knowledge active, fresh, and valid

MIT SIMILE

Semantic Interoperability of Metadata and Information in unLike Environments (SIMILE) is a joint project led by MIT with participation from W3C and Hewlett Packard. Like the subject of this book, the SIMILE project aims to create interoperability among digital assets, schemas, metadata, and services. A key aim is to enable this interoperability of assets between disparate individuals, communities, and institutions.

SIMILE makes use of RDF, with help from Hewlett Packard's Jena toolkit, to create an infrastructure that will provide different views of digital assets and bind those views to services. The initial deployment and use of SIMILE is to be focused in the research library domain space.

Insider Insight—Rick Hayes-Roth on Next-Generation Battlespace Awareness

The modern military, like other modern enterprises, aims (1) to exploit superb information (2) to achieve unprecedented levels of effectiveness through (3) agile, coordinated control of resources. These new enterprises form virtual organizations on an ad hoc basis, quickly assembling resources with needed capabilities and integrating them into a unified operational federation. To succeed, these organizations must collaboratively construct and consistently maintain a shared understanding of several important things: mission intent; alternative plans under consideration and those being executed; and the evolving situation, which includes the past, current, and future expected position and status of all relevant entities in the battlespace.

Continued

[2] For more information, see: www.aktors.org.

The immense potential of emerging semantic technologies to revolutionize collaboration networks—and thereby offer information superiority—is staggering. Dramatic new capabilities will result from innovative collaboration architectures and readily available technologies such as ontologies and logic systems, semantic editors, plan analyzers, machine-aided learning, and automated reasoning. The essential capability required is for human planners and plan executors to be able to receive *just the right information, at the right time* to ensure that their planned actions remain continuously justified in light of current situation data. In short, we need a just-in-time system for delivery of valued information. This generic capability provides personalized, pinpoint delivery of relevant information, thereby improving the recipients' performance and raising the productivity of the entire enterprise.

Important research is under way to define what that next-generation, semantically aware, collaboration network looks like. It is clear that information must be real time and easily updated by humans in the process. Information keyed in from one part of the globe may impact assumptions and tactics in another part of the world. Likewise, each node (human or machine) in the collaboration network represents a unique context, view, and operational scope. Ripe with possibilities, the use of a real-time ontology as a central dissemination point for localized narrow views on dynamic situations could change all the rules. Once an infrastructure of continuously shared, adaptive, information models becomes a reality—a new era of high fidelity situational awareness will be upon us.

Rick Hayes-Roth
Professor, Naval Postgraduate School, and
Former CTO/Software, Hewlett Packard

University of Maryland MINDlab

The University of Maryland established the Maryland Information and Network Dynamics (MIND) Laboratory[3] as a place to explore new information technologies. MINDlab partners with commercial companies, collaborates with other universities, and cooperates with federal agencies.

MINDlab's charter includes broad-based objectives such as:

- Wireless Networking
- Networking Infrastructure and Services
- Information Services and Information-centric Applications
- Information Assurance

A number of ongoing research programs at MINDlab make use of semantic technology, but their key research project is Maryland Information and Network Dynamics Lab Semantic Web Agents Project[4] (MIND SWAP). MIND SWAP is

[3] For more information, see: www.mindlab.umd.edu.

[4] For more information, see: www.mindswap.org.

undertaking a range of projects that include domain modeling and software creation for semantic web services.

IBM Research's SNOBASE

IBM Semantic Network Ontology Base (SNOBASE) is a framework for loading, creating, modifying, querying, and storing ontologies in RDF and OWL formats. Internally, the system uses an inference engine, an ontology persistent store, an ontology directory, and ontology source connectors. Applications can query against the created ontology models, and the inference engine deduces the answers and returns results sets similar to JDBC (Java Data Base Connectivity) result sets.

Hewlett Packard Semantic Web Research Labs

The HP labs[5] investment in the Semantic Web consists of the development of Semantic Web tools and associated technology, complemented by basic research and application-driven research. HP is also part of several collaborative ventures, including involvement in W3C initiatives and European projects.

Hewlett Packard's most interesting work in the Semantic Web area is in the use of Jena—a Semantic Web toolkit. Jena provides ontology management capabilities in addition to basic services such as querying, OWL support, RDF support, and simple reasoning capabilities.

Fujitsu Labs

The American research arm of Fujitsu Labs,[6] in College Park, Maryland, is intensely focused on the application of semantic technology to Web Services infrastructures. The Semantic Web Services focus derives partially in support of MINDlab and partially in support of an overall directive to advance pervasive computing technologies.

TECHNOLOGIES TO WATCH

The array of technologies and architectures covered in this book can be confusing. Like all new innovations, some supporting technologies will survive while others sometimes die out. The following set of technologies appears to have the staying power to propel semantic interoperability solutions forward into a bright future.

[5] For more information, see: www.hpl.hp.com/semweb.

[6] For more information, see: www.flacp.fujitsulabs.com.

Framework—Model-Driven Architecture (MDA)

The Object Management Group's Model-Driven Architecture aims to provide an end-to-end framework for model-driven application development and, more recently, for application interoperability. Complete coverage of the MDA was provided in Chapter 6, Metadata Archetypes.

Semantic Web Services—OWL-S

The Semantic Web Services Initiative aims to create an OWL-based services model for applied use within the Web Services framework. Service descriptions and profiles are better represented in an expressive ontology language such as OWL. However, it is worth noting that other Semantic Web Services initiatives do not rely on OWL-based data representation.

Policy and Rules—SWRL

The Defense Advanced Research Projects Agency (DARPA) DARPA Agent Markup Language (DAML) program, in conjunction with the RuleML activity, has created the Semantic Web Rules Language (SWRL). SWRL is the product of RuleML's extensive support for a wide range of logical operations and OWL's strong utility as an ontology language.

Ontology Language—OWL

The World Wide Web Consortium's Web Ontology Language has emerged as the best-in-class ontology markup language. OWL is a sound semantic model in XML that is geared specifically toward ontology development. OWL-DL is a version of OWL that includes support for Description Logics—allowing for reliable ontology queries. Complete coverage of OWL was provided in Chapter 7, Ontology Design Patterns.

Ontology Visual Models—UML/ODM

The Object Management Group's Unified Modeling Language is the de facto standard for visual modeling in software development. Recent collaboration between key OWL visionaries and the OMG standards body has resulted in the formation of an Ontology Definition Metamodel (ODM) working group—whose charter is to create a framework for the use of UML in the OWL ontology creation process. Complete coverage of the ODM was provided in Chapter 6, Metadata Archetypes.

Ontology Bridges—Topic Maps

Although Topic Maps are envisioned as a mechanism to link topics among XML documents, they have shown great promise in providing context-sensitive bridges between disparate ontology. Topic Maps specify relationships among topics and provide groupings of addressable information objects. Currently, Topic Maps are a standard within the International Standards Organization—ISO 13250.

BARRIERS TO WIDESPREAD ADOPTION

To conclude that semantic interoperability is a sure thing—because industry has begun to adopt it, scientists are researching it, and advanced technology exists to support it—would be folly. The truth is that the road to semantic interoperability may still contain roadblocks that must be overcome. Possible barriers to further adoption could include problems with standards, marketplace entrenchment, and misinformation about actual strengths and capabilities.

Vocabulary Standards Hyperbole

Standards bodies that develop and promote vocabulary standards have a vested interest in self-preservation. As semantic interoperability technology continues to mature, it will become increasingly apparent that vocabulary standards, in fact, are not needed. This will create a rift between some industry consortia and standards bodies developing information-sharing infrastructures that are not dependent on shared vocabularies. One possible avenue to overcome this barrier will be when the standards groups who create and maintain the vocabulary standards realize that they can still add value by supplying domain ontology—without requiring strict vocabulary adherence.

Gaps in Metadata Standards

As much progress has been made across the industry in support of Semantic Web standards—RDF, OWL, OWL-S, MDA, and so on—there is more to do. A robust semantic interoperability infrastructure will require much more flexible mechanisms for bridging and transforming ontologies. Without these bridging and transforming abilities, a domain ontology simply becomes another definitional vocabulary standard—requiring brittle, hard-coded adapters to bind local data to community data. Future work on OWL might be planned to support Topic Map concepts of scope and context, which would be a step in the right direction of making ontologies more adaptive and easier to work with in complicated domains.

Complexity and Risk-Averse Companies

Although this book has shown that semantic technologies are sufficiently mature to offer significant new capabilities and savings to early adopters, the widespread adoption of these technologies will be greatly hindered by the sector's vast complexity. Current modeling and mediation specifications are far too complex for typical software engineers to grasp quickly. Without simple user interfaces and easy-to-understand specifications, the value proposition will not be compelling enough for large organizations to commit significant budgets. In the future, vendor solutions will have to mask the underlying complexities of OWL, RuleML, and Topic Maps in well-conceived enterprise tools. Only then will semantics-based solutions become pervasive.

CROSSING THE CHASM

It is appropriate to conclude this book with a section on "crossing the chasm." Geoffrey Moore popularized this concept in the book of the same name.[7] Crossing the chasm refers to the distinction between visionaries who adopt new—potentially disruptive—technologies first, and the pragmatist early majority who follow. Clearly, Semantic Interoperability is a set of disruptive capabilities that have yet to cross the chasm.

Moore observes that innovators tend to have different expectations about success and failure than pragmatists. Typically, innovators have a strong desire to be seen as leaders in their markets, whereas pragmatists just want to stay ahead. This distinction leads to a much higher buying threshold for pragmatists—new products have to be easy to buy. From a marketing standpoint, these impact product positioning, market analysis, pricing, and distribution. For semantic technologies to cross the chasm they will have to be easy to buy.

For semantic technologies to become easier to buy, semantic technology vendors will have to crystallize their value proposition in terms that new customers and industry analysts can relate to. But more importantly, the software tools provided by semantic technology vendors will have to be simplified to a degree that makes them about as easy to use as desktop office applications (for modeling) and as easy to set up as a Web server (for the run time environment). When these ease-of-use barriers are overcome—and indeed they will be—semantic interoperability will be a dream realized by the masses.

[7] Moore, G.: *Crossing the Chasm*. Harper Business, revised edition, 2002.

Appendix A

Vendor Profiles, Tools, and Scorecards

KEY TAKEAWAYS

- Current tool maturity is acceptable for most problem cases.
- To become widely popular, greater usability must be built into these tools.
- Support for collaborative design environments is still weak.
- Complete end-to-end interoperability support is nonexistent in any tool suite.
- A custom-tailored scorecard should be leveraged to choose vendors.

Although the authors don't expect this appendix to remain accurate indefinitely (business and technology change far too rapidly for that), it should provide an accurate snapshot of commercially available platform capabilities of leading semantic vendors in the 2004 and 2005 fiscal years. Furthermore, this appendix goes beyond a typical vendor overview and provides both a structure and set of metrics for measuring tool set strengths and weaknesses into the future.

The appendix is organized in four main parts. First, a summary of vendor positions and tool highlights is presented. Second, an overview of the ontology tool market and lifecycle requirements is provided. Third, interoperability platforms are examined and their design and run time capabilities are compared. Finally, a sample set of metrics for vendor comparison is provided to assist the tool buyer.

As with all statements about proprietary tools in this book, the authors recommend that the reader contact the tool vendor for any clarification or detailed specifications.

VENDOR OVERVIEWS

Two groups of enterprising vendors are paving the way in this emerging sector. Rambunctious start-ups with innovative, and usually patented, software are making a bid to own the mind share of the market and deliver outstanding value to their customers.

Adaptive Information, by Jeffrey T. Pollock and Ralph Hodgson
ISBN 0-471-48854-2

Tried and true giants of software are also making their bids, attempting to leverage innovations from their research labs and remain nimble in the marketplace. The following two sections provide an overview of key vendors to watch.

Innovative Small and Mid-Sized Vendors

Celcorp

Celcorp is a privately held company founded in 1990 and based in Washington, DC. The company was originally established in Alberta, Canada and has been providing business integration solutions since its inception. It has a number of successful reference clients for its product suite named Celware.

The Celware engine features a Server and Real-Time Planner to integrate applications by streamlining users' work flow when multiple systems must be accessed to perform a single task. The software uses intelligent agent technology based on proprietary extensions to the "Plan Domain Model and the Graph Plan Algorithm."

Running Celware Recorder, a design time tool that watches and records a user's interaction with various systems, automatically generates models to be used later at run time on its server environment.

Website: http://www.celcorp.com

Contivo

Contivo is a privately held company founded in 1998 with offices in Palo Alto, CA. Contivo's corporate investors include industry leaders BEA Systems, TIBCO Software, and webMethods. Venture capital investors include BA Venture Partners, Voyager Capital, and MSD Capital LP. It received series C funding in January 2003.

Contivo's innovative Vocabulary Management System is targeted toward ERP and application integration problem sets. The Server includes a Semantic Dictionary containing enterprise vocabularies, such as various XML, EDI, and ERP standards; a Thesaurus with synonyms that match business concepts; and a Rules Dictionary that governs the field-level data transformation.

Modeling and mapping are performed with the Contivo Analyst tool. Many prebuilt maps for popular data standards are available.

Website: http://www.contivo.com

Modulant

Modulant was founded in 2000 and subsequently merged with Product Data Integration Technologies (founded in 1989) to develop commercially deployable software based on PDIT's proprietary technology and methodology. Modulant is a private, venture-backed company whose existing investors include Sandler Capital Management, Guardian Partners, and First Lexington Capital. Modulant's headquarters are in Charleston, SC.

Modulant's product is called Contextia. The engine, Contextia Dynamic Mediation, uses a central description of enterprise data called Abstract Conceptual Model (ACM) to enable disparate applications exchange information by transforming messages at run time. It reconciles semantic conflicts among disparate applications and data sources.

Modeling is done with Contextia Interoperability Workbench by capturing the meaning, relationships, and context of data elements of all source and target applications and mapping them to the ACM. The mapping specifications and ACM are then used by the Modulant Contextia Dynamic Mediation to transform data from source to target at run time. The Interoperability Workbench accepts a variety of inputs for mapping and modeling, including XML schemas, native schemas, database tables, and delimited files.

Website: http://www.modulant.com

Network Inference

Network Inference was founded in 2000 to commercialize a highly scaleable description logic-based inference engine originally developed at the University of Manchester. A privately held company, Network Inference is headquartered in San Diego, CA and operates worldwide. Network Inference's capital partners include Nokia Ventures and Palomar Ventures.

Network Inference has many reference customers for its two complementary software products, Cerebra Server™ and Construct™. Cerebra Server™ is an enterprise-strength software platform that provides business logic inferencing and processing capabilities for developing dynamic policy-driven applications. Construct™ is a visual modeling tool designed for easy, efficient, and collaborative building of enterprise-ready ontologies in the W3C's OWL language.

Website: http://www.networkinference.com

OntologyWorks

OntologyWorks is privately held and has offices in Maryland and Arkansas. In the first quarter of 2000, it completed development of an initial version (V 1.0) of its tool set and secured its first customer.

The Integration Ontology Development Environment (IODE) utilizes a central description of enterprise data to determine answers to complex queries. Each link in the enterprise ontology is mapped to a query in the "ontology database"; this can either be a data warehouse created as part of the ontology engineering process or a mediated connection to a legacy database. Solutions to queries in the ontology are built with the rules and relations in the ontology, so that the "proof" of the result can be translated in a simple fashion into a program that runs over the databases, to determine the correct answer. Ontology modeling can be done with UML tools, which is then translated into a proprietary OntologyWorks ontology language.

Website: http://www.ontologyworks.com

Ontoprise

Ontoprise® GmbH, formed in 1999, is venture capital backed; it achieved break-even status in 2002. The company is headquartered in Germany. Ontoprise was founded as a spin off of the University of Karlsruhe, which implemented the first version of its technology in 1992.

The main product is OntoBroker. Data integration is done via a several-step process that includes importing data schemas from existing databases and using OntoMap to map concepts and relations from one ontology to the next. These mappings are translated into F-Logic statements, so that OntoBroker can reason over the combined ontology results in data references in the original data sources.

Modeling is done with OntoEdit and OntoMap. Two more tools are needed to complete this picture, which are a rule editor and a rule debugger. The rules state the actual connections between the newly merged concepts and are susceptible to bugs; hence they must be viewable and debuggable.

Website: http://www.ontoprise.com

Semagix

Semagix was founded in 2002 as a result of a merger of Protégé, a UK-based technology incubator and management company, with Voquette, a US-based metadata management company. Semagix is headquartered in London, with offices in Georgia and Washington, DC. It offers semantic-based enterprise information integration and knowledge discovery technology.

Semagix Freedom includes ontology management and metadata extraction components as well as a semantic application development toolkit. Several solutions have been built with the Freedom platform including the award-winning Semagix CIRAS (Customer Identification and Risk Assessment), an anti-money laundering solution. CIRAS builds and maintains a relevant knowledge base containing information about:

- Organizations and their interrelationships (legal/ownership structures, location, shareholders, nature of business activity, key financials, key management)
- Relevant individuals (key information for identity verification, directorships and other key corporate relationships, known criminal and derogatory activity)
- Customized derogatory terms

The knowledge can be automatically drawn from trusted sources to which the CIRAS user has access. Such sources may typically include Dunn & Bradstreet, ICC, Hoovers, Bloomberg, and Corporate 192, as well as government watchlists and any trusted internal customer databases. Content sources are continuously monitored for updates and changes.

Website: http://www.semagix.com

Sandpiper Software

Sandpiper Software provides business semantics infrastructure solutions for context-driven search, collaborative applications, and cross-organizational content interoperability. The company develops semantically aware, knowledge-based products and provides corporate education, training, and ontology-based context development services that facilitate business information interoperability, terminology normalization, and conflict resolution in content across Web-based and enterprise information systems.

Sandpiper's products include the Visual Ontology Modeler™ (VOM). An add-in to IBM Rational Rose®, the VOM v1.1 is a multiuser, UML-based graphical tool for creating, maintaining, composing, and deploying component-based conceptual models, supporting the DARPA Agent Mark-up Language and Ontology Inference Layer (DAML+OIL), the W3C's draft Web Ontology Language (OWL), and other logic-based knowledge representation languages. VOM v1.5, scheduled for release in the US in Q2 2004, will also include an XMI-based interface with Adaptive Foundation™ repository for enterprise metadata management, providing interoperability with other MDA- and MOF-based applications. Sandpiper also sells Standard and Domain-Specific Ontology Libraries. Libraries of general and domain-specific ontology components, including ontology components representing international and US standards for metadata and technical data representation that jump-start the development process. Finally, Medius® Rationalization Server is a persistent, deductive knowledge base that supports graphical diagnostics for ontology comparison, analysis, and construction of large, composite application models from ontology components. The Rationalization Server will be available for beta customers in late 2004.

Website: http://www.sandsoft.com

SchemaLogic

SchemaLogic was formed in 2001 by ex-Microsoft employees and is a privately held company whose investors include Phoenix Partners and other private investors. SchemaLogic is located in Redmond, WA.

SchemaLogic's primary product, SchemaServer, captures and communicates data definitions (enterprise schema and metadata) used across multiple applications and languages, which simplifies information exchange and retrieval. To help create a common model of semantics and syntax, the software imports existing schema, vocabularies, taxonomies, and reference data into an active repository of schema and metadata. Sources can include databases, applications, content management systems, and XML Schema.

SchemaServer manages the associations and dependencies among the separate schema by providing tools to model, map, and describe the contextual meanings across different vocabularies and the structure of various schema. This process involves reconciliation, impact analysis, change management, governance, and synchronization of new or updated metadata (vocabulary, taxonomy, or schema) shared across subscribing systems, using XML, SOAP, and Web Services.

Website: http://www.schemalogic.com

Unicorn Solutions

The company, founded in 2001, is privately held. It is headquartered in New York City with R&D in Israel. Unicorn's investors include Jerusalem Global Ventures, Bank of America Equity Partners, Intel Capital, Israel Seed Partners, Tecc-IS, and Apropos.

The Unicorn System is a design time tool and a script generator for integration with a third-party engine such as WebMethods. Unicorn imports schemas from multiple data sources including XML, RDBMS, COBOL, IMS, and EDI. They are then mapped to a central enterprise model (ontology). Mapping supports creation of data transformation rules. Unicorn can generate transformation scripts as executable SQL, XSLT, and Java Bean code.

Website: http://www.unicorn.com

Large Vendors Pursuing Semantics-Based Solutions

Adobe

Adobe's eXtensible Metadata Platform (XMP) is a specification for document and media metadata that can be seamlessly attached to binary files. Adobe support the specification with a cross-product toolkit that leverages RDF and XML for digital resource management.

Website: http://www.adobe.com/products/xmp/

IBM

IBM Research is home to a large number of projects focusing on aspects of semantic computing. IBM separates the concerns of Data Management, Distributed Systems, Intelligent Information Systems, and Knowledge Management—although the research subjects seem to cross all boundaries. Significant programs include Clio, OptimalGrid, SemanticCalendars, eClassifier, Xperanto, and Active Technologies.

Clio: http://www.almaden.ibm.com/software/km/clio/
OptimalGrid: http://www.almaden.ibm.com/software/ds/OptimalGrid/
SemanticCalendars: http://www.almaden.ibm.com/software/km/cal/
eClassifier: http://www.almaden.ibm.com/software/km/eClassifier/
Xperanto: http://www.almaden.ibm.com/software/dm/Xperanto/
Active Technologies:
http://www.haifa.il.ibm.com/projects/software/amt.html

Sun Microsystems

As with all large organizations, Sun Microsystems employees are faced with information overload on a daily basis as they attempt to use the Web as a tool to find the knowledge needed to do their jobs effectively. While Sun's employees and business partners create and share new knowledge every day during the course of business, because this knowledge is challenging to capture and reuse much is lost. By capturing this knowledge and sharing it effectively with partners and employees, Sun can greatly increase business efficiencies, drive revenues and further its competitive edge in the marketplace. With these benefits in mind, Sun has placed great emphasis on the discipline of knowledge management to further its business goals.

This focus on knowledge management led to the creation of a Global Knowledge Architecture for Sun. The architecture, conceived by Sun's Global Knowledge Engineering Group (GKE) in Sun Services, provides a business and technical framework for sharing knowledge worldwide. Of primary concern to the GKE group is the ability to capture and share the knowledge assets required to provide best-in-class service to its customers.

A key foundation piece of the Global Knowledge Architecture is enabling consistent semantic markup of knowledge assets to allow for personalized, localized and dynamic delivery of content from Sun's vast knowledge bases across the organization. The Sun Metadata Initiative was launched to provide a framework to share a common set of metadata elements, taxonomies, and vocabularies for use across the enterprise. Prior to the launch of the initiative, the use of semantic markup was inconsistent, making it impossible to share and repurpose content in an automated fashion. To achieve this vision, the Sun Metadata Team has leveraged the work of the W3C's Semantic Web which provides technologies and standards to support machine processing of content based on its semantic meaning.

As cited from W3C.org
Website: http://www.w3.org/2001/sw/EO/usecases/byProject.html

Notable Mentions

Callixa

Callixa is a solution provider for integrating disparate data sources across platforms and locations. The Callixa Integration Server is a distributed, high-performance, enterprise data integration platform that helps companies leverage the value of their enterprise information.

Website: http://www.callixa.com/

IGS

IGS provides data transformation and universal adapter tools and solutions designed to solve enterprise application integration challenges. Leveraging middleware and common object models, IGS addresses the unique communication needs of the overlapping application domains while optimizing the end user's investment in the integration infrastructure.

Website: http://www.igs.com/

MetaMatrix

MetaMatrix provides Enterprise Information Integration and Management (EIIM) solutions. The MetaMatrix solution acts as a virtual database to unify and deliver information on demand, across the entire enterprise. Headquartered in New York City, MetaMatrix is privately held.

Website: http://www.metamatrix.com/

Miosoft

MioSoft Corporation is an enterprise software company developing next-generation Customer Data Integration solutions. The miocon suite, MioSoft's initial product, enables firms to start every customer interaction with a complete picture of the customer, regardless of who initiated the contact, the touchpoint or channel, the customer-facing application vendor, or the back-end systems involved.

Website: http://www.miosoft.com/

ONTOLOGY DEVELOPMENT TOOL SUPPORT

Taking a step deeper, this section will explore the use and value of software tools that enable analysts to create ontology and mange them throughout their lifecycle. Ontology tools are defined as software programs that support all, or part, of the processes involved with using conceptual models in enterprise computing environments.

Summary Conclusions

Knowledge models, or ontologies, are a necessary precondition to most semantic interoperability applications. Therefore, the state of ontology development tooling is a key factor in adoption of semantic technology. Today some tool support is available for all stages of the ontology lifecycle. Many tools are still offered as research prototypes, but many others have begun to be commercialized (they are often commercial versions of their direct research counterparts). Standard compliance and support for RDF(S) and OWL is growing. However, because of the different forms of RDF(S), ontology tools still do not interoperate well.

Currently, ontology creation tools require their users to be trained in knowledge representation and predicate logic. The user interface and development paradigms of these tools are different from the standard application development tools. Support for semiautomation (e.g., term extraction) is maturing. Support for collaborative authoring is still weak. To scale ontology-based applications from the pilot/prototype stage to enterprise-level implementations, a new generation of tools is needed that will:

- Improve user experience for ontology building by leveraging familiar interfaces of widely used application development tools or Microsoft Office applications
- Offer a server-based environment with support for consistency checking of interconnected ontologies
- Offer a collaborative environment for model review and refinement that does not require reviewers to be expert modelers
- Feature Web Services-ready interfaces for ease of integration

Ontology Lifecycle and Tools

The ontology lifecycle spans from creation to evolution as shown in Figure A.1.
Tool support is available for all stages of the lifecycle:

Creating

This can be done from scratch, using a tool for editing and creating class structures (usually with an interface that is similar to a file system directory structure or bookmark folder interface). However, there is also a good deal of assistance available at this stage:

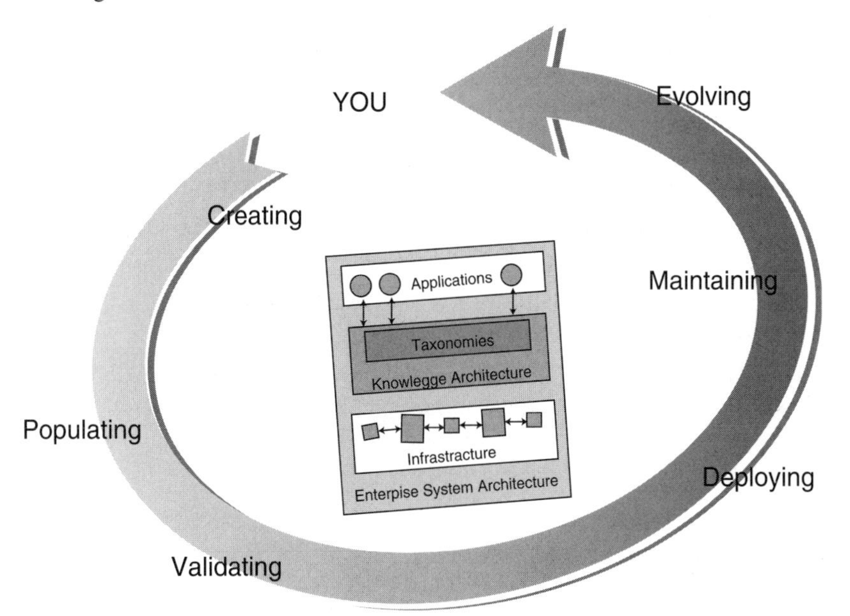

Figure A.1 Ontology lifecycle components

- Text mining can be used to extract terminology from texts, providing a starting point for ontology creation. Also associated with text mining, machine learning algorithms may also be useful on database instance data when attempting to derive ontology from large quantities of denormalized database (or any semistructured source) data.
- Often, ontology information is available in legacy forms, such as database schemas, product catalogs, and yellow pages listings. Many of the recently released ontology editors import database schemas and other legacy formats, such as, Cobol copybooks.
- It is also possible to reuse, in whole or in part, ontologies that have already been developed in the creation of a new ontology. This brings the advantage of being able to leverage detailed work that has already been done by another ontology engineer.

Populating

This refers to the process of creating instances of the concepts in an ontology and linking them to external sources:

- Ordinary web pages are a good source of instance information; so many tools for populating ontologies are based on annotation of web pages.
- Legacy sources of instances are also often available; product catalogs, parts lists, white pages, database tables, etc. can all be mined while populating an ontology.
- Population can be done manually or be semiautomated. Semiautomation is highly recommended when a large number of knowledge sources exist.

Deploying

There are many ways to deploy an ontology once it has been created and populated:

- The ontology provides a natural index of the instances described in it and hence can be used as a navigational aid while browsing those instances.
- More sophisticated methods, such as case-based reasoning, can use the ontology to drive similarity measures for case-based retrieval.
- DAML+OIL and OWL have capabilities for expressing axioms and constraints on the concepts in the ontology; hence, powerful logical reasoning engines can be used to draw conclusions about the instances in an ontology.

Validating, Evolving, and Maintaining

Ontologies, like any other component of a complex system, will need to change as their environment changes. Some changes might be simple responses to errors or omissions in the original ontology; others might be in response to a change in the environment. There are many ways in which an ontology can be validated in order to improve and evolve it. The most effective critiques are based on strict formal semantics of what the class structure means:

- Extensive logical frameworks that support this sort of reasoning have been developed, and are called Description Logics.
- A few advanced tools use automated description logic engines to determine when an ontology has contradictions, or when a particular concept in an ontology can be classified differently, according to its description and that of other concepts.
- These critiques can be used to identify gaps in the knowledge represented in the ontology, or they can be used to automatically modify the ontology, consolidating the information contained within it.
- The task of ontology maintenance may require merging ontologies from diverse provenance. When this is the case tool support is important.
- Some tools provide human-centered capabilities for searching through ontologies for similar concepts (usually by name) and provisions for merging the concepts.
- Others perform more elaborate matching, based on common instances or patterns of related concepts.

Ontology Tools Survey

Many tools are still offered as research prototypes, but many others have begun to be commercialized (they are often commercial versions of their direct research counterparts). Current ontology tools require their users to be trained in knowledge representation and predicate logic.

Typically, highly trained knowledge engineers working with domain specialists or subject matter experts build ontologies. To scale this approach in a large enterprise, a wider group of people must be able to perform some of the ontology lifecycle activities independently. They need to be able to create and modify knowledge models directly and easily, without the requirement for specialized training in knowledge representation, acquisition, or manipulation. A new generation of tools is needed that will:

- Improve user interface for ontology building by leveraging familiar interfaces of widely used application development tools or Microsoft Office applications
- Offer a server-based environment with support for consistency checking of interconnected ontologies
- Offer a collaborative environment for model review and refinement that does not require reviewers to be expert modelers
- Feature SOAP interfaces for ease of integration

In the tables below a list of representative tools and a brief description of their capabilities is provided. A number of commercial tools address multiple, sometimes all, stages of ontology lifecycle. When this is the case, we have placed the tools in the category(s) in which their capabilities are the strongest.

There has been a significant growth in the number of ontology technology products; this report doesn't cover all the available tools. In composing this list we have selected the tools that:

- Support all or some of RDF(S)-DAML+OIL/OWL standard (or have committed to support in the very near future)
- Have strong technical vision for ontology-based solutions
- Are robust and ready to be used

In our research we have also identified several powerful and mature products that have strong value proposition, but currently do not offer standards compliance. These are discussed at the end of this section.

Ontology Tool Capabilities and Standards Support

Lifecycle Phase: Creating

Tool	Protégé-2000
Vendor	Stanford KSL
Lifecycle Phase	Creating
Capabilities	Create concept hierarchies, create instances, and view in several formats. Typically used as a single-user tool. Multiuser support is becoming available.
Standards Compliance and Comments	Open Source (Mozilla); plug-in architecture. Supports RDF, DAML+OIL and OWL

Tool	OntoEdit
Vendor	Ontoprise
Lifecycle Phase	Creating
Capabilities	Create concept hierarchies, create instances. Integrate with common databases. Single-user tool.
Standards Compliance and Comments	Claims to be RDF, DAML+OIL compliant, plug-in architecture. Our tests uncovered some compatibility issues.

Tool	OilEd
Vendor	University of Manchester
Lifecycle Phase	Creating
Capabilities	Create concept hierarchies, create instances, analyze semantic consistency (according to description logics). Single-user tool.
Standards Compliance and Comments	RDF, DAML+OIL support. From the creators of OIL. Free download, integrates with description logics reasoner.

Tool	Medius Visual Ontology Modeler
Vendor	Sandpiper Software
Lifecycle Phase	Creating
Capabilities	Ontology creation. Some support for collaboration.
Standards Compliance and Comments	A limited beta started in January 2002. Extends UML, requires Enterprise Edition of Rational Rose. Supports RDF and DAML+OIL. Includes a library of ontologies that represent the IEEE Standard Upper Ontology (SUO).

Tool	Cerebra Construct
Vendor	Network Inference
Lifecycle Phase	Creating
Capabilities	An advanced ontology construction tool set (with semi-real-time reasoning support). Enables ontology seeding—the absorption of existing ontologies, taxonomies, database schemas, and wrapping with ontology.
Standards Compliance and Comments	Commercial version of OilED extended to integrate with a commercially available graphical editor and enable collaborative authoring. First release in March, 2003. New releases will be delivered in fourth quarter 2003. Supports RDF, DAML+OIL, OWL, SOAP interfaces.

Tool	LinKFactory Workbench
Vendor	Language and Computing
Lifecycle Phase	Creating
Capabilities	Collaborative authoring environment. Originally designed for very large medical ontologies. Has a Java beans API and optional Application Generators for semantic indexing, automatic coding, and information extraction. Compares and links ontologies via a core ontology; related concepts matched on formal relationships and lexical information.
Standards Compliance and Comments	Supports RDF(S); DAML+OIL/OWL. Some support for population and maintenance.

Tool	K-Infinity
Vendor	Intelligent Views
Lifecycle Phase	Creating
Capabilities	Collaborative authoring environment.
Standards Compliance and Comments	Modularized tools supporting all stages of life cycle. Supports RDF and Topic Maps.

Lifecycle Phase: Populating

Tool	OntoAnnotate
Vendor	Ontoprise
Lifecycle Phase	Populating
Capabilities	Copy items from web pages. Create markup.
Standards Compliance and Comments	Integrated with Microsoft Internet Explorer. RDF, DAML+OIL compliant.

Tool	OntoMat
Vendor	AIFB University of Karlsruhe
Lifecycle Phase	Populating
Capabilities	Copy items from web pages. Create markup.
Standards Compliance and Comments	Free download. Adopted by DARPA On-To-Agents project. JAVA-based and provides a plug-in interface for extensions. DAML+OIL compliant.

Tool	AeroDAML
Vendor	Lockheed-Martin
Lifecycle Phase	Populating
Capabilities	Natural language parse of documents to create markup.
Standards Compliance and Comments	Demo available as web service. RDF, DAML+OIL compliant.

Tool	CORPORUM OntoBuilder
Vendor	CognIT AS
Lifecycle Phase	Populating
Capabilities	Basic concepts and relations are represented with single inheritance. Representation of concepts and relations extracted from content may be extended with WordNet information.
Standards Compliance and Comments	Spun off from On-To-Knowledge. Supports RDF, DAML+OIL support. Requires Sesame RDF repository. Focuses on generating editable ontologies automatically from natural language documents. Also supports creation.

Tool	Freedom Enterprise Semantic Platform
Vendor	Semagix
Lifecycle Phase	Populating
Capabilities	Semantically enhancing metadata with associations and concepts unique to the language, structure, and needs of an industry. Automatic ontology-directed classification and semantic annotation of heterogeneous content.
Standards Compliance and Comments	Supports XML with RDF planned for 2004. Also supports deployment and creation. It includes the Knowledge Toolkit for building ontologies.

Lifecycle Phase: Deploying

Tool	Orenge
Vendor	Empolis
Lifecycle Phase	Deploying
Capabilities	Performs natural language search using ontologies. Supported Capability: Concept-based Search
Standards Compliance and Comments	Also supports ontology creation. XML and Topic Map compliant.

Tool	Freedom Enterprise Semantic Platform
Vendor	Semagix
Lifecycle Phase	Deploying
Capabilities	Categorization and search using ontologies. Aggregating and normalizing content from a wide variety of content sources. ESP is an application platform for semantic integration of heterogeneous content including media and enterprise databases. Supported Capability: Automated Concept Tagger, Navigational Search, Concept-based Search.
Standards Compliance and Comments	Also supports ontology creation and population. XML-based, RDF support planned for 2004.

Tool	OntoBroker
Vendor	Ontoprise
Lifecycle Phase	Deploying
Capabilities	Provides framework for processing rules organized by an ontology. Supported Capability: Product Design Assistant
Standards Compliance and Comments	RDF, DAML+OIL compliant.

Tool	Semantic Miner
Vendor	Ontoprise
Lifecycle Phase	Deploying
Capabilities	Constructs semantically meaningful queries from natural language queries.
Standards Compliance and Comments	RDF, DAML+OIL compliant.

Tool	OntoOffice
Vendor	Ontoprise
Lifecycle Phase	Deploying
Capabilities	Just-in-time content delivery based on ontologies. Supported Capability: Context-aware Retriever.
Standards Compliance and Comments	Integrates with Microsoft Office. RDF, DAML+OIL compliant.

Tool	Unicorn Workbench™
Vendor	Unicorn Solutions
Lifecycle Phase	Deploying
Capabilities	Ontology-based Enterprise Data Management. Supports multiuser ontology modeling as well as semantic mapping of data schemas such as XSD, relational and legacy to the ontology.
Standards Compliance and Comments	Download with registration. RDFS/DAML+OIL support. OWL support, plus a proprietary rules language for more expressiveness. Also supports ontology creation.

Tool	Contextia
Vendor	Modulant
Lifecycle Phase	Deploying
Capabilities	Enterprise Data and Application Integration. Accepts a variety of inputs for mapping and modeling, including XML schemas, native schemas, database tables, and delimited files. Can work stand-alone or in conjunction with existing IT infrastructures—such as EAI, B2Bi, business process management, message brokers and off-the-shelf connectors. Can handle data transformations involving complex data elements, nested structures, and incompatible or conflicting semantics. Supported Capability: Semantic Data Integrator, Semantic Application Integration.
Standards Compliance and Comments	Ontology creation supported by FirstStep XG included with Contextia. Express model (ISO 10303) is used for validation; cross-ontology consistencies. Supports XML, Web Services ready, and supports SOAP.

Tool	Cerebra Server
Vendor	Network Inference
Lifecycle Phase	Deploying
Capabilities	Semantic integration of Enterprise Data and Applications leveraging Cerebra's inference engine. This is a commercial version of OilED semantic engine. Entire platform of tools based on Cerebra's engine is planned. Supported Capability: Semantic Data Integrator, Semantic Form Generator and Results Classifier, Knowledge Pulse
Standards Compliance and Comments	Also supports population. Supports maintenance and evolving of ontologies with "hot" addition of new axioms/relationships without system downtime. Supports RDF, DAML+OIL, OWL, SOAP interfaces.

Tool	Tucana KnowledgeStore
Vendor	Tucana Technologies
Lifecycle Phase	Deploying
Capabilities	Distributed database designed especially for metadata and metadata management. The database has been architected to persist and retrieve metadata with extremely fast performance levels while maintaining permanent integrity and secure access.
Standards Compliance and Comments	Also supports creation and population with Tucana Metadata Extractor™. Supports RDF, has SOAP, COM, and Java Interfaces.

Lifecycle Phase: Maintaining and Evolving (Critiquing Activity)

Tool	OntoClean (ODE)
Vendor	University of Madrid
Lifecycle Phase	Maintaining and Evolving (Critiquing Activity)
Capabilities	Checks for consistency of ontologies
Standards Compliance and Comments	Research prototype, DAML+OIL compliant.

Tool	OilED
Vendor	Analyze consistency of ontologies according to description logics
Lifecycle Phase	Maintaining and Evolving (Critiquing Activity)
Capabilities	University of Manchester
Standards Compliance and Comments	Free download. Integrates with Microsoft Office. RDF, DAML+OIL compliant.

Tool	Cerebra Inference Engine
Vendor	Network Inference
Lifecycle Phase	Maintaining and Evolving (Critiquing Activity)
Capabilities	Analyze consistency and draw conclusions based on DL
Standards Compliance and Comments	Commercial version of OilED. Supports deployment. Also supports maintenance and evolving of ontologies with "hot" addition of new axioms/relationships without system downtime. Supports RDF, DAML+OIL, OWL, SOAP interfaces.

Lifecycle Phase: Maintaining and Evolving (Merging Activity)

Tool	PROMPT
Vendor	Stanford KSL
Lifecycle Phase	Maintaining and Evolving (Merging Activity)
Capabilities	Supports merging two or more ontologies
Standards Compliance and Comments	Plug-in for Protege2000.

Tool	Chimera
Vendor	Stanford KSL
Lifecycle Phase	Maintaining and Evolving (Merging Activity)
Capabilities	Allows multiple ontologies to be processed together, provides analysis to find merges
Standards Compliance and Comments	Planned support for RDF and OWL.

Tool	FCA-Merge
Vendor	AIFB, University of Karlsruhe
Lifecycle Phase	Maintaining and Evolving (Merging Activity)
Capabilities	Merges ontologies bottom-up based on common instances
Standards Compliance and Comments	Research prototype.

Semantic Interoperability Platform Capabilities

We have identified the following as key capabilities offered by semantic integration solutions:

Management of Enterprise Data Schemas

- Creating and publishing shared vocabularies of business concepts
- Cataloging data assets, including their schemas and other metadata
- Formally capturing the semantics of corporate data by mapping database and message schemas to the ontology
- Importing a variety of standard data definition formats
- Supporting model management and evolution

Dynamic Data Transformation

- On-the-fly transformations of data based on the available mappings—code generation may be leveraged, but it is transparent to the users because the "work" is done as part of an execution environment and not during design.

Code and Script Generation

- Generating executable code such as SQL, XSLT, and Java
- Generating "wrappers" for data sources
- Embedding of business rules in models
- Automatic updates after change in the model and schemas

Table A.1 Data Table—Comparison of Capabilities Offered by Vendors

	Enterprise Data Management	Dynamic Data Transformation	Code and Script Generation	Semantic Data Validation	Run time Support	Web Services Orchestration
Celcorp Celware	—	—	Yes	—	Yes	—
Contivo EIM Server	Yes	—	Yes	—	—	—
enLeague Semantic Broker	Yes	—	—	—	Yes	Yes
Modulant Contextia Product Suite	—	Yes	—	—	Yes	—
Network Inference Cerebra Platform	—	—	—	Yes	Yes	—
Ontology Works IODE	—	—	Yes	Yes	Yes	—
Ontoprise Ontobroker	—	—	—	Yes	Yes	—
SchemaLogic SchemaServer	Yes	—	Yes	—	—	—
Unicorn System	Yes	—	Yes	Limited	Yes	—

Semantic Data Validation

- Using inference rules to validate integrity of the data based on a set of restrictions. The inference rules will automatically identify inconsistencies when querying for information.

Run Time Support

- Scaleable semantic engine that supports high volume of real-time queries

Web Services Orchestration

- Integration broker
- Intelligent discovery and orchestration (composition and chaining) of Web Services

Table A.1 compares capabilities currently offered by each of the vendors.

The following tables provide a detailed look at each product and its support for open standards.

Product	Celcorp Celware
Adoption and Usage	Mature product, offers a unique approach to application integration. Has a number of reference customers in the financial services industry.
Knowledge Representation	Proprietary, planning to go to RDF in 2004.
Reasoning Capabilities	Based on proprietary extensions to the "Plan Domain Model and the Graph Plan Algorithm."
Interfaces	Import: Screen scraping, SQL statements
Web Services Support	

Product	Contivo Vocabulary Management Systeem
Adoption and Usage	Relatively mature, has a number of reference customers. Focused on complementing webMethods and Tibco.
Knowledge Representation	Proprietary on top of relational database, evaluating RDF reasoning capabilities.
Reasoning Capabilities	Automated processing to create transformation code using relationships and transformation rules maintained in custom ontology.
Interfaces	Import: XML Schema, XML DTD, RDB, SAP, flat files Export: XML Stylesheets (XSLT), EAI (WebMethods,TIBCO), Java, Web Services Support
Web Services Support	XML, SOAP, WSDL

Product	Modulant Contextia Product Suite
Adoption and Usage	Relatively mature, has a number of reference customers. Focused on government, STEP customers.
Knowledge Representation	XML, proprietary
Reasoning Capabilities	None evident
Interfaces	Import: XML, RDB, flat files, STEP 21 files Export: XML
Web Services Support	XML, SOAP

Product	Network Inference Cerebra Platform
Adoption and Usage	New, currently in beta. Initial focus on biotechnology.
Knowledge Representation	RDF, DAML+OIL, OWL
Reasoning Capabilities	Description Logic
Interfaces	Import: XML Schema, RDB (JDBC), RDF/S, DAML+OIL Export: XML Schema (XSLT), RDF/S, DAML+OIL
Web Services Support	XML, SOAP, WSDL

Product	Ontology Works IODE
Adoption and Usage	Relatively mature, has a number of reference customers in government.
Knowledge Representation	Proprietary
Reasoning Capabilities	Robust, based on a proprietary Ontology language OWL (a variant of KIF, not related to w3c standard by the same name)
Interfaces	Import: UML, RDF/S Export: RDB (Oracle, DB2), DDB, RDF/S, XML
Web Services Support	XML

Product	Ontoprise Ontobroker
Adoption and Usage	Relatively mature semantic engine, has a number of reference customers. New to the integration market.
Knowledge Representation	RDF, DAML+OIL, OWL support planned
Reasoning Capabilities	F-Logic
Interfaces	Import: RDB, RDF/S, DAML+OIL, XML Schema Export: RDF/S, DAML+OIL
Web Services Support	XML

Product	SchemaLogic SchemaServer
Adoption and Usage	New. The product can unify structured and unstructured data management. Focuses on helping existing customers of Portal and Content Management products.
Knowledge Representation	XML, Proprietary
Reasoning Capabilities	No
Interfaces	Import: RDF, XML Schema Export:
Web Services Support	XML, SOAP

Product	Unicorn System
Adoption and Usage	Relatively new, focused on enterprise data management. First customer implementations are in progress.
Knowledge Representation	RDF, DAML+OIL, OWL support planned
Reasoning Capabilities	A third-party reasoning engine could be integrated with this standards-based tool.
Interfaces	Import: RDB (Oracle 7i/8i/9i, MS SQL Server 7/2000, DB2), XML Schema, UML (via adopter), ERWin, RDF/S, DAML+OIL Export: RDF/S, DAML+OIL, SQL Transformation Scripts, XSLT
Web Services Support	XML

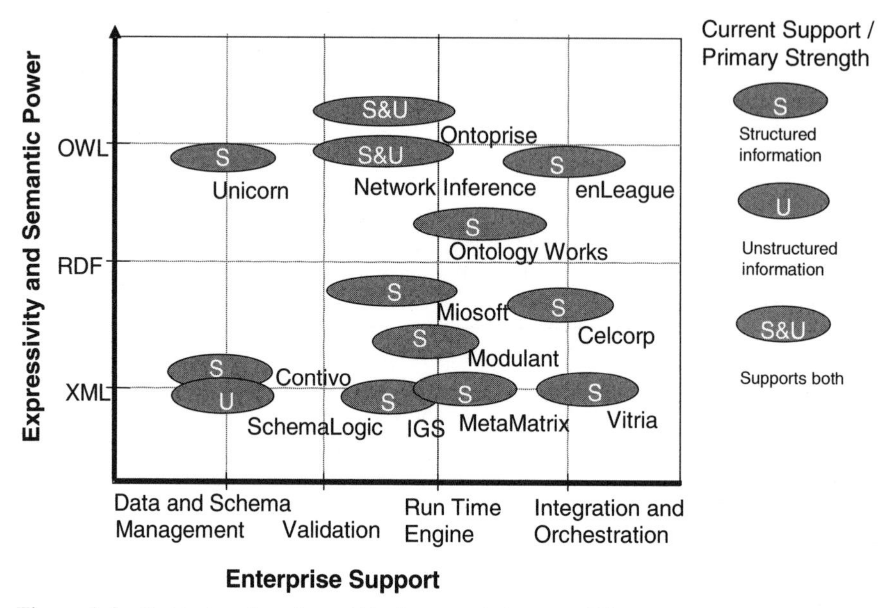

Figure A.2 Positioning of vendors within the semantic interoperability space

Figure A.2 compares how these solutions are positioned within the semantic integration space. The vertical axis represents a vendor's ability to integrate disparate information based on semantics. The horizontal positioning represents a vendor's solution focus. The vertical axis represents a progression—higher positioning indicates more powerful semantic capabilities. The horizontal line doesn't end with an arrow because, unlike the vertical axis, it is not intended to represent a progression of capabilities. The rightmost position of a vendor indicates that its major strength is in "Integration and Orchestration." The vendor may also offer some support, but not full functionality, in the areas of "Data and Schema Management," "Validation," or "Run time."

ABOUT VENDOR SELECTION

Vendors covered in this issue have different strengths as well as different industry and problem focus areas. Choosing the right product will depend on:

- How well it integrates with your data and content sources, infrastructure, and applications
- The degree to which you need run time support
- The product's support for industry-specific XML schemas and vocabularies
- The vendor's flexibility and interest in evolving the product to support your requirements

Table A.2 Data Table—Semantic interoperability vendor scorecard

	Semantic Interoperability Vendor Scorecard									
Context Mediation	(1)	(2)	(3)	(4)	(5)	(6)	(7)	(8)	(9)	(10)
Conflict Accommodation	(1)	(2)	(3)	(4)	(5)	(6)	(7)	(8)	(9)	(10)
Life Cycle Support	(1)	(2)	(3)	(4)	(5)	(6)	(7)	(8)	(9)	(10)
Domain Fit	(1)	(2)	(3)	(4)	(5)	(6)	(7)	(8)	(9)	(10)
Horizontal Capabilities	(1)	(2)	(3)	(4)	(5)	(6)	(7)	(8)	(9)	(10)
Data Coupling	(1)	(2)	(3)	(4)	(5)	(6)	(7)	(8)	(9)	(10)
Service Coupling	(1)	(2)	(3)	(4)	(5)	(6)	(7)	(8)	(9)	(10)
Automation	(1)	(2)	(3)	(4)	(5)	(6)	(7)	(8)	(9)	(10)
Simplicity	(1)	(2)	(3)	(4)	(5)	(6)	(7)	(8)	(9)	(10)
Formalized Methodology	(1)	(2)	(3)	(4)	(5)	(6)	(7)	(8)	(9)	(10)

The authors strongly suggest you contact the vendor with any questions and issue a request for a demonstration of its capabilities.

VENDOR SCORECARD ANALYSIS (HOW TO MEASURE YOUR VENDOR)

When it comes time to choose a vendor for your project, how will you compare the strengths and capabilities of your different options? No doubt there will be a number of project-specific must-haves. Requirements for Web Services orchestration or for code generation may trump other factors.

Although we do not expect that the following scorecard analysis will be the end-all of your decision criteria, we wanted to provide the reader with a set of metrics that we feel are important in a robust semantic interoperability solution. The reader may then individually choose what is most important to him or her. It is our hope that we will have provided some insight into the many distinctions that can be made among semantic interoperability vendor tools.

The scorecard may then be used to produce a comparative graph of different vendors. An example of this kind of graph is shown in Figure A.3.

A description of each axis we chose for this scorecard is provided below.

Context Mediation Capabilities

Context mediation capabilities speak to the vendor's capacity to leverage business context cues as a trigger for using various data semantics. For instance, a vendor may implement an ontology-centric mediation environment based on a "shared understanding" of what the ontology terms mean—in this case, the ontology becomes the context. In other implementations, the vendor may still use an ontology-centric mediation environment but allow users to map to the ontology in whatever way they choose—in this case, semantics are kept in alignment through

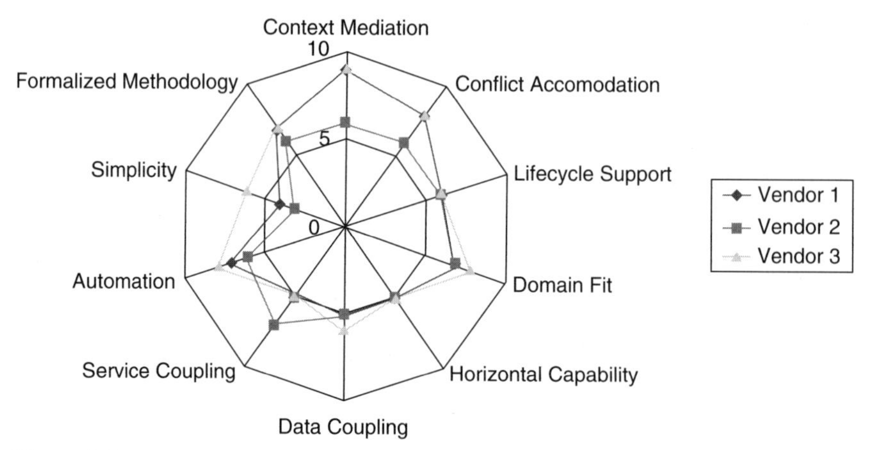

Figure A.3 Sample vendor metrics chart

"context communities" who agree to use the ontology in similar ways because they share context. The scale for measuring context mediation capabilities could look like the following:

- Levels 1–3—no explicit support for context, work-around may be available
- Levels 4–6—basic support for context, slots in the model for simple taxonomy
- Levels 7–10—strong support for context, context is an expressive ontology

Semantic Conflict Accommodation

Semantic conflict accommodation is the ability of the vendor to accommodate each of the semantic conflicts identified in Chapter 5. Table A.3 provides a recap of those conflicts.

When measuring a vendor's tool support for semantic conflict accommodation, the following scale may be used:

- Levels 1–3—not all supported in the modeling environment; of those that are supported, only limited depth and complexity in its support.
- Levels 4–6—all conflicts supported in the modeling environment; only limited depth and complexity in its support, some coding is likely still required.
- Levels 7–10—all conflicts supported in the modeling environment; deep capabilities in each conflict area, very minimal coding or no coding required to accommodate the full range of semantic conflicts.

Table A.3 Data Table—Semantic conflict summary table

Data Type Conflicts	Data type conflicts occur when different primitive system types are used in the representation of data values for a given programming language.
Labeling Conflicts	Labeling conflicts occur when synonyms and antonyms appear with different labels inside the schema specification.
Aggregation Conflicts	Aggregation conflicts arise from different conceptions of relationships among a given data set. This could result in a number of interrelated issues regarding cardinality and structure.
Generalization Conflicts	Generalization conflicts occur when different abstractions and specializations are used to model the same environment.
Value Representation Conflicts	Value representation conflicts occur when a value is derived or inferred in one system but is explicit in another.
Naming Conflicts	A purely semantic conflict, naming issues arise when there are synonyms and antonyms among data values (usually referring to instance values here, as opposed to schema entity names as in the "Labeling Conflict").
Scaling and Unit Conflicts	"Scaling and Units Conflicts" refers to the adoption of different units of measures or scales in reporting."[a]
Confounding Conflicts	Also a purely semantic conflict, confounding conflicts arise when like concepts are exchanged that have different definitions in the source systems—as opposed to simpler "Labeling Conflicts," where the labels are the problem, not the concept.
Domain Conflicts	Domain conflicts are sometimes classified as intensional conflicts. They refer to clashes with the culture and underlying domains of the models inside the software system.
Integrity Conflicts	"Integrity constraint conflicts refer to disparity among the integrity constraints asserted in different systems."[b]
Impedance Mismatch Conflicts	Impedance mismatch is a generalized conflict that refers to the problems of moving data between fundamentally different knowledge representations.

[a]Goh, Cheng Hian, *Representing and Reasoning about Semantic Conflicts in Heterogeneous Information.*

[b]Goh, Cheng Hian, *Representing and Reasoning about Semantic Conflicts in Heterogeneous Information.*

Full Lifecycle Tool Support

This concept was discussed in the ontology tools overview. Full lifecycle support means that the vendor has tool capabilities that span the entire lifecycle of ontology development and run time operations. For ontology development, the lifecycle phases should include Creating, Populating, Deploying, Validating, Evolving, and Maintaining. For the run time environment, the lifecycle phases should include Analysis, Design, Development, Testing, Deploying, and Maintaining.

When measuring a vendor's tool support for semantic conflict accommodation, the following scale may be used:

- Levels 1–3—not all phases are supported
- Levels 4–6—most phases are supported, tools may not be integrated well
- Levels 7–10—all phases are supported, tools are integrated very well

Targeted Domain Fit

Targeted domain fit may be important for some projects. The ontology concepts, work flow, and rules may be specialized enough that a general solution just won't suffice. Some semantic interoperability vendors, such as IGS, do specialize in a particular vertical (telecommunications)—whereas others do not.

When measuring a vendor's tool support for targeted domain fit, the following scale may be used:

- Levels 1–3—minimal to no support for a specific domain
- Levels 4–6—moderate support for the domain (reuse of domain ontology)
- Levels 7–10—tight support for the domain (ontology, rules, process, lingo)

Horizontal Domain Flexibility

Some organizations may seek technologies that can work well across different domains. For instance, an IT organization may wish to begin an enterprise roll-out of semantic technologies to different business units and wants to standardize on the same vendor suite. It is quite possible to have robust capabilities in non-domain-specific tools—the nature of ontology and interoperability work is to operate at an abstraction level above many, if not all, domain-specific logics. A key consideration for horizontal domain support is how complex the models and rules may be. Certain industries, heavy manufacturing for example, have quite complex product structures that can be difficult to accommodate with model-based rules and mappings.

When measuring a vendor's tool support for horizontal domain flexibility, the following scale may be used:

- Levels 1–3—may import models from different domain spaces
- Levels 4–6—supports highly complex domain models and rules

- Levels 7–10—supports a custom-tailored user interface for working with domain models

Formalized Methodology

Because standards development for semantic interoperability is still maturing, many vendors have highly unique approaches to solving semantic conflicts and mediating context. To work efficiently with a vendor's tool set, it should have a well-documented methodology for working through its lifecycle.

When measuring a vendor's tool support for formalized methodology, the following scale may be used:

- Levels 1–3—basic instructional guides are available
- Levels 4–6—formalized methodology documentation is available
- Levels 7–10—very formal documentation and training are available

Level of Data Coupling

Not all semantic interoperability technology approaches have the same degree of loose coupling among data. Some tool sets still require custom application code to link disparate sources—XSL/T and SQL are popular methods leveraged by vendors. Other, more advanced systems rely entirely on model-based associative mappings that require no code and no code generation.

When measuring a vendor's tool support for loose data coupling, the following scale may be used:

- Levels 1–3—much code needs to be written and maintained
- Levels 4–6—some code needs to be written, much may be generated
- Levels 7–10—no code is required, at minimum all code may be generated

Level of Service Coupling

Some customers may want vendor support for advanced Web Services capabilities such as Web Services orchestration, BPEL support, Semantic Web Services, and more. Some vendors go a long way with Web Services support, whereas others simply offer SOAP interfaces to engines, and still others do not offer any comprehensive Web Services support at all.

When measuring a vendor's tool support for loose service coupling, the following scale may be used:

- Levels 1–3—no Web Services support or minimal SOAP support
- Levels 4–6—commitment to Web Services run time capabilities
- Levels 7–10—advanced capabilities already in place (e.g., BPEL, OWL-S)

Automation Support

There are many opportunities for automation in semantic interoperability platforms, from ontology creation and deployment to run time transformations and service configurations. Rather than attempting to categorize the myriad of ways that interoperability support can be automated, a general measure of automation across the lifecycle is considered.

When measuring a vendor's tool support for automation, the following scale may be used:

- Levels 1–3—some aspects of the design lifecycle may automated
- Levels 4–6—some aspects of design and deployment may be automated
- Levels 7–10—advanced automation throughout the lifecycle is supported

Simplicity and Ease of Use

As with all complex processes, the human factors in semantic interoperability are crucial for widespread adoption and support. Innovative GUI approaches to ontology visualization, mapping, low-level testing, and services management will enable semantic interoperability to scale beyond deep technical experts in the field.

When measuring a vendor's tool support for automation, the following scale may be used:

- Levels 1–3—complicated interfaces and tools
- Levels 4–6—easy to use, but nongraphical modeling and management
- Levels 7–10—graphical and innovative UI for full lifecycle of modeling and management of data, services, and processes

Appendix B

Acronyms and Abbreviations

Acronyms and Abbreviation Definitions

.NET	Microsoft's proprietary Web Services solution framework
A2A	Application to Application
AI	Artificial Intelligence
AMS	American Management Systems
ANSI-SPARC	American National Standards Institute—Standards Planning and Requirements Committee
API	Application Programming Interface
ASCII	American Standard Code for Information Interchange
DML	Data Manipulation Language
ASP	Application Service Provision/Provider
B2B	Business to Business
B2Bi	Business to Business Internet
BPEL	Business Process Execution Language
BPM	Business Process Management
BPML	Business Process Modeling Language
CASE	Computer-Aided Software Engineering
CG	Conceptual Graphs
CIA	Central Intelligence Agency
CIIM	Context-Based Information Interoperability Methodology
CIM	Computation-Independent (business) Model
CIO	Chief Information Officer
COBOL	COmmon Business-Oriented Language
COM	Component Object Model
CPC	Collaborative Product Commerce
CPFR	Collaborative Planning, Forecasting, and Replenishment
CPU	Central Processing Unit
CRM	Customer Relationship Management
CSV	Comma Separated Values
CWM	Common Warehouse Metamodel
CYC	A large knowledge-based system—enCYClopedia

Semantic Interoperability, by Jeffrey T. Pollock and Ralph Hodgson
ISBN 0-471-48854-2

DACNOS	A prototype network operating system for multivendor environments, from IBM European Networking Centre Heidelberg and University of Karlsruhe
DAML	DARPA Agent Markup Language
DARPA	Defense Advanced Research Projects Agency
DCM	Demand Chain Management
D-COM	Distributed Common Object Model
DECnet	Proprietary Digital Equipment Corporation network protocol
DIA	Defense Intelligence Agency
FEA	Federal Enterprise Architecture
DL	Description Logics
DNS	Domain Naming Service
DSS	Decision Support System
DTD	Document Type Definition
EAI	Enterprise Application Integration
ebXML	E-Business XML
EDI	Electronic Data Interchange
EDIFACT	EDI For Administration, Commerce, and Transport
EPS	Earnings Per Share
ERD	Entity Relationship Diagram
ERP	Enterprise Resource Planning
ETL	Extract, Transform, and Load
CGEY	Cap Gemini, Ernst & Young
FBI	Federal Bureau of Investigation
FIPA	Foundation for Intelligent Physical Agents
FOL	First-Order Logic
FTP	File Transfer Protocol
GGF	Global Grid Forum
GPA	Grade Point Average
GUI	Graphical User Interface
HAL	Heuristically Programmed ALgorithmic Computer
HL7	Health Level 7
HTML	HyperText Markup Language
HTTP	HyperText Transfer Protocol
I2I	Industry to Industry
IDE	Integrated Development Environment
R&D	Research and Development
IIOP	Internet Inter-ORB Protocol
IP	Internet Protocol
IR	Integrated Resource (in STEP)
ISO	International Standards Organization
IT	Information Technology
J2EE	Java 2 Enterprise Edition
KIF	Knowledge Interchange Format
KM	Knowledge Management

KR	Knowledge Representation
LDAP	Lightweight Directory Access Protocol
LLC	Logical Link Control (in OSI)
M&A	Mergers and Acquisitions
MAC	Media Access Control (in OSI)
MAFRA	Ontology MApping FRAmework
MDA	Model-Driven Architecture
MIT	Massachusetts Institute of Technology
MOF	Meta-Object Facility
MOM	Message-Oriented Middleware
MOMA	Message-Oriented Middleware Association
MRP	Material Requirements Planning
NEON	New Era Of Networks
NSA	National Security Agency
OAG	Open Applications Group
OASIS	Organization for the Advancement of Structured Information Standards
OCL	Object Constraint Language
ODM	Ontology Definition Metamodel
OGSA	Open Grid Service Architecture
OGSI	Open Grid Services Infrastructure
OIL	Ontology Inference Layer
OMG	Object Management Group
OO	Object-Oriented
OSI	Open Source Initiative
OWL	Web Ontology Language
OWL-R	Web Ontology Language—Rules
OWL-S	Web Ontology Language—Services
P/E	Price-to-Earnings Ratio
PDM	Product Data Management
PICS	Platform for Internet Content Selection
PIM	Platform-Independent Model
PLM	Product Lifecycle Management
PSM	Platform-Specific Model
QoS	Quality of Service
RDBMS	Relational Database Management System
RDF	Resource Description Framework
RDF/S	Resource Description Framework Schema
RFID	Radio Frequency ID
RIM	Reference Implementation Models
RISSnet	Regional Information Sharing System Network
RMI	Remote Method Invocation
ROCIC	Regional Organized Crime Information Center
ROI	Return on Investment
RPC	Remote Procedure Call

RuleML	Rule Markup Language
SCM	Supply Chain Management
SMTP	Simple Mail Transfer Protocol
SNMP	Simple Network Management Protocol
SOA	Service-Oriented Architecture
SOAP	Simple Object Access Protocol
SQL	Structured Query Language
STEP	STandard for the Exchange of Product Model Data
SWSI	Semantic Web Services Initiative
SWRL	Semantic Web Rules Language
TCO	Total Cost of Ownership
TCP	Transmission Control Protocol
TDWI	The Data Warehousing Institute
TTM	Tabular Topic Maps
UCC	Uniform Code Council
UDDI	Universal Description, Discovery, and Integration
UML	Unified Modeling Language
URI	Universal Resource Identifier
URL	Uniform Resource Locator (a type of URI)
VIN	Vehicle Identification Number
W3C	World Wide Web Consortium
WiFi	Wireless Fidelity
WSDL	Web Services Description Language
X12	American version of EDIFACT
X-Map	XML Mapping
XMI	XML Metadata Interchange
XML	eXtensible Markup Language
XPath	XML Path Language
XQuery	XML Query Language
XSD	XML Schema Definition
XSL/T	XML Stylesheet Language Transformations

Endnotes

1. *Does IT Matter? An HBR Debate*. Hagel, Seely-Brown, June 2003.
2. *Technology Sending*. CIO Insight, June 2002.
3. US Productivity Growth, 1995–2000. McKinsey Global Institute, October 2001.
4. *Does IT Matter? An HBR Debate*. Hagel, Seely-Brown, June 2003.
5. Rekha Balu, "(Re) Writing Code," Fast Company, April 2001.
6. Information Integration A New Generation of Information Technology, M.A, Roth. *IBM Systems Journal*, Volume 41, No. 4, 2002.
7. The Big Issue, Pollock, *EAI Journal*.
8. *Experience Design*, Nathan Shedroff.
9. The Big Issue, Jeffrey Pollock. *EAI Journal*, April 2002.
10. Smart Data for Smart Business: 10 Ways Semantic Computing will Transform IT, Michael Daconta. *Enterprise Architect*, Winter 2003.
11. Using the G.M.A. Grube translation (*Plato*, *Five Dialogues*, *Euthyphro*, *Apology*, *Crito, Meno*, *Phaedo*, Hackett Publishing Company, 1981, pp. 6–22) [6e, G.M.A. Grube trans., Hackett, 1986].
12. Miller, Barry, "Existence", *The Stanford Encyclopedia of Philosophy* (Summer 2002 Edition), Edward N. Zalta (ed.), URL = <http://plato.stanford.edu/archives/sum2002/entries/existence/>.
13. Sowa, Chapter, 3.6 "Knowledge Representation".
14. Biletzki, Anat, Matar, Anat, "Ludwig Wittgenstein", T*he Stanford Encyclopedia of Philosophy* (Winter 2002 Edition), Edward N. Zalta (ed.), URL = <http://plato.stanford.edu/archives/win2002/entries/wittgenstein/>.
15. Sowa, John F., ed. (1998) *Conceptual Graphs*, draft proposed American National Standard, NCITS.T2/98-003.
16. Joseph Becker: Becker 1975—"The Phrasal Lexicon." In *Proceedings, [Interdisciplinary workshop on] Theoretical issues in natural language processing*, Cambridge, MA June 1975. pp. 70–73.
17. Graeme Hirst, "Semantic Interpretation and the Resolution of Ambiguity".
18. Kirby, Simon, *Language Evolution Without Natural Selection: From Vocabulary to Syntax in a Population of Learners*. Edinburgh Occasional Papers in Linguistics, 1998.
19. *Collected Papers of Charles Sanders Peirce*, volume 2, paragraph 97.
20. *Collected Papers of Charles Sanders Peirce*, volume 7, paragraph 98.
21. Turing A. M. (1936–7), On computable numbers with an application to the Entscheidungsproblem.
22. R. Davis, H. Shrobe, and P. Szolovits. What is a Knowledge Representation? *AI Magazine* 14(1):17–33, 1993.
23. Esther Dyson, Release 1.0.
24. Goh, Cheng Hian, *Representing and Reasoning about Semantic Conflicts in Heterogeneous Information Systems*, 1997.
25. Goh, Cheng Hian, *Representing and Reasoning about Semantic Conflicts in Heterogeneous Information Systems*, 1997.
26. Goh, Cheng Hian, *Representing and Reasoning about Semantic Conflicts in Heterogeneous Information Systems*, 1997.
27. Goh, Cheng Hian, *Representing and Reasoning about Semantic Conflicts in Heterogeneous Information Systems*, 1997.
28. *Financial Information Integration In the Presence of Equational Ontological Conflicts*, Aykut Firat, Stuart Madnick, Benjamin Grosof, MIT Sloan School of Management, Cambridge, MA.

Adaptive Information, by Jeffrey T. Pollock and Ralph Hodgson
ISBN 0-471-48854-2

29. Goh, Cheng Hian, *Representing and Reasoning about Semantic Conflicts in Heterogeneous Information Systems*, 1997.
30. IBM submission to OMG Ontology Working Group, Dan Chang, Yeming Ye, 08/2003.
31. IBM submission to OMG Ontology Working Group, Dan Chang, Yeming Ye, 08/2003.
32. *Thinking in Java*, 2nd ed 2000, Bruce Eckel.
33. *Ontologies: A Silver Bullet for Knowledge Management and E-Commerce*, Dieter Fensel.
34. Release 1.0, February 2003.
35. Knowledge Transformation for the Semantic Web Proceedings 2002.
36. Information Integration: A New Generation of Information Technology, M.A. Roth, *IBM Systems Journal* Volume 41, No. 4, 2002.
37. John Sowa, *Knowledge Representation: Logical, Philosophical, and Computational Foundations*. Brooks/Cole 2000.
38. *Automated Negotiation from Declarative Contract Descriptions*, Ashish Mishra, Michael Ripley, and Amar Gupta MIT, Sloan School of Management.
39. Alexander Maedche, Boris Motik, Nuno Silva, Raphael Volz, *"MAFRA—A MApping FRamework for Distributed Ontologies in the Semantic Web*." Proceedings of the Workshop on Knowledge Transformation for the Semantic Web (KTSW) 2002. Workshop W7 at the 15th European Conference on Artificial Intelligence, July 2002.
40. Wang, David, "*Automated Semantic Correlation between Multiple Schema for Information Exchange*." Massachusetts Institute of Technology Department of Electrical Engineering and Computer Science, master thesis, June 2002.
41. EPM ExpressWay, January 2001 ["PDM Interoperability", Emery Szmrecsanyi].
42. RTI / NIST Report, March 1999 ["Interoperability Cost Analysis of the U.S. Automotive Supply Chain"].
43. *Place to Space*, Peter Weill, HBR 2001.
44. Robert S. Kaplan and David P. Norton, *The Balanced Scorecard: Translating Strategy into Action*; ISBN: 0875846513, Harvard Business School Press, 1996.
45. Andy Neely, et al., *The Performance Prism—the Scorecard for Measuring and Managing Business Success*, FT, Prentice-Hall, Pearson Education, ISBN 0-27365334-2, 2002.
46. Stephen Denning, *The Springboard: How Storytelling Ignites Action in Knowledge-Era Organizations*, Butterworth-Heinemann, October 2000.
47. Andreas Maier, Hans-Peter Schnurr, and York Sure, *Ontology-Based Information Integration in the Automotive Industry*, The Semantic Web, ISWC 2003 Proceedings, D. Fensel et al. (Eds), pp. 897–912, Springer-Verlag Berlin, Heidelberg, 2003.
48. A. Sheth, et al., "*Semantic Association Identification and Knowledge Discovery for National Security Applications*", *Journal of Database Management*, 2004.
49. Technical Memorandum # 03-009, LSDIS Lab, Computer Science, the University of Georgia, August 15, 2003. Prepared for Special Issue of *Journal of Database Management* on Database Technology for Enhancing National Security, Ed. Lina Zhou.
50. Yao Sun, *Information Exchange Between Medical Databases Through Automated Identification of Concept Equivalence*, Ph.D. Thesis in Computer Science, Department of Electrical Engineering and Computer Science, Massachusetts Institute of Technology, February 2002.
51. Bachler, M.S., Buckingham Shum, S., De Roure, D., Michaelides D. and Page, K.: "*Ontological Mediation of Meeting Structure: Argumentation, Annotation, and Navigation*", 1st International Workshop on Hypermedia and the Semantic Web, ACM Hypertext, Nottingham, 2003.
52. BPEL4WS (Business Process Execution Language for Web Services) is a specification co-authored by IBM, Microsoft, BEA, SAP and Siebel Systems. Influenced by Microsoft's XLANG and IBM's WSFL, it is a work flow language with simple control flow constructs such as *if*, *then*, *else*, and *while-loop*.
53. Sheila A. McIlraith, David L. Martin, James Hendler (ed.) "*Bringing Semantics to Web Services*", IEEE Intelligent Systems, 2003.
54. Ryusuke Masoka, Bijan Parsia and Yannis Labrou, "*Task Computing—the Semantic Web meets Pervasive Computing*", The Semantic Web, ISWC 2003 Proceedings, D. Fensel et al. (Eds), pp. 866–881, Springer-Verlag Berlin Heidelberg, 2003.

55. Abraham Bernstein and Mark Klein, "*Discovering Services: Towards High-Precision Service Retrieval*", in *Proceedings of the CaiSE 2002 International Workshop, WES 2002, on Web Services, e-Business, and the Semantic Web*, Springer, pp. 260–275, 2002.
56. T. W. Malone, K. Crowston, J. Lee, B. Pentland, C. Dellarocas, G. Wyner, J. Quimby, C. Osborn, A. Bernstein, G. Herman, M. Klein, and E. O'Donnell, *Tools for inventing organizations: Toward a handbook of organizational processes, Management Science*, vol. 45, pp. 425–443, 1999.
57. Mark Hansen, Stuart Madnick and Michael Siegel, "*Process Aggregation Using Web Services*", Web Services, E-Business and the Semantic Web, CAiSE International Workshop, WES 2002, Toronto, Canada, pp. 12–27, 2002.
58. Gartner Report, "Semantic Web Technologies Take Middleware to the Next Level", 8/2002.
59. Maier, J. Aguado, A., Bernaras, I. Laresgoiti, C. Pedinaci, N. Pena, T. Smithers, Integration with Ontologies, WM 2003, Lucerne, Switzerland.
60. N.F. Noy and M.A. Musen. PROMPT: Algorithm and tool for automated ontology merging and alignment. In *Proceedings of the National Conference on Artificial Intelligence (AAAI)*, 2000.
61. Simos, M. et al., *Organization Domain Modeling (ODM) Guidebook*, 1995.
62. Simos, M.A.: Organization domain modeling (ODM): Formalizing the core domain modeling life cycle. In: *Proceedings of the 1995 Symposium on Software Reusability*, Seattle, Washington (1995).
63. Moore, G.: *Crossing the Chasm*. Harper Business, revised edition, 2002.
64. Moore, G.: *Crossing the Chasm*. Harper Business, revised edition, 2002.
65. "The Semantic Web", Tim Berners-Lee.
66. Grueninger and Fox, The Role of Competency Questions in Enterprise Engineering, *Proceedings of the IFIP WG5.7 Workshop on Benchmarking—Theory and Practice*, Trondheim, 1994.
67. "The Semantic Web", Tim Berners-Lee.
68. Uschold and Grueninger, "Ontologies: Principles, Methods and Applications", *Knowledge Sharing and Review*, 1996.
69. Grueninger and Fox, The Role of Competency Questions in Enterprise Engineering, *Proceedings of the IFIP WG5.7 Workshop on Benchmarking—Theory and Practice. Trondheim*, 1994.
70. Grueninger and Fox, The Role of Competency Questions in Enterprise Engineering, *Proceedings of the IFIP WG5.7 Workshop on Benchmarking—Theory and Practice. Trondheim*, 1994.
71. Fernández-López, Gómez-Pérez, Pazos-Sierra, Pazos-Sierra, Building a Chemical Ontology Using METHONTOLOGY and the Ontology Design Environment, IEEE Intelligent Systems & their applications, January/February 1999.
72. Moore, G.: *Crossing the Chasm*. Harper Business, revised edition, 2002.

Index

Adaptive Information, by Jeffrey T. Pollock and Ralph Hodgson
ISBN 0-471-48854-2